Environmental Studies

(CONCEPTS, IMPACTS, MITIGATION AND MANAGEMENT)

THIRD EDITION

[COVERING UGC SYLLABUS FOR UNDERGRADUATE COURSES IN ALL BRANCHES OF ENGINEERING AND NON-ENGINEERING]

Prof. M.P. Poonia
Vice-Chairman
All India Council for Technical Education (AICTE)

S.C. Sharma
Formerly General Manager
THDC India Limited

Prof. Santosh kumar
Former University Professor
Bihar College of Engineering (Now NIT Patna)
Patna University

KHANNA BOOK PUBLISHING CO. (P) LTD.
Publisher of Engineering and Computer Books
4C/4344, Ansari Road, Darya Ganj, New Delhi-110002
Phone : 011-23244447-48 **Mobile:** +91-9910909320
E-mail : contact@khannabooks.com
Website : www.khannabooks.com

Dear Readers,

To prevent the piracy, this book is secured with HIGH SECURITY HOLOGRAM on the front title cover. In case you don't find the hologram on the front cover title, please write us to at contact@khannabooks.com or whatsapp us at +91-99109 09320 and avail special gift voucher for yourself.

Specimen of Hologram on front Cover title:

Moreover, there is a SPECIAL DISCOUNT COUPON for you with EVERY HOLOGRAM.

How to avail this SPECIAL DISCOUNT:

Step 1: Scratch the hologram
Step 2: Under the scratch area, your "coupon code" is available
Step 3: Logon to www.khannabooks.com
Step 4: Use your "coupon code" in the shopping cart and get your copy at a special discount
Step 5: Enjoy your reading!

Price: ₹ 495.00

ISBN: 978-93-90779-02-4

Environmental Studies
by M.P. Poonia, S.C. Sharma, Santosh Kumar

First Edition: 2017
Second Edition: 2019
Third Edition: 2021
Reprint: 2023, 2024, **2025**

To view complete list of books, Please scan the QR Code:

Published by:
Khanna Book Publishing Co. (P) Ltd.
CIN: U22110DL1998PTC095547

Visit us at: **www.khannabooks.com**
Write us at: **contact@khannabooks.com**

Printed in Bharat.

PREFACE TO THE THIRD EDITION

We are happy to note that the first two editions of the book have been exhausted very fast. This shows that the book have been liked by the students and teachers. The reasons for appreciating the book are; it covers the UGC syllabus for undergraduate courses in all branches of engineering and non-engineering, besides, the book also covers the topics on "Environmental Impact Assessment", "Resettlement and Rehabilitation", Environmental Management", and "Environmental Management Plan." Objective type questions in large numbers have also been added in the book. The book has also been recommended for "Environmental Science" by AICTE.

The book has now been thoroughly revised and large number of examples on various environmental problems highlighting the causes, impacts, mitigation measures etc. have also been added. A chapter for "Field Work (Project Work)" has been included. A large number of questions including objective type questions have also been incorporated.

The book is now more useful for various competitive examinations.

The book further deals the environment from concepts to Impacts to Mitigation and then management in detail. The topics like 'Ecological balance', 'Agricultural solid waste management', 'Sustainable energy and material', and 'Protection of ozone layer' have been added.

We are thankful to Shri Mukul Seth of **M/s Khanna Publishing House** for taking keen interest in the publication and marketing the book. We would welcome suggestions, which may be received from the readers, for further improvement of the book.

–Authors

PREFACE TO THE THIRD EDITION

PREFACE TO THE FIRST EDITION

Environmental degradation has been a major concern since past few decades, because of economic growth and development across the world, it has caused major impacts on the Earth's ecosystems and natural resources to an extent that can limit the well-being of future generations.

The ignorance of people about retaining the ecological balance, distancing from Indian traditions which used to emphasize the need for conservation and sustainable use of resources are also the reasons for degradation of environment. The continuing of large scale urbanisation, population explosion, industrialisation, commercialisation and agriculture have caused such a vast impacts on our environment that the very existence of man has come under a threat. Sustainable development, under these circumstances, has become the prime necessity of today.

All the above factors are responsible for progressive decline in the quality and magnitude of natural resources including clean air, water, healthy soil, forests, biodiversity. This has also caused other environmental problems, such as desertification, hunger, malnutrition, acid rain, ozone depletion, global warming etc.

In view of the above, efforts are being made at international and national levels towards the right to healthy environment. These efforts have created awareness on environmental issues in NGOs, environmentalists, intellectuals etc.

India has recently started to realize the importance of environment and of the environmental education. Following the 2001's Supreme Court directive, the environmental education has been or is being included in the curriculums, right from school stage to the College/University level.

This book covers the AICTE & UGC syllabus for under-graduate courses, and also the syllabii for technical and other universities for different disciplines, may it be in the name of environmental studies, environmental science, ecology, or natural resource management. The book also covers the syllabus of Institution of Engineers (India) on the subject. This book has been written to bring about an awareness of a variety of environmental concerns and deals from concepts through impacts, mitigation to management.

We hope this book will be useful for the students of all the streams as well as professionals in this field. We would welcome suggestions, which may be received from the readers, for further improvement of this book.

–Authors

ACKNOWLEDGEMENT

We feel great pleasure to present the first edition of 'Environmental Studies' in the hands of the readers, the students. Even though, there are many textbooks available in the market, the authors have attempted to bring the material in the best possible manner after realizing the difficulties, generally faced by the students while studying the subject.

The help received from Er. Himmi Gupta, Assistant Professor, National Institute of Technical Teachers Training and Research, Chandigarh for checking and proof reading the manuscript is thankfully acknowledged. The valuable comments led to the betterment of the book.

We are grateful to the entire team of M/s Khanna Book Publishing Company (P) Ltd., Delhi for extending their full co-operation for early release of this first edition.

ABOUT THE AUTHORS

Prof. M.P. Poonia born on 7th July, 1959 in a small village of Rajasthan state in India, Prof. M.P. Poonia is specialized in the field of Mechanical Engineering (IC Engines, Gas Dyanics, Ref. & AC), Renewable Energy and Sustainable Development. He is M.Tech (Mech.) and Ph.D (Thermal Engg) from Indian Institute of Technology, Delhi. He possesses more than 30 years of total experiences, out of which 6 years as Principal of Govt. Engg. College, Bikaner (Rajasthan) and 1 year and two months as Dean, Planning and Development in MNIT, Jaipur, (Rajasthan). He has published 80 papers in National and International Journals and published several books and manuals in the field of Mechanical Engineering. He has published/edited 8 books and 02 Lab Manuals too. He is the Member of IEI, ISTE & Society of Automotive Engineers. Prof. Poonia is also a google scholar. On 20th January, 2017, Prof. Poonia took up the responsibility of Vice-Chairman, All India Council for Technical Education (AICTE), New Delhi. During the period from 20th July, 2012 to 18th January, 2017, say around a period of 4½ years, Prof. Poonia served as a Director, National Institute of Technical Teachers Training and Research (NITTTR), established by Ministry of Human Resource Development, Govt. of India, Chandigarh (India). Prof. Poonia made significant contribution as a Director, NITTTR, Chandigarh for the growth and development of technical education in the northern states of India and to conduct the sub-regional and in-country programmes in collaboration with Colombo Plan Staff College for Technician Education, Manila, Philippines. Under his able leadership, the institute has to its credit, the conduct training programme for Nigeria. He has undertaken the projects sponsored by All India Council for Technical Education, Deptt. of Science and Technology, Delhi and Govt. of Rajasthan and MHRD, Govt. of India. He is the recipient of many Awards for his academic achievements and National Award for the Empowerment of Persons with Disabilities – 2013 by the Hon'ble President of India on 3rd December, 2013 in New Delhi in recognition of outstanding performance in the field of Best Institution for Empowerment of Persons with Disabilities in the country. During his tenure as a Director, NITTTR, Chandigarh, he remained member of Board of Governors of, as many as 20 Government/Autonomous bodies/Universities in the state of Punjab, Haryana, Himachal Pradesh, Delhi and Chandigarh. He has visited many countries like USA, China, Canada, Thailand, Singapore & Sri Lanka.

S.C. Sharma, after completion of graduation in 1966, joined as a Lecturer in Mechanical Engineering. He had been associated for more than 4 decades in various fields including teaching and management of projects in India and abroad in different capacities. While working in hydropower projects for more than 15 years, he was actively associated with various environmental and Rehabilitation & Resettlement issues and successfully resolved various complicated issues. He has also worked as a consultant for matters related to safety, environment, and R&R. He is a Member of IE (India) and AIMA.

He has written about a dozen books on subjects related to engineering and management including management of projects.

Prof. Santosh Kumar, former Professor of Civil Engineering at the Bihar College of Engineering/NIT Patna, obtained his Bachelor's degree in Civil Engineering from Regional Institute of Technology/NIT Jamshedpur and Master's degree from University of Roorkee/IIT Roorkee. Further he obtained Dip. H.E. from Delft University, Netherlands. He completed his Ph.D. from Patna University.“ Prof. Kumar taught water resources related subjects for more than 40 years. During his tenure he successfully guided four Ph.Ds. and a record 28 M.Tech dissertations. He also produced more than 50 research papers. Later he also acquired advanced knowledge in Environmental Science.“Dr. Kumar was picked up by the World Bank as a consultant to its projects in Bihar at FMISC, where he served for four years. He is a Fellow of Institution of Engineers India; Chairman Indian Water Resources Society (Patna Chapter).

Dr. Kumar has authored an important book "River Engineering - Theory and Practice". He is also a visiting faculty to Patna University and Magadh University. Being a social activist he coaches poor children (Class VI to X) in Science and Mathematics at his residence.

CONTENTS

PART II : ECOLOGY, ECOSYSTEM AND BIODIVERSITY

CHAPTER – 17 : ENVIRONMENTAL CHALLENGES 331–343

CHAPTER – 18 : HUMAN POPULATION AND THE ENVIRONMENT 344–358

CHAPTER – 19 : RESETTLEMENT AND REHABILITATION (R&R) 359–384

PART VI : ENVIRONMENTAL PROTECTION

PART VII : ENVIRONMENTAL MANAGEMENT

Chapter – 23 : Environmental Management and Environmental Management Plan 448–470

Chapter – 24 : Field Work (Project Work) 471–482

Exercise (Defining, Distinguishing, Short Answer Questions, Expansions, Fill in the Blanks, MCQs, True-False) 483–506

PART I : GENERAL

1

Chapter

An Introduction—Multi-disciplinary Nature of Environmental Studies

1.1 ENVIRONMENT

Literary meaning of the word **'environment'** is the surroundings of an object. Our immediate concern is the environment in which humans live. In a layman's language, we can say that environment encompasses everything that is around us. Environment is a complex of several things (light, water, soil, temperature, air etc.) which surround an organism. Generally speaking, the environment is equated with nature wherein physical components of the planet earth *viz.* land, air, water, etc., support and affect life in biosphere. In the beginning the environment was considered to be consisted of only physical aspects of the planet earth (land, air and water) and biotic communities, but with the passage of time and advancement of society, man extended his environment through his social, economic and political functions.

As explained above, dictionary meaning of the word 'environment' is a surrounding, external conditions influencing development or growth of the people, animals or plants, living or working conditions etc. Here, following questions arises:

(a) What is surrounded? Answer to this question is living object in general and man in particular.

(b) By what surrounded? If man is taken to be surrounded, answer becomes environment.

(c) Where surrounded? Answer to this question is space or habitat.

Primarily, the concern is with the environment of man. But, since man cannot exist or be understood in isolation from other forms of life and from plant life, therefore environment of all biological population should be the concern. Therefore, '*environment refers to the sum total of conditions which surround man at a given point in space and time*'. The impact of man in shaping and bringing environment to its present state is, however, too dominant to be overlooked. Thus, the '*environment is the representative of phyiscal components of the earth, wherein man is an important factor affecting the environment*'.

Our immediate concern is the quality of space we live in, the air we breathe, the food we eat, the water we drink and the resources we draw from the environment to support our economy. Environment is an inseparable whole and is constituted by the interacting systems of physical, biological and cultural elements which are inter-linked individually as well as collectively in many ways. Here,

(a) Physical elements (space, water bodies, climate, soils, rocks and minerals) determine the variable character of the human habitat, its opportunities as well as limitations.

(b) Biological elements (plants, animals, micro-organism and man) constitute the biosphere.

(c) Cultural elements (economic, social and political) are essential man-made features.

The word **'environment'** is taken from the Greeks word *'Environ'* which means surroundings or outside. The environment supplies the required food, nutrient and energy and other living conditions that may influence the organisms directly or indirectly.

The environment is of two types—*living* or *non-living. Living* or *biotic* environment includes all the living beings (plants and animals) with whom the organism or its population or the entire community interacts. The each of the species that is interacting, is part of the biotic environment for the other species. The physical and chemical components that influence the living organism, make up the *non-living* or *abiotic environment.* These components are the soil, atmosphere, water, sunlight, climate and nutrients.

Thus, the types of plants and animals living in a particular region depends upon the physical environment in that region. A change in physical environment brings about a change in biological environment or the *vice-versa.*

Changes in the physical environment, due to passage of time, and corresponding variations in the biological environment have been occuring as normal phenomena during long history of the earth while some of the changes are due to natural processes, others are caused by human activities.

'Environment' can also be defined as:

(i) Everything collectively that affects an organism during its lifetime.

(ii) The aggregate of all external conditions and influences, affecting the life and development of organisms. The term takes into account all variables which directly or indirectly infringe on it. These include physical, social, cultural factors and conditions influencing the existence of development of organism. The five factors, *i.e.*, air, water, land, fire, flora and fauna, co-evolve, co-adopt and are inter-related and inter-dependent.

(iii) The surrounding in which any organisation or living system operates and includes air, water, land, its resources, humans, flora (plants), fauna (animals) and their inter-relationships.

1.2 CLASSIFICATION OF ENVIRONMENTAL FACTORS

As discussed above, environment is a complex of so many things (light, temperature, soil, water, etc.) which surround an organism. Any external force, substance or condition, which surrounds and affects the life of an organism in any way, is known as *environmental factors.* These factors may be living (biotic) as well as non-living (abiotic). Sum of all these living and non-living factors make the environment of an organism.

Environmental factors can be classified in many ways, some of them are as follows:

1. On the Basis of Direct or Indirect Factors

(a) **Direct factors** are light, temperature, humidity, soil-air, soil-water, soil nutrients etc.

(b) **Indirect factors** are soil structure, soil organisms, altitude, wind, slopes etc.

2. On the Basis of Variable Factors or Causes

(a) **Substances**—soil, water

(b) **Conditions**—temperature, light

(c) **Forces**—wind, gravity

(d) **Organisms**—plants (flora), animals (fauna)

(e) **Time**

3. On the Basis of Ecological Factors

(a) **Climatic or Aerial Factors:** These include:

(i) light

(ii) temperature of air (atmospheric temperature)

(iii) rainfall (precipitation)

(iv) humidity

(v) atmosphere (gases and wind)

(b) **Topographic Factors:** These are the factors related to physical geography of earth, and includes the following:

(i) altitude

(ii) direction of mountain chains and valleys

(iii) steepness and exposure of slopes

(c) **Edaphic Factors:** These are the factors which are related to formation of soil, its physical and chemical properties.

(d) **Biotic Factors:** These are the factors which are related to interaction between different forms of life *i.e.*, plants, animals, micro-organisms etc.

4. On the Basis of Living and Non-living Components

(a) **Physical, Abiotic (Non-living) Components:** These include:

(i) solid (solid earth)

(ii) liquid (hydrosphere)

(iii) gas (atmosphere)

(b) **Biotic (Living) Components:** These include:

(i) plants (flora)

(ii) animals (fauna)

5. On the Basis of Spatial Scales

Physical environment can be classified as follows:

(a) mountain (b) plateau
(c) plain (d) lake
(e) river (f) maritime
(g) glacier (h) desert
(i) coastal

6. On the Basis of Physical, Social and Economical Aspects

(a) **Physical Environment:** This includes:
 (i) habitat (space) (ii) air
 (iii) water (iv) food

(b) **Social Aspects:** All the organisms work to form their social groups at several levels. 'Social-man' establishes institutions, formulates laws and policies to safeguard his existence, interest and social welfare.

(c) **Economic Aspects:** Man drives and utilises resources from physical and biotic environment with his skills and technologies. It is this aspect which makes the man an environmental process and transports, the matter and energy from one component of the ecosystem to the other. When this aspect exceed the critical limit, the equilibrium of the environment/ecosystem is disturbed and several environmental and ecological problems arise, which become detrimental to man himself as well as whole population of species in a given ecosystem.

1.3 NATURE

The dictionary meaning of nature is external world untouched by man. Thus, that part of the planet earth which has not been invaded by man is 'nature' or 'wild area'. But actually there is no area on our earth which has not been approached by man, and therefore, we do not possess any space, having virgin physical environment. Therefore, the qualifying conditions for 'nature' or 'wild area' have to be lowered down and those areas should be termed 'nature' which have been least invaded by 'economic-man'.

When the man appeared first on the earth, the whole of the earth was 'wild area' and the physical environment was in natural state but as the population grew and the skill of man increased, more and more 'natural areas' were converted into farms, villages, cities and towns, highways and several economic establishments resulting into rapid rate of shrinkage of 'natural areas'. A group of conscious people, now known as 'environmentalist' has alarmed the economic man against the devastating impact of unplanned, unscientific and reckless exploitation of 'nature'.

Due to growing popular support for wilderness, almost every nation has attempted to protect a few wild areas as 'national park', 'wild life sanctuaries' etc., but these have also been adversely affected by ever-increasing pressure of population growth and recreation, and tourism industry.

1.4 ENVIRONMENT AND RESOURCES

Natural resources are significant for the successful survival of all types of life on the planet earth, in general and for human being in particular. Therefore, use and misuse, conservation and management of all sorts of resources either natural or cultural, renewable or non-renewable are very significant aspects of environment. The very fundamental principle of resources is that resources, even renewable, are finite. For example, air and water are in abundance and are renewable but when they are polluted and degraded, they become unusable and non-renewable because it becomes very difficult to restore the original quality of air and water, if they have been degraded beyond a certain critical limit.

Withdrawal of resources from the nature, mode of their uses by human being and their disposal have severe adverse effect on the environment. It is, therefore, necessary to study the linkages between resource use and the environment so that proper management of resources and remedial measures of environmental pollution and degradation may be made possible.

Natural resources are of prime concern to the present society and the future generations, because (*i*) Majority of the raw materials are being consumed at an alarming rate, which may result in low supply of certain key minerals and even in complete depletion, (*ii*) Withdrawal and use of minerals have immense impact on environment. For example, extraction of minerals create pits and leads to destruction of biotic community. Similarly, withdrawal of groundwater results in the formation of large cavities beneath the earth's surface, which may cause surface collapse and subsidence. Disposal of certain waste materials into water and soil pollutes, which may then be hazardous to organisms.

1.5 ENVIRONMENT AND SOCIETY

Environment and society are inter-dependent. Different social structures like industrial, agricultural, political, cultural, religious, aesthetic etc., have evolved and developed during various stages of development of human civilization, and these are primarily based on natural environment. The existence and quality of environment depends on the responses of these social structures. Several issues other than the quality of environment, like environment degradation and pollution, ecological imbalance, depletion of resources, disruption of ecosystems etc. have been raised during last three decades, but the solution to any of these problems depend upon the entire society and the society's response towards the 'environment improvement programmes.' For example, mass falling of trees to meet the need of timber and formland for ever-increasing population leads to accelerated rate of soil erosion, causing siltation of river beds and consequently recurring floods. Solution to this problem depends upon the willingness of the society to consume the natural environment.

The society has now awakened towards the environmental problems, but there is a need to translate this into practice and efforts are to be made so that this tempo is sustained for a long time.

1.6 ENVIRONMENT AND MAN

In the beginning of the process of evolution of man and his society, physical elements of the planet earth *e.g.*, terrain, soil, water, climate, flora and fauna formed man's environment and man was basically a *'physical man'* because of his limited wants, requirements and total dependence on nature. As the man became social, economic and technological, he broadened his environment by creating his own environment for better food, shelter, access and comfort.

Man felt that, he is superior to all creatures and everything is created for his use and enjoyment. This ideology instigated the man to exploit natural resources and to subdue nature without considering the after-effects of reckless and uncontrolled plundering of natural resources, this approach led to excessive and rapid rate of exploitation of natural resources.

Role of man in the natural environmental system was fundamentally that of the user of environmental resources. As the skill and technology of man developed with the cultural development, his roles towards natural environment also changed from user through modifier and then to destroyer of the environment. Industrial and scientific revolutions have led to rapid changes in our environment. Materialistic outlook of man have also played significant roles in changing man – environment relationships on a large scale. Man is changing the environment for his vested interests to such an extent that even the very existence of human beings is threatened.

The man's impact on environment falls into two broad categories:

1. **Direct or Intentional Impacts:** These are preplanned, and their consequences, both positive and negative are known. Changes launched for modifying the natural environment for economic development include:
   (a) land use changes e.g., clearing of forests for crop cultivation,
   (b) felling of trees for commercial purposes,
   (c) changes in cropping pattern for adopting new farming techniques, new high yield seeds, irrigation facilities etc.
   (d) construction of dams, reservoirs, canals, construction of roads and bridges, mining, drilling of mineral oils, withdrawal of groundwater etc.,
   (e) increase in urbanisation,
   (f) use of chemical fertilisers, pesticides and insecticides,
   (g) weather modification programmes like, cloud seeding dispersal and clearing of clouds and fogs, checking of hailstorms,
   (h) nuclear programmes etc.

   It is important to point out that the effects of such changes in the natural environment are noticeable, but these effects are reversible because both, before and after studies may enable the man to set the adverse effects right to certain extent.

2. **Indirect or Unintentional Impacts:** These arise from those human activities which are directed to acceleration the pace of economic growth, especially industrial development. These indirect impact are not immediately noticeable because these changes take the long time to cross the sensitivity of the ecosystem. These effects may change the overall natural system and the chain effects sometimes become suicidal for human being. Majority of indirect impacts of such activities are related to pollution and environmental degradation. Examples of this category are:
   (a) Release of toxic elements into the ecosystem through their uses as insecticides, fertilizers etc.

(b) Release of industrial wastes into stagnant water, rivers and seas contaminates water and causes several diseases and deaths of organisms and thus, disturbs ecological balance.

(c) Urbanisation, industrial expansion and long use changes very often change weather and climate though in long-time perspective.

(d) Burning of hydrocarbon fuels has increased the concentration of carbon dioxide in the atmosphere. This may change heat balance of the atmosphere, because CO_2 allows more absorption of incoming short wave solar radiation and outgoing longwave terrestrial radiation.

(e) Depletion in ozone layer. Ozone layer, concentrated mostly at the height of 15 km to 36 km, absorbs ultraviolet rays of the Sun and thus, protects the earth from becoming too hot. Thus, depletion in ozone layer means increase in the temperature at the earth's surface. This also causes skin cancer, immuno-suppression in human body, decrease in water use efficiency and decrease in crop yield.

1.7 ENVIRONMENTAL STUDIES

Definition

Environmental studies provide an approach towards understanding the environment of our planet earth and the impact of human life upon that environment. It also provides a search for solutions to the environmental problems.

Environmental science deals with environmental components—air, water, soil and living beings, their characteristics and inter-relationships. It includes not only the study of physical, chemical and biological characteristics of the environment but also the social and cultural factors and impact of man on the environment. The components of environment are used and protected by adopting technological measures which are studied under *environmental engineering*.

1.7.1 Importance

Environment belongs to all and is for all. Every body is affected by environment and the environment is also affected by human activity. Knowledge of ecology and environment is very important to learn about the nature and the values of living in tune with the nature which is very necessary for leading a peaceful, happy, satisfied and a healthy life.

For existence of life on the earth, there are complex interactions between and within the environmental components, such as atmosphere, hydrosphere, lithosphere, biosphere and social components namely population, technology, economics and politics. The study of environmental science helps in refining the lifestyle of people towards protecting the environment.

Effective utilisation of nature's capital that contains air, water, soil, energy, mineral resources and solar energy with services, such as pollution control, nutrient recycling, climate control, pollution control, waste management, pest control, disease control and biodiversity conservation is necessary to maintain sustainability.

Environmental studies deal with the most important issues, like safe and clean drinking water, hygienic living conditions, clean and fresh air, fertile land, healthy food and development that is sustainable.

The main aim of environmental studies is to reduce or minimise the impact of human activities on the environment by making efforts to avoid the overuse, misuse or abuse of environmental resources. The main basis of drawing an environmental strategy should be *(i)* environmental planning, *(ii)* environmental status evaluation, *(iii)* environmental impact assessment, and *(iv)* environmental legislation and administration.

Environmental concerns have been an integral part of Indian traditions and recognised in the religious books, like *Vedas* and *Upanishads*, which stressed the need for conservation and sustainable use of natural resources. Further, importance of these concerns are evident in the constitution of India under the Directive Principles which provide that the state shall endeavour to protect the environment and safeguard the forests and wildlife of the country. The constitution also provides that it shall be the fundamental duty of every citizen to protect and improve the natural environment.

The importance of environmental studies have been felt by realising environmental problems of local, regional and global nature. There are some major issues, like global warming, depletion of ozone layer, greenhouse effect, acid rains, dwindling forests and energy resources, loss of global biodiversity etc., that are going to affect the mankind as a whole.

1.7.2 Scope

Earlier, Ecology (as a branch of biology with detailed study of zoology and botany) and Geography were studied for understanding the environment. During the last three decades, importance of environmental sciences have been recognised as an indispensable tool for creating and maintaining the quality of human environment.

Study of natural ecosystems alongwith human ecosystems is considered as relevant, since it helps in understanding the environmental problems created by humans. Ecology deals with living and non-living entities in relation to their surroundings in the total environment. Living environment consists of men, animals, birds, fish, insects, micro-organisms, fungi, algae, plants etc. Soil, water, air and nutrients constitute the non-living components. The Sun is the chief source of energy in the entire system. These are the essential resources for the living environment.

From a human perspective, environmental issues involve concerns about science, nature, health, employment, economics etc. Besides life sciences, environmental studies require knowledge of physical sciences, such as physics, chemistry, geography, geology, meteorology, economics, commerce, sociology and psychology. Applied sciences, like technology, especially biotechnology, genetic engineering. Management studies, especially resource management, environment management, pollution management, wildlife management, remote sensing, soil science, hydrology, oceanology etc.

Environmental issues have today drawn the attention of common people, since environmental degradation and pollution have assumed global dimension and are even threatening the very existence and survival of mankind. This has cautioned the scientists, administrators, academicians, social workers and common people alike against environmental hazards and disasters.

It is, therefore, desirable to understand interactions between technologically advanced 'economic man' and the environment and the processes, leading to the environmental degradation and pollution, to evaluate ecological resources, to carry out successfully the environmental impact assessment is successfully carried out, pollution abatement programmes are scientifically formulated and implemented, and environmental planning and management programmes are properly executed to maintain ecological balance and ecosystem stability at local, regional and global levels.

From the above, we can conclude that comprehensive coverage of subjects that underlines conservation of natural resources and strategies for sustainable development is essential to developing countries as well as to prosperous industrialised nations.

1.8 NEED FOR PUBLIC AWARENESS

As discussed earlier, environment is the life support system. The nature is the valuable asset that provides all the basic requirements, such as food, air, water and shelter for the livelihood of living beings. All these components of the environment are the resources which are being continuously utilised by the living organisms for their survival. These resources are replenished, some at slower rate and some at relatively faster rate.

In the earlier stages of civilisation, the natural resources were plenty and in their purest form. During the last over more than a century, man, however, instead of learning to live in tune with nature, has been distancing himself from nature in the name of scientific development and under an urge to conquer nature. This has resulted in a large scale adverse impact on the ambient environment, causing severe adverse effects on human health. The continuing large scale urbanisation, industrialisation, commercialisation and agricultural activities has caused such vast local, regional and global adverse impacts on our environment that the very existence of man has come under threat.

The magnitude of the demand made by man in pursuit of satisfying both, his basic needs and greed is increasing. The living conditions, though have surely improved today, to a great deal due to industrialisation and urbanisation, but only at the cost of environmental degradation. The resources are not only getting depleted, but are also getting polluted and wastes are being generated. With the growing concerns about the environment, issues regarding hazardous waste, global climate change, ozone depletion, groundwater contamination and disaster mitigation have become the focus of environmental attention in recent years. This has been attributed to the lack of awareness amongst masses about the problems and to the unwise and exploitative use of natural resources. Therefore, attempts are being made to create greater awareness by disseminating the knowledge and information on how to protect our environment through formalising various courses for students of different standards.

In this context, sustainable development assumes a great significance, which can be defined as **"the development that meets the needs of the present without compromising the ability of the future generations to meet their own needs."** This can be done only when there is a public pressure on the industry, trade and pressure, on the other hand, will be developed only when the public knows the values and the importance of the environment, and of the harmful effects being caused by the environmental pollution. It is, therefore, extremely necessary to import atleast basic knowledge of the environmental studies to all people, and more particularly to the coming generation of the student community.

The pivotal role of legal instruments, mass media for promoting public awareness and education is, therefore, now fully recognised. The public has to be educated about the fact that if we are degrading our environment, we are actually harming our ownselves. To ensure a meaningful public participation and involvement, it is necessary to create awareness about the environment, pollution and other related implications and adverse effects on health. We have to feel that the rivers, jungles and the entire environment is ours and for us and can be preserved by us only. Thus, it will result in a feeling of owness and it is our duty to preserve the environment.

The civic development activities concentrated only on the economic and social aspects, which degraded the quality of the environment. This is referred to as unsustainable development, where no concern was given to maintain the quantity and quality of resources. This trend has now reached at such a damaging state that we are facing a scarcity of resources and they would not be sufficiently available for use by the future generation.

The environmental education is, therefore, necessary to increase the public awareness of the problems in this field as well as possible solutions, and to lay the foundation for a fully informed and active participation of the individual in the protection of environment and the prudent and national use of resources.

The U.N. conference on the Human Environment (Stockholm, 1972) took the resolve to defend and improve the environment for present and future generations and emphasized the role of education in combating the threats to the human environment.

Thereafter, First Inter-governmental Conference on Environmental Education (Tbilisi, 1977) discussed broad environmental issues and recommended Environmental Education to cater to all ages and socio-professional groups in the population.

Similarly, the United Nations Conference on Environment and Development held at Rio de Janerio (Earth Summit) in 1992; and World Summit on Sustainable Development (WSSD) at Johannesburg in 2002, have made it clear that no citizen of the entire world can afford to be ignorant of the existing environmental issues and ever increasing problems of environment and health-care, threatening the very survival of human race.

Thus, keeping such factors into consideration, the Honourable Supreme Court of India, through one of the landmark judgement delivered in 2003, has directed the University Grants Commission (UGC) to prescribe a course on the subject of "Environmental Studies" for all the Colleges of India and to consider the feasibility of making it a compulsory subject at every level of school and college education. In compliance with the said order, the UGC has designed a course for a new subject on "Environmental Studies", which shall compulsorily be taught at every under-graduate level of the college education.

1.9 ENVIRONMENTAL EDUCATION

The issue of "Environmental Education" has been thoroughly discussed at several national and international seminars, workshops and conferences since 1970s, may it be U.N. Conference on Human Environment at Stockholm in 1972, International Workshop on Environmental Education at Belgrade (Yugoslavia) in 1975, organised by UNESCO or International Conference on Environmental Education by UNESCO at Tbilisi (USSR) in 1977 or International Conference on Environmental Education held in New Delhi in 1980 and again in 1985.

Most relevant saying on the subject is "If you plan for one year, plant rice, if you plan for ten years, plant trees, but if you plan for one hundred years, educate the people." In fact education is to be regarded as investment of the highest order, which in time to come, becomes an asset.

Goals of Environmental Education as decided at the UNESCO, 1975, are to develop a world population that is aware of and concerned about total environment and its associated problems, and commitment to work individually and collectively towards solution of current problems and the prevention of new ones.

The Main Objectives of Environmental Education as formulated at the Tbilisi Conference in 1977 are: individual and social groups acquire awareness and knowledge, develop attitudes, skills and abilities and participate in solving real-life environmental problems. The perspective should be integrated, inter-disciplinary and holistic in character.

There is need for education about environment to the public in rural, tribal, slum and urban areas, women, students and teachers in schools, colleges and universities as well as planners and decision-makers, programme implementors and R&D workers.

There are a number of government as well as non-governmental organisations that are engaged in environmental studies. These are being controlled/co-ordinated by the Ministry of Environment and Forest (MoEF), Government of India through Department of Environment, Forests and Wildlife.

Following are the *important recommendations of the Tbilisi Conference* on Environmental Education:

1. Environmental Education should develop understanding of the complex relation between socio-economic development and the improvement of the environment.
2. It should aim basically to succeed in making individuals and communities to understand the natural and built environment resulting from the interaction of the biological, physical, social, economic and cultural aspects, and acquire the knowledge, values, attitudes and practical skills to participate in a responsible way in solving environmental problems and in the management of the quality of the environment.
3. It should consider the environment in its totality, be a continuous life-long process, be inter-disciplinary in its approach, examine major environmental issues from local, regional, national and international points of view, focus on current and potential environmental solutions, relate the studies to plans for development and help to discover the symptoms and real causes of environmental problems.

For Environmental Education, *central environmental* issues are:

(a) The impact of emissions from production units carried through air or water with the humans, animals and plants at the receiving end,

(b) The safety and health risks connected with the use of agricultural and industrial products and their impact on environment, when used; and

(c) The effect of exploitation of natural resources.

The most generalised issue of environmental protection involves:

(i) Waste disposal, and

(ii) Recuperation, replenishment and recycling.

1.10 ENVIRONMENTAL PROBLEMS

Environmental problems arise as negative effects in the process of development and massive deforestation, which threatens the sustenance of healthy environment. Poverty, inadequate resources and environmental illiteracy are the basic sources, paving the way for massive deforestation, which threatens the sustenance of the forest base. The basic human needs, food, fuel, shelter and employment for a large section of our population impinge on scarce resources which ultimately lead to greater poverty. Systematic planning also plays a crucial role in protecting the environment. Education has a great role for generating awareness among the masses and removing the problem of environmental illiteracy.

In the aftermath of Rio (United Nations Conference cn Environment and Development, better known as *Earth Summit* or simply *Rio Summit* was organised in June, 1992 in Rio De Janario city of Brazil for the protection of the earth and its environment, maintenance of ecological balance and to enrich biodiversity), the concept of sustainable environment has become the prime focal point of global environmental debate.

Growing population, human habitation, loss of forest resources have changed the monsoon and weather routine into an abnormal tendency, hence, by threatening human habitation. Sustainable development is a goal that invites the attention of policy-makers and scientists to investigate various aspects for checking the environmental problem. It refers to increasing economic activity, awareness of mass, purposive co-operation and involvement to feel and realise the true essence of global environmental change by concealing socio-economic barriers. Acid rain and growing forest fires would be the major concern for our country. It causes ozone depletion and global warming, engulfing the globe.

Industrial world is responsible for emission that led to global warming, the effect of which will be felt all over the globe by declining growth in harvesting, growing water shortage, drought and rising sea–level. Environmental pollution deteriorates working conditions by affecting air and water base which ultimately brings drought, famine and cyclone.

Presently, the requirement for facing the dangerous consequence of environmental degradation would be by promoting social forestry, plantation to protect the sound ecological base in our country.

Environmental problems being faced in India can be divided into following categories:

1. *Type I — Environmental Problems:* These are the problems arising out of lack of appropriate development, such as deforestation due to fuelwood extraction, unhygienic slums, soil erosion, lack of sanitation and polluted water supply etc. If the pace of development is slow, the Type I problems will intensify.
2. *Type II — Environmental Problems:* These are those problems, which are caused by lack of environmental safeguards in development projects, such as air and water pollution, hazardous waste disposal, water–logging in irrigation projects, pesticide pollution, fertilizer run off into streams etc. If sufficient investments are not made in development projects to compensate for environmental degradation, these problems will increase.
3. *Type III — Environmental Probems:* These are global environmental problems, such as global warming, ozone depletion, marine pollution, loss of genetic diversity, automobile pollution etc. These problems are created by the current pattern of development, life-style and intensive energy use, etc.

The existence of the first two types of environmental problems make the situation a different one to handle. If we do not develop rapidly, the Type I problems will overtake us. If we do care for Type II problems, they will overwhelm us. Environmental management issues have to be seen in this context. Hence, the options open for India, have to be examined in terms of two parameters *i.e.*, economic growth and degree of environmental protection.

QUESTIONS

1. What is an environment? Define and explain.
2. Discuss the different components of environment. How will you classify various environmental factors?
3. (a) Distinguish between 'nature' and 'environment'.
   (b) Discuss the importance of environment.
4. Discuss the relationship between:
   (a) Environment and resources.
   (b) Environment and society.
5. Write an essay on 'Environment and Man'. Also, state the impact of man on environment.
6. What is 'environmental science'? Explain the importance of 'environmental studies.'
7. Discuss the scope of environmental studies in detail.
8. Explain the need for 'public awareness'. What are the ways to fulfill this need?
9. What is environmental education? What are its activities?
10. Discuss the historical background of 'environmental awareness.'
11. Discuss 'goals' and 'objectives' of environmental education.
12. Narrate the causes of environmental degradation.
13. What is sustainable development? Discuss about its principles.
14. Describe different environmental problems.
15. Discuss the following environmental problems in brief:
    (a) Over–population (b) Pollution
    (c) Depletion of resources (d) Global warming
    (e) Green house effect (f) Depletion of ozone

2
Chapter

THE EARTH AND THE BIOSPHERE (THE EARTH SCIENCES)

2.1 INTRODUCTION

The Earth is a unique planet of the solar system, as it has conditions favourable for the evolution and survival of various forms of life, including humans. This is possible because of optimum distance of the earth from the Sun, availability of water and air. The temperature on the earth is neither too high, as on Venus and Mercury nor too low, as on Jupiter and other distant planets. The moderate temperature conditions enable water to be present in large quantities on the earth. The temperature variations due to the heating of the Sun enable water to be present in solid and gaseous state also, enabling its circulation. The layer of air which surrounds the earth, contains oxygen which is essential for all forms of life. The air envelope also moderates the temperature on the surface of the earth. This enables the temperature to remain within moderate limits between day and night or between summer and winter. The circulation of air and water brings about changes in the climate conditions in different seasons.

The science which deals the study of the earth is known as Geology. The earth is one of the nine planets of the solar system and revolves around the Sun through space. The earth is the only home for human beings, animals, plants and other organisms. It is the fifth biggest planet in the solar system, and is neither the heaviest nor the fastest member of the solar system—the universe. Besides Sun, these eight major planets—Mercury, Venus, Earth, Mars, Jupiter, Saturn, Uranus and Naptune, there are 32 moons or satellites. These eight major planets and their moons are revolving nearly on the same plane around the Sun. In addition to this movement around the Sun, these planets are undergoing at constant rotation about their axis.

2.2 ORIGIN OF THE EARTH

There are two schools of thoughts about the origin of the earth, or strictly speaking the origin of the solar system. First school, also known as *single body hypothesis,* suggests that the planets were formed without intervention from forces outside the solar system. This hypothesis assumes the existence of a large globular mass of gas, slowly rotating through space. This gaseous mass, called a *nebula*, gradually become smaller in response to its inherent gravitational pull, and as it contracted, its speed of rotation increased. The outermost part of the nebula rotated so rapidly that rings of gas got separated from the shrinking nebular body. These rings then slowly got condensed to form the planets, including the earth and the central mass became the Sun.

Second school of thought assumes that the solar system originated as a result of forces created by the accidental interference of another celestial object. This is also known as second body hypothesis or the two-star theories. According to this at sometime in the remote past, an 'intruder' star passed by the sun, thereby exerting a gravitational force on the sun, strong enough to pull masses of solar material from opposite sides of the sun. This detached material was dragged along, in the direction of movement of the passing star, and later got cooled and condensed into solid particles, each persuing its own orbit around the Sun.

Thus, we see that before coming into present shape, it was a fire ball, surrounded by the clouds of hot gases. It took millions of years to cool down and the envelope of hot gases turned into clouds. These clouds rained for a long time and the rain water accumulated in the low lying areas of the earth, which turned into oceans.

The universe began to form over 15,000 million years ago. The scientists with the help of radioactivity and fossil records have calculated the age of the earth to be about 4.5 to 5 billion (4500 to 5000 million) years.

In the beginning, the earth was a giant landmass, called 'Pangaea'. This landmass gradually split into different landmasses, which are called *'continents'*. In the process of cooling, the upper surface of the earth became solid which we call as the earth's crust. About 570 million years ago, the life began on the earth in the form of micro-organism. The first 345 million years saw the development of acquatic life. For the next about 160 million years, reptiles came into existence and subsequent 65 million years saw the development of mammals. The development of man is only an event of one million years old.

The life of the earth is dependent on the sun. It will probably end, as when the sun's nuclear fuel gets exhasuted and it stops radiating energy. It has been predicted that the life on the earth will further continue to flourish for atleast 5 billion years. This time period has been based on the assumption, the Sun is likely to radiate for atleast 10 billion years, before the hydrogen fuel of its core gets exhausted.

2.3 CHARACTERISTICS OF OUR SOLAR SYSTEM

Following are the main known characteristics of our solar system:

(i) All the planets revolve around the sun in the same direction, in nearly circular orbits, all of which lie almost in the same plane.

(ii) The moons (satellites) of the different planets exhibit the same regularities, except for a few outer satellites that move in the opposite direction.

(iii) The sun is also rotating in the same direction as the planetary orbits.

(iv) The velocity of rotation of the sun is very small as compared to those of other planets. Infact, inspite of possessing a mass more than 99% of the mass of the entire solar system, it possesses an angular momentum (angular momentum of a rotating body is the product of its mass m, velocity v and radius of orbit R *i.e.* mvR) equal to only 2% of the total angular momentum of the entire solar system.

(v) Although the different planets are uniformly related to the sun, yet the comets are not uniformly connected. Comets move in highly elliptical orbits that are strongly inclined to the orbital plane of the planets.

Do you Know?

1. The Earth's rotational velocity at the equator	1690 km/hr
2. A solar day is greater than a sidereal day by	4 minutes
3. Each degree of latitude equals	111 km (approx.)
4. A person crossing International Date Line from East to West	loses one day
5. The sun is vertically overhead at the equator on	March 21
6. Indian Standard Time accepted Meridian	82.5° E, passing through Allahabad
7. The Earth's tilted axis causes	seasons
8. IST is ahead of GMT by	5½ hours
9. The Earth is round, was first theorised by	Plato

2.4 THE EARTH'S SHAPE

Years of accumulated knowledge, experience and observations have conclusively established that the earth is spherical in shape. However, it is not a perfect sphere, since its equatorial circumference is greater than its polar circumference. The true form of the earth resembles a sphere that has been compressed along the polar axis and made to bulge slightly around the equator, just as in an orange which is a little flattened at both ends. This form is known as an *oblate ellipsoid*. A cross-section through poles resembles an ellipse rather than a circle. The oblateness is caused by the centrifugal force of the Earth's rotation, which deforms the somewhat plastic the Earth into a form that is in equilibrium with respect to the forces of gravity and rotation.

Some of the dimensions of the Earth are:

- Polar diameter — 12714 km
- Equatorial diameter — 12756 km
- Circumference — 40,075 km
- Surface area — about 510 million square km
- Volume — about 1042 billion Cu km
- Mass — about 5.97×10^{21} tonnes
- Relative Density — 5.52

2.5 TILT OF THE EARTH'S AXIS

The axis of the Earth is inclined to the plane of *ediptic* (the plane in which the earth orbits round the sun) at an angle of 66.5°, giving rise to different seasons and varying lengths of day and night. If the axis were perpendicular to this plane, all parts of the globe would have had equal days and nights at all times of the year. During one half of the year, each hemisphere is tilted towards the sun, while during the remainder of the year, each hemisphere is tilted away from the sun. The sun is vertically overhead at the equator on two days each year *i.e.*, on March 21st and September 23rd, and on these two days all places on the Earth have equal days and nights. On June 21st the northern hemisphere have its longest day and shortest night. On December 22nd, the southern hemisphere will have its longest day and shortest night.

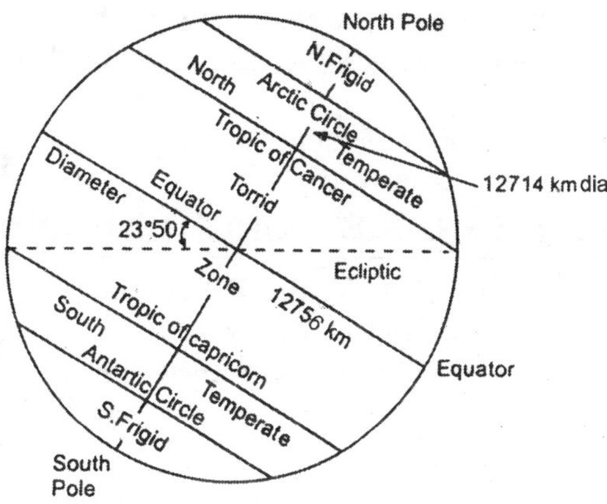

Fig. 2.1: Spheroidal earth.

2.6 THE EARTH'S MOVEMENTS

The Earth rotates on its own axis that is tilted at an angle of 23.5° to the vertical. It spins from west to east once in every 24 hours (23 hrs. 56 min./sec), on its own axis, causing day and night. The linear velocity of rotation varies, achieving a maximum of 1690 km/hr at the equator, reducing gradually to zero at the poles.

The Earth revolves round the sun once in every 365¼ days. This, coupled with the tilt of the Earth's axis, causes seasons. Since, it is not possible to show a quarter of a day in the calender, a normal year is taken to be 365 days, and an extra day is added every four years in a Leap year. The Earth revolves on an elliptical orbit at a speed of 106,560 km/hr. When viewed from a point in space, the Earth moves counter-clockwise.

Besides rotation and revolution, these are three periodic deviations in the relative position of the Earth and the Sun, which affect the seasonal distribution of solar radiation on the Earth. These are : *(i)* eccentricity of the Earth's orbit about the sun, *(ii)* changes in obliquity of the Earth's axis, and *(iii)* precession *i.e.*, wobbling of the Earth's axis.

(i) **Eccentricity:** The Earth's orbit undergoes a change of shape during a cycle of 90,000 to 100,000 years. The orbit, sometimes, forms a longer ellipse and at other times, it returns to a more circular shape. This results in variation of distance from the Earth to the Sun affecting terrestrial temperature. Ice Ages and Global Warming are sometimes argued as effects of this eccentricity.

(ii) **Obliquity:** Presently, the Earth's axis is inclined at an angle of 23.5° to the vertical. Obliquity is the variation in the angle of inclination of the Earth's axis from 22.1° to 24.5° during a cycle of about 41,000 years. As a result of the variation in the angle between the Earth's and the ecliptic, the Tropics and Polar circles are displaced. Changes in the inclination of this angle influence temperature differences between the winter and summer.

(iii) **Precession:** Like a spinning top, the rotating Earth also wobbles on its axis. At present, the axis of Earth points towards the North Star. But by the year 14000 AD the axis will be pointing towards the star Vega. The period of precession is about 26000 years. This cyclic wobbling of the axis will trigger the climatic change of great magnitude. The tilt of the axis towards will be about 12,000 years. The northern hemisphere will experience winter when the sun is farthest from Earth and summers when the Earth is nearest to the sun. Thus, summers will be warmer and winters will be colder than they are at present.

2.7 LATITUDES AND LONGITUDES, LONGITUDE AND TIME

In order to precisely locate any place on the Earth's surface, imaginary lines have been drawn on the globe. One set of lines, running east-west, parallel to the equator, is called *lines of latitude*. The other set runs north-south passing through the poles and is called *lines of longitude*. The intersection of latitude and longitude pinpoints any place on the Earth's surface.

Latitude is the angular distance of a point on the earth's surface, measured in degrees from the centre of the Earth. As the earth is slightly flattened at the poles, the linear distance of a degree of latitude at the pole is a little longer than that at the equator. Average distance of about 111 kms is taken per degree, which can be used for calculating approximate distances between the places.

Longitude is the angular distance, measured in degrees along the equator, east or west of the *Prime Meridian* (the meridian that passes through *Greenwich* near London).

Longitude have very important function to play. They help in determining the local time in relation to *Greenwich Mean Time* (GMT). The Indian Government has accepted the meridian of 82.5° east, which passes through Allahabad as the standard meridian, which is 5 hours 30 minutes ahead of GMT.

2.7.1 Longitude and Time

Since the Earth makes one complete revolution of 360° in 24 hours, it passes through 15° in one hour or 1° in 4 minutes. Since the Earth rotates from west to east, every 15° we go eastwards, local time advances by 1 hour. Conversely, if we go westwards, local time is retarded by 1 hour. Thus, the places east of Greenwich see the sun earlier and gain time, whereas the places west of Greenwich see the sun later and lose time. Most countries adopt their standard time from the central meridian of their countries. Larger countries like USA, Canada and Russia having greater east west stretch have to adopt several time zones.

A traveller, going eastwards, gains time from Greenwich until he reaches 180°E meridian, where he will be 12 hours ahead of GMT. Similarly going westwards, he loses 12 hours till he reaches 180° W. There is, thus, a total difference of 24 hours or a whole day between the two sides of the 180° meridian. This is the International Dateline, where the date changes by exactly one day when it is crossed. Therefore, a traveller, crossing this line from east to west, loses a day and while crossing the dateline from west to east, gains a day. The International Dateline is in the mid-pacific curves from the normal 180° meridian at the Bering strait, Fiji, Tonga and other islands to prevent confusion of day and date in some of the island groups that are cut through by the meridian.

2.8 COMPONENTS OF ENVIRONMENT

The environment as a whole includes the lithosphere, hydrosphere, atmosphere and biosphere. These environmental components have different physical, chemical and biological characteristics and they interact with each other and maintain a dynamic balance for sustainability. Through the natural cycles, the components are continuously transformed into various forms to support life.

The environment consists of three components, *viz.* (*i*) abiotic or inorganic or physical component, (*ii*) energy component, and (*iii*) biotic or organic component. In this chapter, we are concentrating on abiotic component, which includes mainly the lithosphere, the hydrosphere, and the atmosphere. Generally, these are considered as land or soil, water and air respectively.

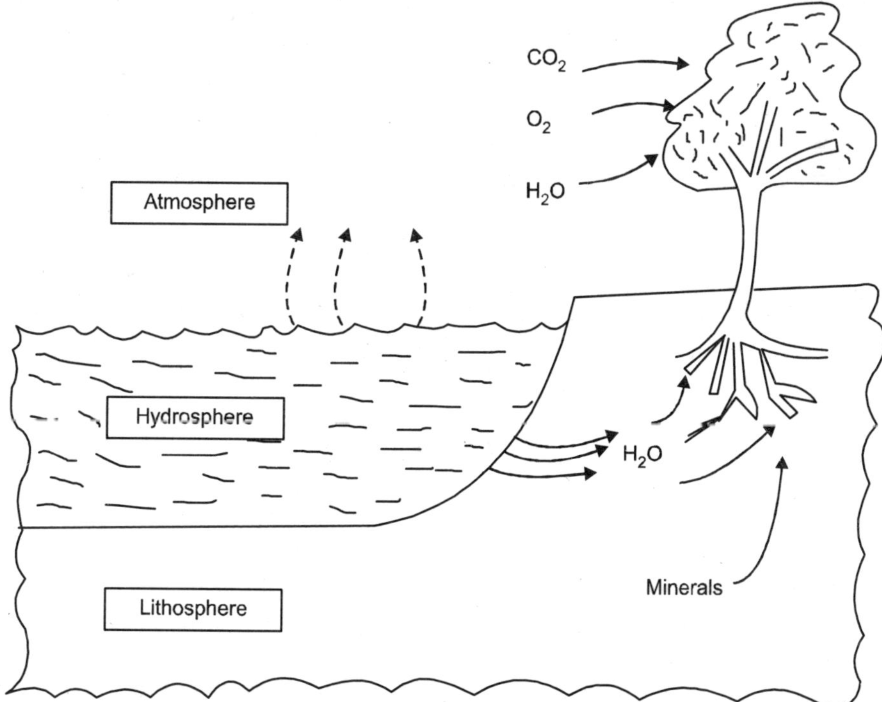

Fig. 2.2: Contribution of each of the three components of environment for the maintenance of living matter.

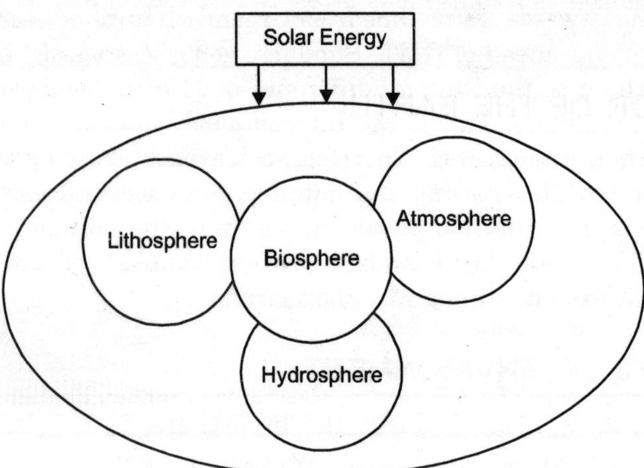

Fig. 2.3: Interaction of environmental components.

2.9 LITHOSPHERE

Lithosphere literally means a sphere or a zone of rocks. Thus, **lithosphere** refers to the solid layers of rock material on the earth's surface, both, on the continents and ocean floors. It is made up of rocks in great variety and on the land, it is commonly covered by a blanket of soil or other loose deposits, such as desert sands. Its average thickness is 30 to 40 km under the continents, 5 to 10 km under the ocean floor and upto 60 km under the mountains.

The crustal layer is of the lighter density as compared to the interior layers. The average density of the whole Earth is 5.5, whereas that for the rocks making most of the upper crust of the earth is 3, which infers that the materials beneath the surface is of a much higher density. The temperature increases with the depth. The hot *lava* erupted in the *volcanoes* also indicate that the deep interior parts of the earth are very hot. The pressure has also been observed to be increasing with depth. Experiments have shown that the temperature increases at the rate of 10°C for every 32 meters. With this rate of increase, all the matter at the core should have been in a molten state, but since the pressure at such great depths is also very high, the melting point is higher than the prevalent temperature and the inner core is in a solid state. In the outer core, however, the pressure is lesser and the matter is in a liquid state.

Rocks occurring in the crust fall into two groups:

(a) **Light rocks**, which include granite and related types. These rocks have a specific gravity of about 2.7. These rocks are very rich in silica, while alumina is the most abundant of the remaining constituents. Such rocks are known by the term ***SIAL*** (derived from first two letters of silica and alumina respectively).

(b) **Dark and heavy rocks**, which include basalt and related types having density ranging from 2.9 to 3.4. In these rocks, silica (40 to 50 percent) is leading constituent, while magnesia takes up the second place. Thus, the whole group is called *SIMA* (from silica and magnesium).

The SIAL is the dominant material of the continent crust, down to a depth of several kilometers. The SIMA forms the foundations of the ocean floor. It extends beneath the continents and is visualised as a continuous layer. In fact the continents are visualised to be floating on the SIMA below.

2.10 THE INTERIOR OF THE EARTH

The entire 6370 km thickness of the earth can be divided into the following major zones:

(a) the crust,

(b) the mantle, and

(c) the core

These different zones are discussed below:

(a) The Crust

As discussed above, the topmost thin layer of the earth's body, having a solid thickness of about 30 to 40 km in continents and about 5 to 6 km in the oceans, is known as the **crust**. Below the crust, there is *M–discontinuity* (Mohorovicic discontinuity) boundary between the crust and the M–discontinuity.

If in the continents, the total depth is about 35 km, out of which the bottom 5 km depth consists of denser Basalt rock (density = 3.0), the top 30 km consists of lighter Granite rock (density 2.7). There is no such granite beneath the oceans, and only 5 to 6 km Basalt rocks occur. These Granitic rocks of the continents and the Basaltic rocks of the oceans are generally covered by a top layer of unconsolidated sediments (about 1 km thick) deposited since the formation of these continents or oceans.

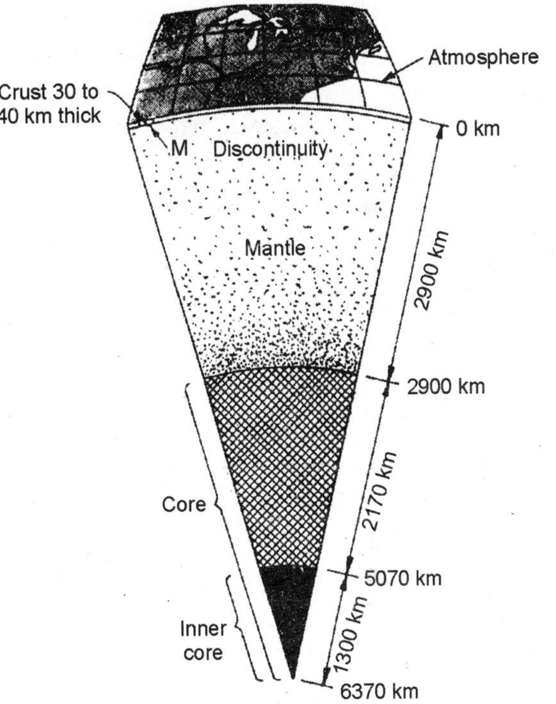

Fig. 2.4: Wedge of the Earth, showing the major zones and boundaries.

The Earth's crust, thus, provides us very useful hard and soft rocks, which are classified as—Igneous, Sedimentary and Metamorphic rocks.

The upper Granitic and the lower Basaltic layers of the crust are separated by a seismically determinable boundary called the *Conrad discontinuity*. The uppermost portion of the mantle on which the crust rests, is rigid. The crust and this solid uppermost mantle are together called *lithosphere*.

(b) The Mantle

Beneath the M – discontinuity, still denser rocks (average density of 3.3) known as *ultra basic rocks* are existing. This portion of the earth, having thickness of about 2860 km is known as **mantle**. Upper mantle, below the lithosphere, becomes so hot that it resembles hot plastic, it can be made to change shape and can be moulded. The soft plastic layer in the upper mantle is called the *asthenosphere.* The hot plastic character of the asthenosphere and the convective heat cells facilitate the movement of lithospheric plates overlying this layer.

It is not possible for us to make any direct examination and testing of the material, which lies at a depth of more than 10 km, therefore, much of the information what lies below the surface of the earth down to its centre has come from indirect geophysical methods. The most important evidence about the nature of the earth's interior, comes from the study of the earthquake waves. The energy released during disturbance in the crustal rocks is transmitted through the rocks in the form of *seismic* (earthquake) waves.

The mantle is suggested to have a mixture of iron – magnesium – silicates, much of which is probably a mineral. The mantle is supposed to be in a rigid, solid condition due to very high temperature and pressures. The mantle behaves as a slowly flowing plastic body. Its density is between 3.1 to 5.00. The mantle continues upto a depth of about 2860 km, where another boundary of abrupt change in behaviour of the seismic waves is met with. This boundary is known as *Gutenburg discontinuity*. Beyond this boundary lies the core of the earth.

(c) The Core

The **core of the earth** consists of metals in liquid or plastic state because of high temperature and pressure. The core of the earth has a radius of about 3400 km. As nickel (Ni) and iron (Fe) are dominant in the core, it is called *Nife*. As one goes down into the interior of the earth, there is progressive increase in temperature. The increase is not uniform at all depths and at all places. The average rate of increase of temperature is about 1°C per 32 metres of depth. However, the temperature increases at the following rate:

(a) Upto the depth of 100 km — 12°C per km

(b) From 100 to 300 km depth — 2°C per km

(c) Below 300 km depth — 1°C per km

From the above, we see that the core of the earth should be liquid or gaseous. But, the pressure of overlying layers on the earth's interior has forced to acquire the properties of a solid. Therefore, the earth's interior is in a plastic state. It is rigid and behaves mostly as a solid down to a depth of 2900 km. Whenever the pressure is released, the rocks become molten. The materials come out through volcanic eruptions proved that the core is in liquid form: Volume of these layers of the earth are : Crust – 0.5%; Mantle – 16%; and Core – 83%. Its average thickness is 2900 to about 6400 km and density is 5.1 to 13.0.s

Because of this core, the earth is having magnetic field, and earth's magnetic field has a north pole and a south pole.

2.11 CONTINENTS

There are seven large continuous landmasses, called **continents**. They are : Asia, Europe, Africa, North America, South America, Australia and Antarctica. According to the German Geologist, Alfred Wegener, once, the earth was a single landmass, known as *Pangaea*.

Nearly 200 million years ago, this landmass broke into two parts—Laurasia and Gondwana land. According to Wegener's *Continental Drift Theory*, these two large landmasses continued to drift away from each other. Present North America and Eurasia (Europe and Asia) emerged out of Laurasia, while the South America, Africa, Australia and Antarctica owe their existence to the Gondwana land. According to this theory, these solid continental masses of the Earth's crust are assumed to be moving over the underlying viscous fluid in the mantle. These masses, thus, act like floating bodies, which may drift from one place to another under the influence of forces, like tidal forces.

Since east coast of South America fits like a jigsaw puzzle into the west coast of Africa, which clearly exhibits that both those continents were together in the ancient past.

Continental shelf extend from land to about 300 km or so. The floor shows a gentle slope, the sediments deposited in this zone are well-sorted with coarser particles in the initial shallower zone, and finer particles towards the deeper portions. The maximum depth in this zone is about 180 metres or so. The upper part of this zone is liable to be affected by the sunlight, seasonal temperature changes, and current and wave action. A major part of sea, plant life and animals grow here in this zone. The lower part shows very little penetration of sunlight and is very less affected by temperature changes. The sea animals like *Corals* and brachiopods grow in this zone.

2.12 ROCKS

Rocks are the aggregates of minerals, including hard as well as soft materials, like stones, sands, days etc. Rocks constitute the significant part of the Earth's crust, and are very very important to mankind. A rock can be defined as a definite unit of the earth's crust, which itself is made up of a certain mineral assembly. Depending upon the mode of formation, rocks are divided into three main classes, *viz.*, *Igneous, Sedimentary* and *Metamorphic*.

• Igneous Rocks

Rocks that have solidified from very hot, molten silicate material, known as magma, are termed as igneous rocks. Such rocks are formed by gradual cooling of molten rock, called magma which gets erupted by volcanoes. As the earth cooled to a solid crust, the original crust of the earth consisted of igneous rocks. Igneous rocks are, therefore, known as *primary rocks*. Basalt and Granite are the examples of igneous rocks.

As discussed earlier that at a certain depth (about 30 to 35 km), the temperature becomes so high that all the rocks will melt, and will exit in a molten state. This natural hot molten material or liquid rock is called the *magma*. When this hot magma erupts out on the earth's surface due to volcanic activities, it is called *lava*. The magma is a complex high temperature solution of silicates, containing water and other gases. The gases may escape when magma erupts out as lava.

• Sedimentary Rocks

Sedimentary rocks are formed by the deposition and consolidation of 'new sediments', in layers, over the pre-existing rocks. New sediments may consist of particles of gravel, sand, silt or clay. These are eroded away from some old rocks by weathering and denudation. The physical disintegration and chemical decomposition of rocks is responsible for formation of finer fragments and dissolved matter, which is transported by different agents, like rivers, wind and glaciers and deposited at some suitable place over some existing rocks, which on consolidation, well result in the formation of sedimentary rocks. Since the sediment is generally deposited and consolidated in layers, such rocks are, therefore, sometimes called as *stratified rocks*. Further, since these are formed from some existing primary rocks, they are also called as *secondary rocks*.

In secondary rocks, the loose sediments get compacted into rocks by pressure of overlying sediments and presence of cementing materials, such as lime. Limestone and sandstone are sedimentary rocks. These rocks are distinguished by the presence of stratification (layering) and fossils (organic remains) *i.e.*, remains of plant and animal organism. The marks indicative of their transport and deposition through the agency of water or air are also characteristics of sedimentary rocks. Sediments are accumulated in variety of places, like shores, deltas, flood plains and deserts etc., since physical, chemical and biological characteristics are different at different sites, the resulting sedimentary rocks show different characters depending upon the environment of the site where they are deposited.

In these rocks, different layers can be identified on the basis of differences in composition, texture, hardness, cohesion or colour. These layers are disposed in approximately parallel bands, and each layer is called a *bed*. Thickness of which ranges from a few metres to fraction of a centimeter. The plane which separates two beds is called as a *bedding plane* and the rocks break easily along the bedding planes.

• Metamorphic Rocks

Metamorphic rocks are those rocks which are formed by the change of form of pre-existing igneous or sedimentary rocks, by the process of metamorphism. Metamorphism is a natural process whereby the existing rocks are altered under conditions of high pressure and temperature, associated with depths of several thousand metres beneath the surface. The formation of these metamorphic rocks takes place in the solid state. The minerals present in rocks get metamorphosed or altered, owing to high temperature or pressure and new minerals are formed. Limestone gets converted to marble, sandstone to quartzite, and shale to slate.

The igneous and sedimentary rocks, when subjected to metamorphism, undergo changes that are either physical, chemical or both. Physical changes are reflected by the changes in the texture of old rocks, whereas chemical changes produce new minerals. A metamorphic rock may further undergo metamorphism, resulting in the formation of a different metamorphic rock. For example, *Slate* (a metamorphic rock) formed from *Shale* (a sedimentary rock) when further metamorphosed, give rise to a new metamorphic rock called *Phyllite*; which in turn, when again subjected to metamorphism, gives rise to *Schist* (another metamorphic rock).

Rock Cycle

Once a rock is formed in the crust of the earth, it does not become a permanent feature there, since when it is exposed at the surface, it is subjected to the weathery giving rise to a new sedimentary rock. This rock may be subjected to crustal movements, which carry them down to depths within the crust where effect of high temperature and great pressure transforms it to a metamorphic rock. Ultimately, the effect of heat and pressure is so pronounced that the rocks subjected to them are actually melted at a greater depths giving rise to new magma. When conditions are suitable, the magma tries to come near the surface of the earth and coals to form a new igneous rock. This phenomenon is described as *rock* cycle.

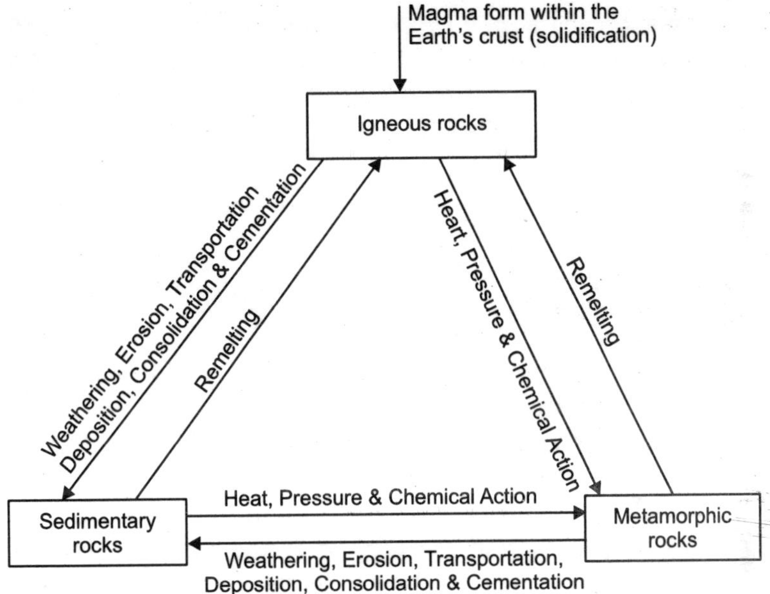

Fig. 2.5: Rock cycle.

2.13 MINERALS OF THE EARTH'S CRUST

A mineral is defined as a naturally occuring homogeneous solid, inorganically formed, having a definite chemical composition or a definite range of composition and an ordered internal atomic arrangement. As pointed out earlier, the Earth's crust contains various types of rocks, which are very useful to man in his building activity. Since these rocks are made up of minerals, we must first study these minerals. Further, we shall, first of all, discuss them.

An element is a substance that cannot be separated into simpler forms of matter by ordinary chemical means. It is, thus, the smallest unit in nature. When two or more elements combine together chemically, they form a compound. Presently, about 100 elements are known, out of which only 8 elements occur in the earth's crust it amounts greater than 2 per cent and constituets more than 98 per cent. All the other elements are collectively found to be less than 2 per cent. These elements of the earth's crust rarely occur individually in nature, but are generally found in the form of compounds. Specific compounds always contain a definite proportion of its constituent elements, irrespective of its place of occurrence.

Table 2.1: The Common Elements of the Earth's Crust.

S.No.	*Name of element*	*Chemical formula*	*Percentage amount present (by weight)*
1.	Oxygen	O	46.60
2.	Silicon	Si	27.72
3.	Aluminium	Al	8.13
4.	Iron	Fe	5.00
5.	Calcium	Ca	3.63
6.	Sodium	Na	2.83
7.	Potassium	K	2.59
8.	Magnesium	Mg	2.09
	All other materials		1.41
	Total		100.00

2.14 MOUNTAIN FORMATION

The surface of the land is not uniform in height or appearance over large areas. Mountains, plateaus and plains are major landforms on the continents. Landforms vary in size and shape from place to place, and also change with the passage of time. Such changes may occur suddenly (due to earthquakes or volcanic eruption or floods) or gradually. Landforms are caused by the action of two types of natural processes on the surface of the land, namely external processes (caused due to changes taking place in the atmosphere and hydrosphere), and internal processes (takes place in the interior of the earth). These internal processes cause movements of the earth's crust, leading to formation of mountains and plateaus.

Mountains may be defined as the raised or elevated portions of the Earth's crust. In general literature, *hills* and *mounds* progressively represent relatively smaller sizes of mountains (less than 900 meter high), but in general term, all such raised portions are termed as mountains. A long and narrow part of a mountain is called a *ridge*; and a number of parallel or sub-parallel ridges formed during the same mountain building process is known as a *mountain range*. A number of mountain ranges formed during the same geological age, are called a *mountain system*. However, a number of mountain ranges formed during different geological periods are termed as *mountain chain*.

Mountains are classified in the following four categories, based on the process of their formation:

(a) **Fold Mountains:** These are made up of rocks, which are generally jointed, folded and faulted. These are made up of many layers of rocks and are formed due to violant contractions and pressure inside the earth. Because of this, the rocks turn away and overlap each other and their layers get folded to rise above the earth's surface to form mountains. The folds usually occur in long parallel rows of mountain ridges separated by valleys. The Himalayas is one of the mountains came into existence as a result of folding process.

(b) **Fault Mountains or Black Mountains:** Faulting in crustal rocks always result in displacement of the blocks along the rupture plane. This displacement is lateral or vertical or oblique.

(c) **Residual Mountains:** Residual mountains are formed due to denudation and erosion by which high plateaus are gradually shaped into peaks and ridges. Denudation and erosion is caused by natural agents, such as wind, water, snow etc. Infact, due to differential erosion, the weaker and the less resistant rocks are worm down and lowered, while hard and more resistant rocks remain standing above the ground surface, making hills and hillocks, called residual mountains. Most of the medium-sized 4 Indian mountains, including the 'Vindhyas', the 'Satpuras', the Eastern and the Western 'Ghats', but excluding the 'Himalayas' and the 'Aravallis' all belong to this category.

(d) **Volcanic Mountains:** These are formed due to accumulation and solidification of lava, ash and debris erupted from the earth's interior. They are basically cone-shaped with a creator at the top. Volcanic mountains are generally small or medium-sized mountains.

2.15 SURFACE FEATURES (GEOGRAPHICAL ZONES) OF INDIA

India may be divided into three physio-graphic (Geo-graphical) zones. They are:

1. The Himalayan Mountains.
2. The Great Plains.
3. The Peninsular Plateaus.

1. The Himalayan Mountains

The Himalayas standing on the northern frontier of India are the highest mountains of the world. The great Indus and the mighty Brahmaputra rivers divide them into three sections: The main Himalayas, the north-west Himalayas, and the south-west Himalayas. The Himalayan mountain ranges are much younger as compared to the Peninsular Plateau and are called the young fold mountains. Till about 70 million years ago the entire Himalayan area as well as the plain area remained under sea-water in a geosyncline of *Tethys Sea* as sediments. To the north and south of the *Tethys sea* were the landmasses of *Angarland* and *Gondwana land* respectively. The sediments in the *Tethys sea* were deposited by the rivers flowing into it from the two landmasses. As a result of northwards drift of the Peninsular Plateau during the mountain building process, these sediments were folded and the present Himalayas were formed.

The Himalayan ranges were formed when the Indian plate was driven northwards and pushed beneath the Eurasian plate. About 30-60 million years ago, the two plates came closer and the *Tethys* sea crust began to fracture into thrust edges. About 20-30 million years ago, the Himalayan ranges started emerging. The Himalayan ranges run for about 2400 kms from the *Indus* river in west to *Brahmaputra* river in the east. The average width of the Himalayas is 160-400 kms. The Himalayan mountain has a group of three mountain, ranges running almost parallel to one another along the longitudinal axis. These ranges are:

(a) **The Great Himalayas or the Inner Himalayas:** This is the highest mountain range of the Himalayan system. This always remain snow covered. Its average altitude exceeds 6000 meters and its average width is 25 kms. Almost all the important peaks are located in this range *e.g.*, Mount Everest (Nepal, 8848 m), Kanchanjunga (Sikkim, India 8598 m), Makalu (8481 m), Dhaulagiri (8172 m), Naga Parbat (8126 m), Annapurna (8078 m), Nanda Devi (7817 m).

(b) **Middle Himalayas:** This range is located in the south of Great Himalayas and runs parallel to the east-west direction. Its average height is from 3700 to 4500 meters and is nearly 80 kms wide on an average. In between Great Himalayas and Middle Himalayas, there are longitudinal valleys, like Kashmir Valley and the Kathmandu Valley. Most of the hill stations, such as Shimla, Mussourie, Dalhausie, Nainital, Darjeeling etc. are located in this range.

(c) **Outer Himalayas or the Shivalik Range:** The southern-most range of Himalayas is the *Shivalik* range, whose height varies from 900 to 1200 metres and width varies from 10 to 50 kms. In between the 'Middle Himalayas' and the 'Outer Himalayas' there are longitudinal valleys, known as 'dunes' in the west and 'duars' in the east. Dehradun is situated in one of such valleys.

Most of the Himalayan ranges have steep slope on the southern side while they have comparatively gentle slope on the northern side. The northern side has larger accumulation of 'glaciers'.

2. The Great Plains

The Great plains of India lies in between the Himalayas in the north and Peninsular plateau in the south. Geologists believe that there was large depression, caused due to Himalayan uplift out of the 'Tethys sea' and the subsidence of the northern flank of the peninsular plateau. Rivers flowing into this depression from the lofty mountains of the north and the plateau of the south deposited sediment in the depression and in course of time, it turned into a flat land of alluvium deposits.

Half of the Great plains lies in Uttar Pradesh and balance in the states of Bihar, West Bengal, Assam, Punjab and Haryana. The older alluvium occupying higher grounds is known as the *'bangar'* and the newer alluvium covering the low lying flood plains adjacent to the river banks, is known as the *'khadar'* in Uttar Pradesh and the *'bet'* in Punjab. The two big depressions in Bihar are known as the *'jala'* in Patna and *'tal'* near Mokamah. These get flooded every year during the rainy season. The Assam Valley is a continuation of the Ganga plain. Because of very gentle slope of the valley, sands and silts are deposited in and around the Brahmaputra river courses. This causes wide-spread flood. The plain contains a large number of alluvial lakes locally known as the *'bils'*.

The Punjab plains are drained by the Satluj and Beas. The Ganga and its tributaries drain the control part of the north India, and the tributaries of the Brahmaputra, the eastern part of the plain.

The plains are those large areas of the earth's surface which are flat and low-lying land surface. On the basis of their formation, the plains are divided as under:

(a) **Structural Plains:** These are the plains which are formed by the uplift of a part of the sea-floor or by depression of a large part of the earth's surface due to its crustal movements.

(b) **Erosional Plains:** These plains are formed by the erosive action of running water, wind etc., on the raised parts of the earth's surface. Pebbles and stones can be seen in them.

(c) **Depositional Plains:** These are the plains which are formed by filling up sediments into depressions along the foothills, lakes and seas. The huge amount of sediments are transported by natural agents, like running water, waves, wind and ice and deposited into the depressions.

Thar Desert

The Arid plains of Rajasthan are the continuation of the Great plains of India. Lying to the west of the Aravalli range, these plains with low rainfall have become arid. This area was also under the sea and was uplifted. In greater part of the area, the groundwater is impregnated with salt. There are numerous salt lakes of which the sambhar lake is the largest.

3. The Peninsular Plateau

A plateau is an elevated area and has a larger area on the top. It has an extensively even or undulating surface and has steep sides.

The southern part of India consists of an old plateau, which is known as the 'Southern Plateau'. This is also called the 'Peninsular Plateau because it is surrounded by sea on three sides. According to the Geologists, it is the oldest landform of the Indian sub-continent. At the time of uplift of the Himalayas, the western flank of the plateau subsided and the *Indian Ocean* advanced landwards leading to the formation of the *Arabian Sea*. The Peninsular Plateau has four physiographic units; namely Central Highlands, Peninsular Plateaus, Coastal Plains and Islands.

(a) **Central Highlands:** A wide belt of hilly country bordered on the west by the Aravalli range and on the east by the Vindhya scarps is known as the 'Central Highlands'. Its southern limit is the right valley of the Narmada. One-half of Madhya Pradesh, one-third of Rajasthan and a small part of Uttar Pradesh are included in the highlands.

(b) **Peninsular Plateaus:** Facing the bay of Bengal and the Arabian sea, the Peninsular plateau forms the largest physiographic division. It stretches from the Satpura range in the north to Kanyakumari in the south (1500 km nearly) and from the Western Ghats in the West to the Rajmahal hills in the east (about 1400 km). Triangular in shape, it has four physiographic sub-units: The Western Ghats, the Deccan Plateau, the Eastern Ghats and the Eastern Plateau.

The Godavari, the Krishna and the Kaveri rivers, have their sources on the Western Ghats and flow into the Bay of Bengal. The Deccan plateau in the northern section comprises the Satpura range and the Maharashtra plateau. The plateaus of Andhra Pradesh and Karnataka form the sourthern part of the Deccan. The Baghelkhand plateau (Son river basin), the Chhotanagpur, the Mahanadi basin and the Dandakaranya form the eastern plateaus (also locally known as Chhatisgarh basin).

The Eastern Ghats are non-contiguous, contrary to the Western Ghats. In Odisha the Eastern Ghats are known as the Mahendragiri. Southwards, they are low detached hills. The Nilgiri, for south in Tamil Nadu, is the meeting place of the Western Ghats, the Eastern Ghats and the Southern hills.

(c) **Coastal Plains:** The Peninsular plateaus are bordered on the east and the west by the coastal plains. The coastal plain of the east is wider and drier than the west coastal plain. The east coast has a number of deltas, namely the Mahanadi, the Godavari, the Krishna, and the Kaveri deltas. The west coast has no delta. The west coast is more or less straight but the east coast is curved. The coastal tract of Odisha is known as the Utkal plain. In Andhra Pradesh and Tamil Nadu, the east coast is known as the 'Payan Ghat' (coromandal coast) and extends from the delta of Godavari to Kanyakumari. It has a width varying between 100 to 130 km.

About 1500 km long between Kanyakumari and Surat, the west coastal plain ranges in width between 10 and 25 km. The coastal plain in Maharashtra is known as 'Konkan', in Karnataka, the 'Kanara plain', and in Kerala, the 'Malabar plain'.

(d) **Islands:** There are a number of islands in the Bay of Bengal and the Arabian sea. Islands in the Bay are much larger and more habitable. These are elevated portions of submarine mountains. Andman and Nicobar are a group of islands. The great Andman is a collection of three islands. Portblair is located in the South Andman. The Nicobar contains nineteen islands.

The islands in the Arabian sea are entirely built of coral deposits. Lakshadweep contains a number of islands. Of these, eleven are comparatively larger and have fringing reefs close to their shores.

2.16 THE HYDROSPHERE

Hydrosphere, as the name suggests, includes all the natural waters of the outer part of the earth in the form of oceans, lakes, rivers and other water bodies. Water covers 71 per cent of the total surface area of the earth. Water occurs in the form of ice-sheets in polar regions and on high mountains. Water also occurs below surface of the land in the form of underground water. Water occurs in the form of water vapour in the lower layers of the atmosphere.

Hydrological Cycle

Various bodies of water in the hydrosphere, atmosphere and lithosphere are linked by a single cycle, known as the hydrologic cycle. Differential heating by the sun is responsible for the circulation of water in the hydrosphere. When the surface water in the oceans, lakes etc. gets heated by the sun's rays, *evaporation* takes place and water vapour is added onto lower layers of the atmosphere. This water vapour in the atmosphere may get cooled leading to *Condensation* of water into tiny droplets which form clouds. Such clouds may cause *precipitation* of water in the form of rainfall or snowfall on the surface of earth. Rainfall on the land leads to surface *run-off* in the form of rivers which reach the oceans ultimately.

In this process of circulation, some water is consumed by plants and animals in the biosphere. Water may get temporarily stored on the land in the form of lakes, ice-sheets or as underground water below the land surface. This circulation of water between hydrosphere, atmosphere and lithosphere is called the 'hydrological or water cycle'.

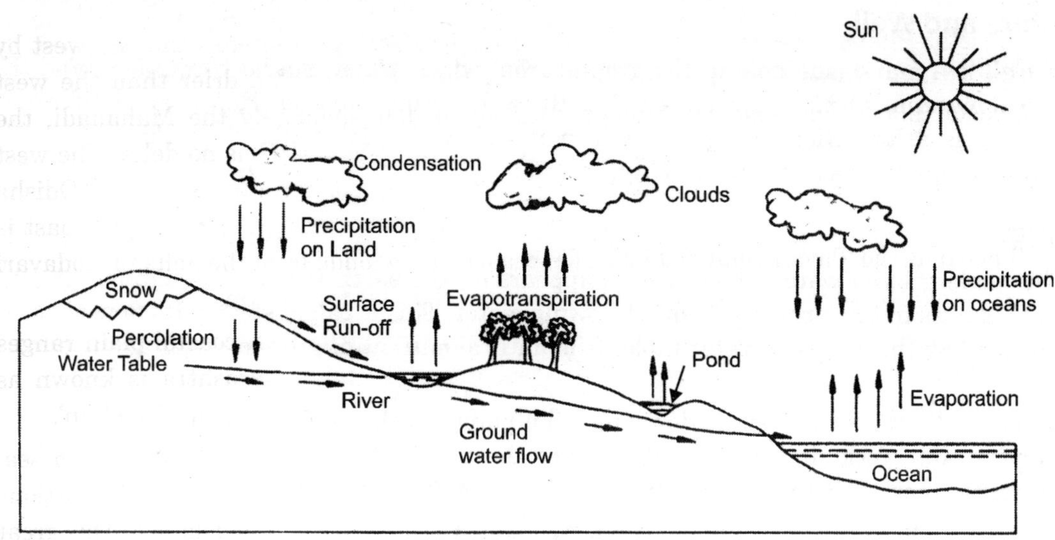

Fig. 2.6: The Hydrologic Cycle.

Water component of the abiotic or inorganic or physical components consist of: (*i*) Ground water, (*ii*) Surface water, and (*iii*) Oceanic water.

Underground Water

The water which infiltrates through the faults, fractures and joints is held in the porous of the bedrocks is termed as *ground water*. Thus, ground water is the under-ground water that occurs in the saturated zone of variable thickness and depth below the Earth's surface. Greeks and pores in the existing rocks and unconsolidated crystal layers, make up a large underground reservoir, where part of precipitation is stored.

The rainfall that percolates below the ground surface, passes through the voids of the rocks and joins the water table. These voids are generally inter-connected, permitting the movements of the groundwater. But in some rocks, they may be isolated, preventing the movement of water. Therefore, occurrence of groundwater depends largely upon the geology of the area. The volume of water contained in the groundwater reservoir in any localised area is dependent upon:

(i) the porosity and permeability of rocks;

(ii) the rate at which water is added to it by infiltration; and

(iii) the rate at which water lost from it by evaporation, transportation, seapage to surface courses, and withdrawal by man.

The underground water is held in the pores and fissures of the rocks making them saturated. The upper surface of this saturated zone is called *water table*. Water table can also be defined as the static level of water in wells penetrating the zone of saturation. It is generally higher under the hills and lower under the valleys. The water table is, thus, the surface of a water body, which is constantly adjusting itself towards an equilibrium conditions, with water moving from higher points to lower points.

The water is drained from the groundwater reservoir, either under some natural phenomena like springs, or artificially by constructing well and lifting water through them for fulfilling domestic, rural, municipal or industrial demands.

Aquifers and Well

A *well* is a man-made hole in the ground from which water can be withdrawn; and the geological material which yields the water to the well is called *aquifer*. Hence, an aquifer may be defined as a formation of a permeable material, which is capable to yield appreciable quantities of groundwater under gravity.

Springs

The underground water sometimes re-appears on the earth surface through an opening in the rockmass in the form of current of flowing water. This is called spring. Its flow of water is controlled by the slopes of water table. It is natural outflow of ground water at the earth's surface.

Some of the rain water sinks down into the porous rocks, and when this water reaches an impermeable rock, it cannot sink further down, but flows along the surface of the impermeable rock. If it finds way out and this water flows out, it forms a spring.

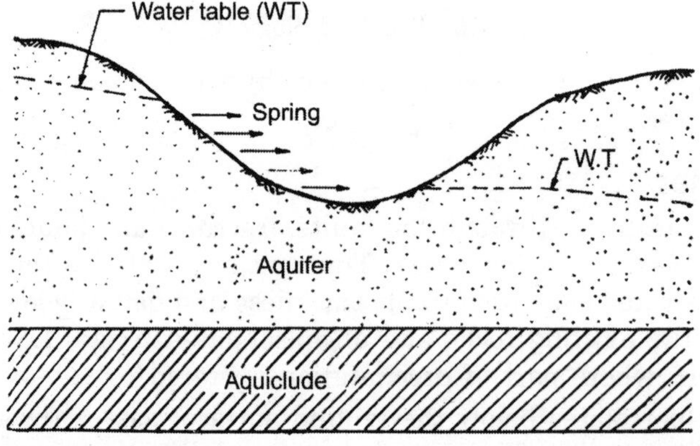

Fig. 2.7: Spring formed due to favourable slope in a valley.

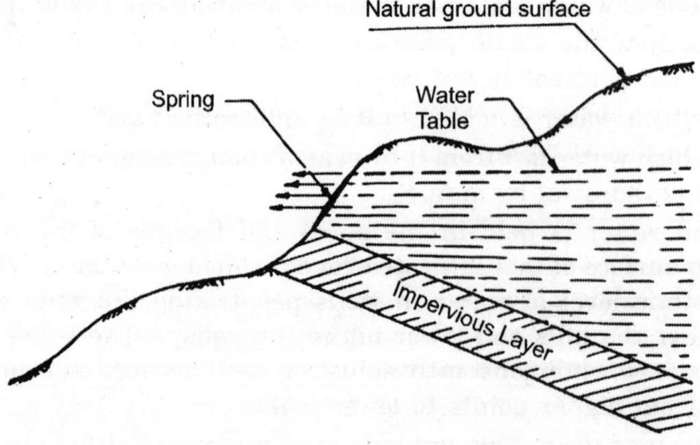

Fig. 2.8: Spring formed due to the presence of an impervious obstruction against an aquifer or water table.

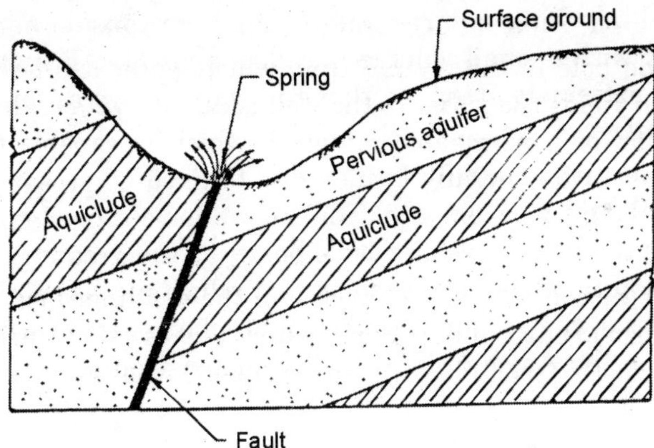

Fig. 2.9: Spring formed due to the presence of a fault across a water bearing formation.

Hot Water Springs

The temperature of water of a hot spring is higher than that of ordinary springs. As we have already studied that groundwater temperature rises with increasing depth at the rate of about 1°C for each 25 metres. When underground hot water flows out regularly through some crack or fissure, it is called a hot spring. In some volcanic regions, the steam formed by the heat of the molten rocks is forced out in the form of hot water.

Geysers

The Geysers are special types of hot springs from which hot water is erupted out at regular intervals with enough force, so that the water rises to height which may vary from a few metres to 100 metres or so. The cause of the geyser is the internal heat of the earth which converts underground water into steam. This steam forces its way out of a weak spot in the crust of the earth. It is then converted into boiling water and sprouts up like a fountain.

Thus, a geyser is a jet of boiling water and steam issuing from the earth in a few volcanic regions. Beneath the geyser, there is some hot rock and a narrow fissure, connects the rock to the mouth of the geyser. Water reaches the rock through this fissure, where it is heated to about 100° celsius. Now the steam pressure inside, forces the hot water to come out to the surface of the earth. The flow of hot water continues till such time, as cold water does not return to the fissure. The cold water is again converted into steam and as the pressure rises, the hot water is thrown out.

Crater Lakes

Due to volcanic eruptions, very broad and shallow craters are formed. They are surrounded by a well-like structure made up of lava. When there is no eruption for a long time, many craters become filled with rain water, forming lakes. These are called crater lakes. Crater lakes are normally on the top of the mountains.

Water Falls

Water falling from the height of mountains is called *waterfall*. Waterfalls are the important features of a river profile, and are generally found in the mountainous reaches of the river.

If the volume of falling water is large and it falls from considerable height, it is known as *coteract*. A waterfall with a small volume is called a *cascade*. If instead of a vertical fall, the river bed becomes extremely steep, so that the water flows at an excessive slope, then the fall is called a *rapid*.

Waterfalls and rapids are generally formed due to unequal erosion of different types of bedrocks encountered along the river path. Thus, if we have a hard stratum outcropping in the bed of a river, with softer and weaker beds below it; the greater wear of the latter will develop sufficient inequality of bed levels, so as to produce rapids. With progressive erosion and increasing steepness of the stream bed, the rapids change into vertical falls. The highest waterfall in the world is the Angel Falls with 979 meter drop in Venezuela.

Caves

A cave is a natural opening in the earth that is big enough for a person or animal to enter. Most caves are formed by the erosive action or rain water or underground water on rocks such as limestone or dolomite. That contains dissolved carbondioxide which forms dilute carbonic acid. This acid dissolves the rock, forming passages and large open spaces. The other types of caves are *lava caves* (found near the base of a volcanic mountain), *Ice caves* (formed within glaciers); and *sea caves* (formed in the coastal rocks by the action of sea waves).

The early man lived in caves. *Amarnath* cave and *Vaishnodevi* cave are the famous caves in India.

2.17 GLACIERS

Glaciers are the moving masses of ice. They generally occur in polar region and on high mountain ranges having a permanent cover of ice and snow. Glaciers cover about 10% of the earth's land surface.

Glaciers are of two major types:

(a) Continental Glaciers

(b) Mountain Glaciers

(a) **Continental Glaciers:** These occur in polar regions in the form of extensive and thick ice-sheets covering the entire land surface. The large continent of Antarctica is covered by thick ice-sheets. The extent of continental glaciers is not always constant, because when the earth gets warmer, they melt away along the margins reducing their extent. About one million years ago, when the earth was much cooler continental glaciers, extended over northern parts of North America, Europe and Asia.

In the coastal regions of Greenland and Antarctica large masses of ice broken off from the glaciers float on the sea. They are called *icebergs*. Normally about one-tenth part of an iceberg is above the sea-level and the remaining under the water. The largest glacier in the world is Lambert about 515 km long and 70 km wide. It lies in the Australian Antarctica region.

(b) **Mountain Glaciers:** The height above which there is permanent cover of snow and ice is called the snow line. In the equatorial regions, the snowline is at a height of about 5500 metres above the sea-level. The snowline in the Himalayas region is about

4800 metres. As one moves poleward away from the equator, the height of the snow-line decreases gradually, and in the polar regions, the snowline is at the sea-lead. These mountain glacier occur in high mountain regions, such as the Alps and the Himalayas. In such regions, ice and snow accumulate in depressions and along valley heads near the top. As the movement of solid blocks of ice is slowed by friction with the sides and floor of the valley, the average rate of movement of valley glaciers is hardly a few centimetres per day. As the valley glaciers descend down to lower lands which are warmer, they melt away in the form of streams. The Himalayan rivers, such as the Ganga and Yamuna originate from valley glaciers.

Vast areas covered with deep ice and snow for a considerable time are known as *ice cap* or *continental ice-sheet*.

Arctic Regions

The Arctic ocean lies on the northern border of the earth. In its centre is situated North Pole. It is completely frozen and on the three sides, it is flanked by northern boundaries of Asia, Europe and North America, while Greenland and other small islands exist on its fourth side. It is covered with snow all the year round and icebergs float on open sea. Very few people live in the region and hardly any vegetation exists.

2.18 EROSION

Erosion is a process of scraping, scratching, grinding and transportation of the soil and rocks on the earth's surface by running water, moving ice or glacier, waves and currents of the sea, rain and wind. Lakes, water-falls, caverns, valleys and natural bridges are created by erosion. Thus, erosion is the destructive mechanism caused by the agents of transportation. Erosion is widely exhibited in the form of resultant landscapes, such as formation of gullies on hillslides, landslides etc.

Agents of Erosion

As pointed out above, the erosion is caused mainly by following agents:

1. Water
2. Wind
3. Ice
4. Marine

1. Water Erosion

Water is the most important agent of erosion and creates maximum havoc in the form of erosional damages to the land surface:

(a) **Erosion caused by falling rains:** Water causes erosion by detaching the soil particles, and then transporting them usually downhill. The water erosion process starts as soon as a rain starts, and the following rain erodes the surface material in two ways: Firstly, the impact caused by the falling raindrop on the soil, detaches the soil particles. The loosened or detached material is separated and splashed into the air under this process, the amount of material eroded from a hilly land is more than that from a flat land. Secondly, the rainwater that does not percolate to the underground, will flow as run off and will cause erosion which depend upon the slope of the land surface, and the nature and extent of vegetative covers.

(b) **Downslope movement of Weathered material:** The materials dislodged by weathering, tends to move downslope under the influence of gravity and reach upto the river or the stream, from where it is carried away upto the sea. The downslope movements are generally slow processes, but, however, when slopes are disturbed by construction of engineering structures, such as railways, roads, buildings etc., more rapid movements may be triggered off, causing much damages.

Massive downslope movements, called *mass movements* may be triggered on in many ways, such as by:

(i) *Undercutting of the slope,*

(ii) *Overloading of the slope*, so that it is unable to support its new weight and therefore slides out,

(iii) *Vibrations* from earthquakes or explosions that break the bond holding the slope in place, and

(iv) *Additional water*, which reduces the internal cohesion of the overburden by filling the void space, acts as lubricant for the movement, and also by adding more weight to the overburden.

(c) **Erosion caused by Rivers:** Rivers cause erosion from their beds, receive and transport the products of weathering that are fed into them by run off from precipitation as well as by downslope movement, and finally carry the entire denundated material into the sea for deposition.

Erosion processes in the river depend upon the nature of the bed rock quality of water contained in the river.

Some of the important processes are:

(i) The bedrock is slowly and steadily get eroded away chemically in the water by the process, called erosion.

(ii) Hard and resistant material of the bedrock is eroded away slowly by abrasive action caused by sands, gravels, and other fragments present in the river water.

(iii) Softer bedrock and loose material may be eroded away by the flowing water.

The rate of stream bed erosion may depend upon: (*i*) velocity of river water, (*ii*) nature and characteristics of the bedrock, and (*iii*) sediment load present in the stream and its sediment carrying capacity.

The process of development of a *river valley* starts when the water after rainfall concentrates its movement along the naturally available ground slope, slowly cutting the ground by erosion in the form of small gullies, which on further development with the passage of time acquire the shape of a valley. The physical processes involved in the formation of a river valley are: (*i*) *deepening* of the river valley is caused by down-cutting of the river bed, which depends considerably on the velocity of flow; (*ii*) *widening* of river valley is done, partly by the lateral erosion caused by running water, and partly by the processes, like weathering of banks, rainwash, creep landslides etc. taking place from the sides of the valley.

Sediment load carried by a river tends to settle down as and when the flow velocity of the river reduces. Depending on, as to where and when the river sediments are dropped and deposited, different types of features, like river delta, alluvial fans and cones (formed where stream flows out mountains into broad valley in plains), flood, plains etc., may develop.

2. Wind Erosion

As movement of the atmosphere in a direction parallel to the earth's surface, is *wind*. Like water, wind is also an agent of erosion, transportation and deposition. It is an effective agent is generally caused by two main erosion processes; namely, deflation and abrasion.

Deflation: It is a process of removing loose sand and dust-sized particles by fast moving winds. Deflation may result in the formation of depressions or hollows, generally a few meters deep and several hectares in area.

Abrasion: The blowing winds generally pick up sand particles from the Earth's surface, and then move with them as sand laiden winds. When these sand laiden winds strike the portending rocks or other structures in their way, they will cause them to erode by the rubbing and grinding action of the moving sand particles, maximum erosion is caused in the lower parts of the obstruction *i.e.*, within 30 to 60 cm from the Earth's surface.

When the velocity of a fast flowing sediment laiden wind is checked by some obstruction or otherwise, the sediments get dropped and deposited. There are two types of such deposits, namely sand dunes and loess.

Sand dunes are formed by the deposition of bed load (particles moved by winds within 30 to 60 cm of the earth's surface) from sand storms. Sand dunes are the huge heaps of sand formed by the natural deposition of wind blown sand.

Loess are formed by the deposition of suspended load from the dust storms (wind blown material finer than 1/16 mm, and are lifted by the moving winds by a distance of hundreds of metres above the Earth's surface). Loess are the sheet like deposits of silt and clay formed, under the action of falling rains.

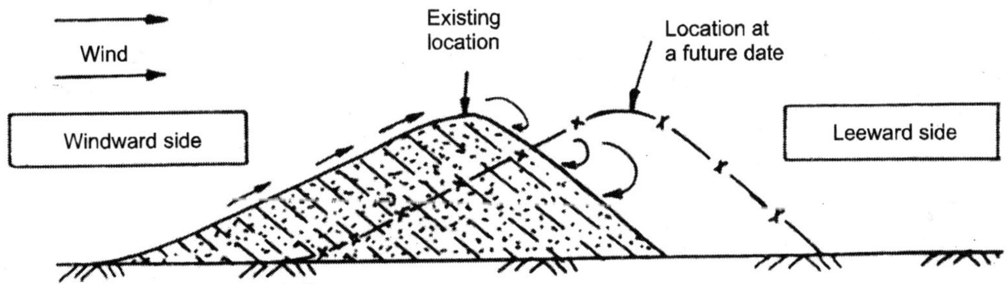

Fig. 2.10: The drifting movement of an active (or unstabilised) sand dune with time.

3. Ice Erosion

A *glacier* is a mass of moving ice (moving naturally under the action of gravity), which causes erosion of the surface over which moves. They are slow erosion agents. Glaciers are of following two types:

(a) **Valley Glaciers:** These are often called *mountain glaciers*, may be defined as the rivers of ice, developed in previously formed stream valleys. Accumulation of snow in such pre-existingg river valleys when gets converted into ice, the glacier begins to move down the valley. A glacier, thus, moves down the valley under the influence of increasing accumulation of snow. As the glacier moves down the valley, the rate of melting increases due to the warmer climates of the lower altitude regions. The valley glacier, thus, further erodes its pre-existing stream valley, transports the eroded material, and finally deposits the same at some places downstream.

(b) **Continent Glaciers:** These are often called *ice-sheets* or *ice-caps*, consists of extensively large masses of glacial ice, covering enormous area of the existing topography. Major ice-sheets are: Greenland ice-sheet, and Antarctic ice-sheet. Erosive effects of continent glaciers are much less, and are generally limited to rounding the topography, and rerapping off much of the soil and over-burden.

4. Marine Erosion

Marine water in the form of seawaves and currents, considerably erodes the land near the sea-coast, and may cause extensive damage to an unprotected seashore, or to the man-made structures situated near the shore. The erosion by wores is generally takes place due to abrasion or solution (*i.e.*, corrosion). *Abrasion* is the process of constant rubbing and grinding action of sandy sea-water, continuously striking with the loosely packed sediments or rocks existing on the seashore. The *solvent action* of sea-water, particularly erodes the rocks, which are soluble or easily attacked by marine water, and this process is called *corrosion*. The coastal rocks are broken or loosen at the weak portion under the influence of pressure created by seawaves.

The sediments entering seawater either from rivers or by erosion by seawaves, are carried in suspension by the waves, and are deposited on the floor of the sea at some place or the other. Coarser and heavier sediments are generally left near the coast, and the finer sediments are taken to far-off places into the deep sea.

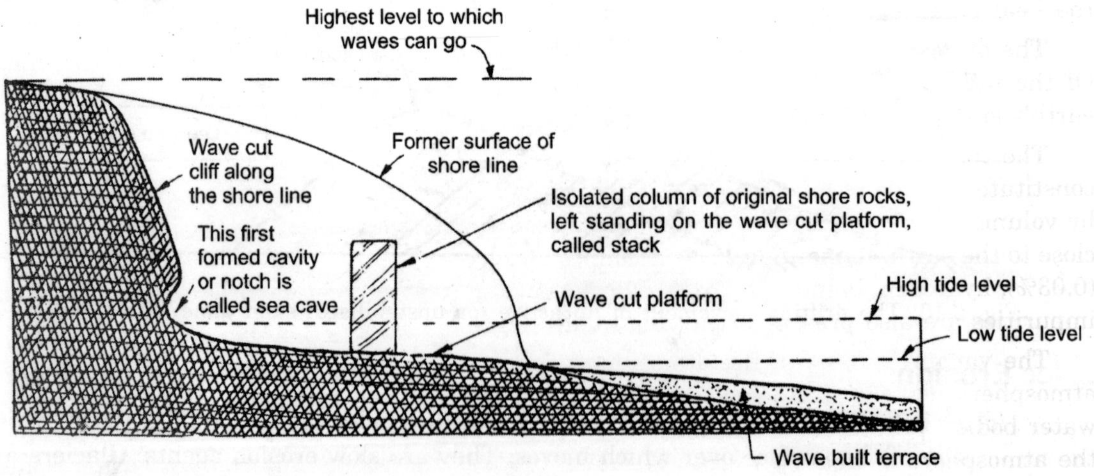

Fig. 2.11: Structural features of marine erosion.

2.19 DESERT

A desert is a barren and dry (arid) regions that has vast area of sand. It has usually dry and hot high temperature and an extremely low rainfall (less than 12.5 cm). In deserts, only a few varieties of plants and animals can exist. These are those species that require little water. In most deserts, days are very hot but the nights are cold. In deserts, movement of soluble salts is restricted to the upper few centimetres. Owing to high evaporation, salts usually travel upwards and accumulate in the surface layer.

The best known deserts are the *Sahara* and the *Kalahari* in Africa, the south-west America in desert, the Great Australian desert, the *Chilean* desert in South America, Arab Thar desert in the south-west Asia and the deserts of Turkistan and Gobi. In India, desert soils are found in large parts of Rajasthan and the *Rann of Kutch*. Thin and sandy desert soil of Rajasthan state looks yellow, because it has been weathered from a yellowish rock, called sandstone.

Sand dunes are formed due to the action of strong winds. They may be upto 500 metres high and 900 metres long. The highest sand dunes are found in *Sahara* desert in Algeria.

2.20 THE ATMOSPHERE

Our planet is enveloped by a blanket of gases extending several kilometres above its surface. Like the lithosphere and hydrosphere, the atmosphere too is an integral part of a gigantic system....the earth. The right kind of temperature coupled with the force of gravity, has provided the earth an atmosphere that makes the Earth unique in the solar system. **Atmosphere, is inseparable from the earth** due to force of gravity, and is more commonly known as air. It is mobile, elastic and compressible.

Atmosphere contains life-giving gases, like oxygen for man and animal, and carbon dioxide for plants. It also acts like a greenhouse and, therefore, keeps the Earth warmer than it would otherwise be. Thus, the atmosphere acts like a blanket for the earth and regulates the heat balance. It also protects us from the harmful ultraviolet radiation of the sun.

The atmosphere serves as a storehouse for water vapour, which leads to precipitation. Of the total mass of the atmosphere, 99 per cent is within a highest of 32 kms from the earth's surface.

The composition of the atmosphere is taken as the composition of pure dry air, which constitutes mainly of nitrogen (78%) and oxygen (21%) that together make up 99% of the air by volume. Their proportion remains almost constant within lower layers of the atmosphere close to the earth. Remaining 1% is accounted for by gases, like argon (0.93%), carbon dioxide (0.03%), hydrogen, helium and ozone. Besides water vapour, dust particles, smoke, salts and impurities are also present in varying quantities.

The vapour content in the atmosphere ranges between 0 to 5 percent by volume. The atmospheric vapour is received through the evaporation of moisture and water from the water bodies, vegetation and soil covers of the lithosphere and seas and oceans. Vapour in the atmosphere decreases from the equator towards the poles. This trend of distributional pattern of atmospheric vapour alongwith temperature affects vegetation. The moisture content in the atmosphere is responsible for the creation of several forms of condensation and precipitation (clouds, fogs, frost, rainfall, ice, hailstorm, snowfall etc.).

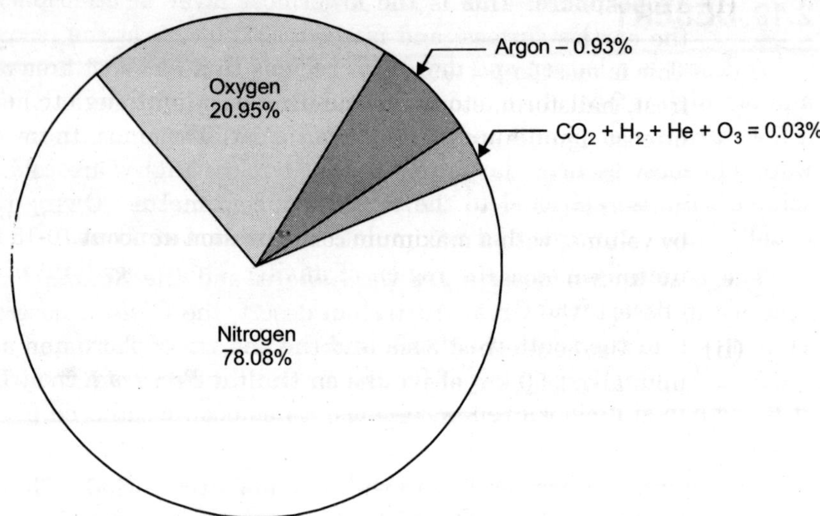

Fig. 2.12: Composition of Atmosphere.

Structure

The atmosphere has a layered structure, because of density stratification, as a result of which lighter gases move up and denser ones settle down. There are four definite layers, having varying characteristics of temperatures and combinations of gases in the atmosphere. From the lower atmosphere to the upper atmosphere, these are as follows:

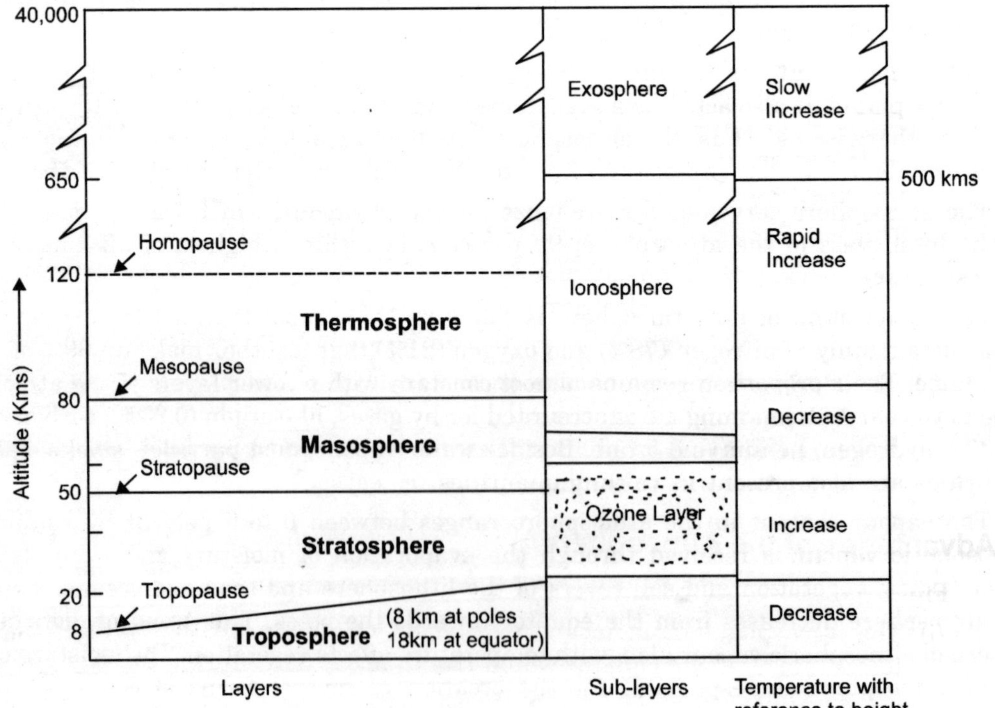

Fig. 2.13: Layers of Atmosphere.

(i) **Troposphere:** This is the lowermost layer of atmosphere which is in contact with the earth's surface, and is about 8 km high at the poles and 16 km at the equator. This is most important layer because almost all of the weather agents (*e.g.*, fog, cloud, frost, hailstorm, storms, cloud–thunder, lightning etc.) occur in this layer. This is of utmost significance for all life-forms including man. In this layer, temperature decreases with increase in height at the average rate of about 6.5°C per 1000 m (known as normal lapse rate). Troposphere also contains varying quantities of water of 1 to 4% by volume, with a maximum concentration at about 10-15 km altitude. The troposphere literally means the region of mixing. The upper limit of the troposphere is called as 'tropopause'.

(ii) **Stratosphere:** This layer is above the troposphere and extends to an average height upto about 50 km above the earth. It differs from the troposphere in that it contains very little water vapours and no clouds. Though its gaseous composition is similar, but the gaseous mass is only 15% of the total atmosphere, and it contains more ozone. In this layer, there is gradual rise in temperature and there is 0°C temperature at the height of 50 km, the upper limit of the stratosphere which is known as 'stratospause'. Because of the concentration of ozone in this layer, this layer is very important. The ozone acts as a protective cover for the biological communities, as it absorbs almost all of the harmful ultraviolet rays of solar radiation and protects the earth surface from becoming too hot.

(iii) **Mesosphere:** This layer extends between 50 km and 80 km and temperature again decreases with increase in height. At the upper limit of mesosphere *i.e.*, at 80 km (known as mesopause) temperature becomes – 80°C. This layer consists of less gaseous mass, no water vapour and ozone is also present in it.

(iv) **Thermosphere:** The part of the atmosphere beyond mesopause (beyond 80 km) is known as thermosphere, wherein temperature increases rapidly with increasing height. From 80 to 120 km oxygen and nitrogen are abundant, but at 500 km helium and hydrogen are the main gases. Thermosphere (also known as heterosphere) contains no water vapour or ozone.

Thermosphere is further divided into two layers, *i.e.*:

(a) *Ionosphere,* extends from 80 km to about 650 km. This lower part of thermosphere contains electrically charged particles called 'ions'. These particles reflect radiowaves back to the earth's surface and enable wireless communication.

(b) *Exosphere,* represents the uppermost layer of the atmosphere extending beyond 650 km height from the earth's surface. The exosphere has very high temperature due to solar radiation. There is no distinct upper limit to the exosphere and this gives way to the inter-planetary space.

Advantages of the Atmosphere

From the above, we have seen that:

(i) The atmosphere is an insulating blanket of air around the earth. Without it, the temperature at the equator would rise to about 80°C during the day and drop to as low as –140C° at night.

(ii) It burns up meteors that would bombard the surface of the earth from the space.

(iii) It is a reservoir of oxygen needed by the animals and of carbon-dioxide essential for plants.

(iv) It serves as a distribution system for moisture.

(v) Without the atmosphere, there would be no sound and no light.

(vi) There would be no conventional long-distance radio communication, because this is dependent on the electrons in the upper atmosphere.

(vii) The atmosphere shields the earth from lethal concentrations of ultraviolet relations.

2.21 BIOSPHERE

Biosphere or *Ecosphere* is the largest biological system, in which several ecosystems can operate together. In simplest term, it is that part of the earth in which life exists. The biosphere is a life supporting layer which surrounds the earth and makes plant and animal life possible.

The moderate temperature conditions and presence of water favours growth and evolution of various species of plant and animal life, including humans. Thus, the earth is unique in having a life-bearing layer *i.e.*, biosphere. The biosphere consists of all the living or biological (biotic component), energy (energy component) and physical environment (the abiotic component) and there are continuous interactions between living organisms and physical environment and among the living organisms themselves.

The average thickness of the biosphere or life supporting layer consisting of air, water, soil and rock is about 30 km. The upper limit of the biosphere is determined by the availability of oxygen, moisture, temperature and air pressure. Though certain bacteria have been found upto the height of 15 km in the atmosphere but the lower layer of the atmosphere upto a few hundred metres accounts for most of the living organisms because of the favourable environmental conditions for the growth and development of living organisms.

The lower limit of the biosphere is determined by the availability of the required amount of oxygen and light which can sustain life. Thus, the depth of the biosphere over the land is upto the depth of the deepest roots of the trees or the depth upto which can live the burrowing organisms. The biosphere extends upto greater depth in the oceans. The existence of life has been detected upto a depth of 9000 metres in the deep oceanic trenches and deep sea plains.

Biosphere is, therefore, that particular zone on the Earth, where the lithosphere, the hydrosphere, and the atmosphere come into contact with one another. It is in fact that portion on the Earth, where alone, life is in existence. There is a continuous exchange of matter between these three elements (*i.e.*, lithosphere, hydrosphere, and atmosphere). For example, plants draw their food from the nutrients and moisture found in the soil layers of the lithosphere. Similarly, plants use carbon dioxide from the atmosphere and sunlight for their growth. Dead plants and animals are decomposed and become soil nutrients. The evaporation of water from hydrosphere, and subsequent condensation of water vapour in the atmosphere and consequent rainfall on the Earth, provide water supply for the organisms in the biosphere.

Table 2.2: Biosphere Reserves of India.

S. No.	*Name*	*Area in sq. km.*	*Location (States)*
1.	Nilgiri	5,520	Tamil Nadu, Kerala and Karnataka
2.	Nanda Devi	5,861	Uttarakhand
3.	Nokrerk	820	Meghalaya
4.	Manas	2,837	Assam
5.	Sunderbans	9,630	West Bengal
6.	Gulf of Mannar	10,500	Tamil Nadu
7.	Great Nicobar	885	Andaman Nicobar Islands
8.	Similpal	4,374	Odisha
9.	Dibru-Saikhowa	765	Assam
10.	Dehang Debang	5,112	Arunachal Pradesh
11.	Pachmarhi	4,926	Madhya Pradesh
12.	Kanchanjanga	2,620	Sikkim

The organisms in the biosphere provide a variety of food for us and also raw materials which provide clothing and shelter and other needs of man. Our existence and survival on the earth depends on the biosphere.

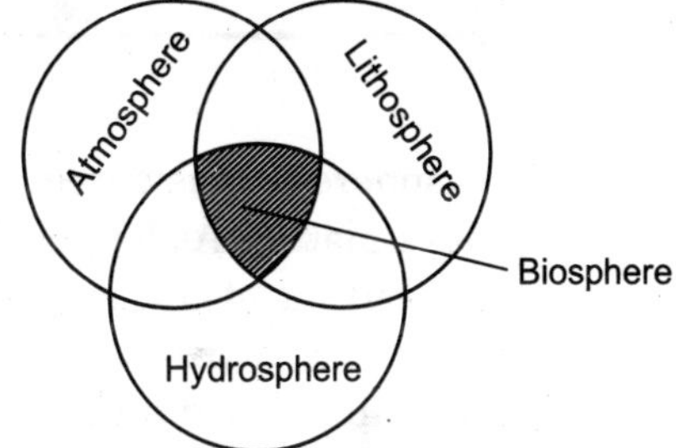

Fig. 2.14: Biosphere

Any change in the physical environment brings about suitable adaptation by various organisms. The abiotic components of the physical environment influence the types of organisms living in an area. For example, the organisms living on the landmasses are different from those in sea-water. Climatic conditions bring about variations in the types of plants and animals found in different landmasses.

The non-living components consist of chemical substances found in the soil, water and atmosphere. These chemical may be *inorganic* substances, like water, oxygen, carbon dioxide and minerals; or *organic* materials, like carbohydrates, fats, proteins and vitamins. Other abiotic elements of the climate are temperature, rainfall, duration of sunlight, winds, nature of soil, slope of the land, composition of water bodies etc.

The biotic components are broadly divided into two major groups *viz.*, the *producers* and the *consumers*. The producers are organisms which produce their own food from the physical environment. Green plants are primary producers. All other organisms are called consumers or *heterotrophic organisms,* as they depend on other organisms for their food. An organism that feeds only on plants is called a *herbivore* or *primary consumer*. A consumer that feeds on animals only is called *carnivore* or *secondary consumer*. A rabbit is a herbivore, while a lion is a carnivore. The humans are *omnivore,* as they eat both plants and animals. A fourth group of consumers feed on dead or decomposed tissues of plants and animals and are called *decomposers* or *detritus feeders e.g.*, bacteria and fungi, termites.

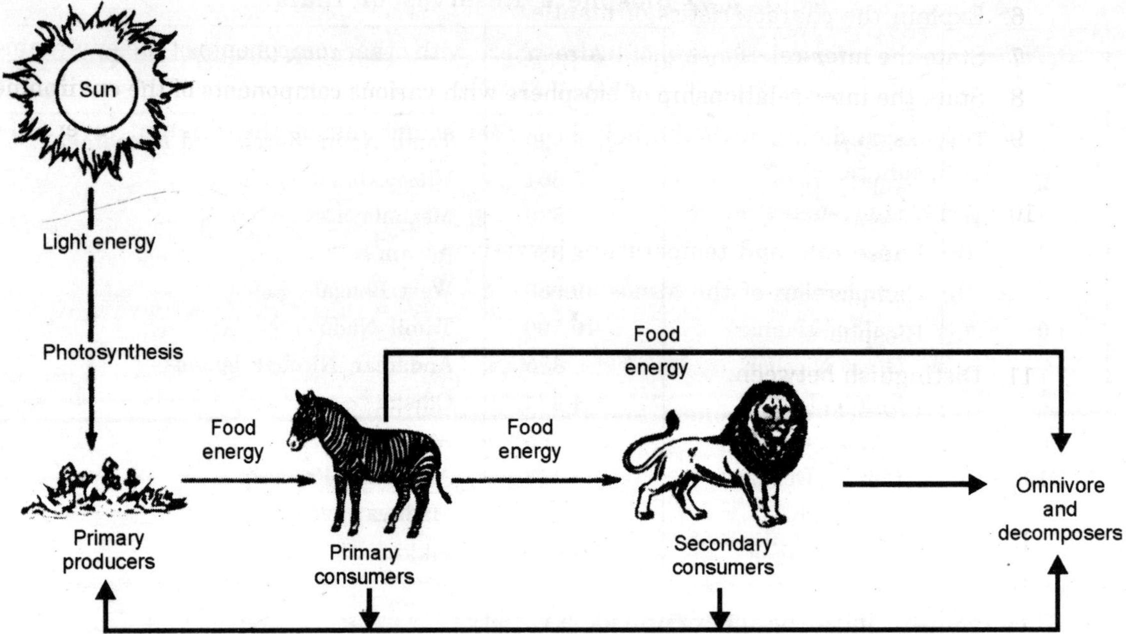

Fig. 2.15: Food chain.

The biosphere is considered as an open system because there is continuous flow of energy and output of matter. As long as the balance is maintained between the input of energy and matter, and output of matter, the biospheric system remains in equilibrium condition but when only of the attributes of the biospheric system crosses critical limit, the system equilibrium is disturbed and several environmental and ecological problems crop up.

It may be pointed out that the equilibrium state of the biosphere system which is in natural state, self-sustaining and ecologically efficient, depends on the intimate relationship between various components of the biosphere and a series of large-scale cycle mechanisms (energy cycle, hydrological cycle, sediment cycle, nutrient cycle etc. – collectively known as biogeochemical cycle), as these cycles influence the biotic, abiotic and energy components of the biosphere, whereas in turn these components also influence the cycle mechanisms involving the movement and transfer of energy, water, sediments and nutrients throughout the biospheric system.

QUESTIONS

1. **Why is the earth considered a unique planet?**
2. **Write short notes on:**
   (a) **Origin of the earth** (b) **Shape of the earth**
   (c) **The Earth's movement** (d) **Longitude and time**
3. **What is a solar system? Discuss its main characteristics.**
4. **Discuss the different components of environment.**
5. **Discuss about the different layers of lithosphere.**

6. Explain the characteristics of mantle.
7. State the inter-relationship of hydrosphere with other components of the environment.
8. State the inter-relationship of biosphere with various components of the environment.
9. Discuss in detail, with the help of a neat sketch, about the different layers of the atmosphere.
10. Write short notes on:
    (a) Lapse rate and temperature inversion.
    (b) Composition of the atmosphere.
    (c) Biosphere
11. Distinguish between:
    (a) Physical environment and Biological environment
    (b) Troposphere and stratosphere
    (c) Evaporation and condensation
12. Discuss the significance of lithosphere to man.
13. Describe the composition and structure of the atmosphere.
14. Discuss the role of biosphere in the environment.
15. (a) How are the oceans important to us?
    (b) Discuss the importance of atmosphere.
16. Discuss the 'food chain' with the help of a neat sketch.
17. (a) What is a water table?
    (b) Discuss different types of 'springs'.
18. (a) What are the different forms of precipitation?
    (b) Permeable and impermeable rocks.
    (c) Discuss the importance of underground water and its occurrence.
19. Explain 'hydrological cycle' with the help of a sketch.
20. Write short notes on:
    (a) Desert
    (b) Effect of altitude on temperature
    (c) Photosynthesis
    (d) 'Producers and consumers' with reference to 'food chain'.
21. Discuss the interior of the earth.
22. What is a rock? Discuss different types of rocks in brief.
23. (a) What are the main minerals found in the earth's crust?
    (b) Discuss the procedure of mountain formation.
24. Discuss the following in brief:
    (a) Acquifers (b) Geysers (c) Coter lakes
    (d) Water falls (e) Caves (f) Glaciers

25. Write short notes on:

    (a) Erosion (b) Desert (c) Thermosphere

26. Define the term 'Biosphere'. How does biosphere contribute to the maintenance of living matter?
27. What is the role of ozone in protecting the biosphere? Which substances deplete the ozone layer?
28. What is the tilt of the Earth? What is its effect?
29. Differentiate between lines of latitude and lines of longitude.
30. Discuss different classes of rocks, depending upon their mode of formation.
31. List the names of continents of earth.
32. Briefly discuss about the Geographical zones of India.
33. What is 'erosion'? Explain various agents of erosion.

PART II : ECOLOGY, ECOSYSTEM AND BIODIVERSITY

3

Chapter

ECOLOGY

3.1 INTRODUCTION

Ecology is a relatively new science and deals with various principles which govern relationships between the two components of nature *viz.*, organisms and their environment. The term **ecology** was coined by combining two Greek words **oikos** (meaning 'house') and **logos** (meaning 'the study of') to denote relationships between the organism and their environment.

Definitions

Some of the definitions of ecology are given here under:

(i) ".......the knowledge of the sum of the relations of organisms to the surrounding outer world, to organic and inorganic conditions of existence......". —*Ernst Haeckel*

(ii) "the study of organisms in relation to their environment." —*Worming*

(iii) "a science which investigates organisms in relation is their environment, and a philosophy in which the world of life is interpreted in terms of natural processes." —*Woodbury*

(iv) "the science of all relations of all organisms to all their environment." —*Taylor*

(v) "the scientific approach to the study of environment interactions which control the welfare of living things; regulating their distribution, abundance, reproduction and evolution." —*Petrides*

(vi) "ecology is the scientific study of the interactions that determine the distribution in abundance of organisms." —*Krobs*

(vii) "the science of environment." —*Karl Friederichs*

(viii) "the study of the structure and junction of the ecosystem." —*Evgene P. Alum*

(ix) "ecology concerns itself with the inter-relationships of living organisms, plant or animal and their environments, these are studied with a view to discover the principles which govern the relationships. —*A. Macfadijen*

(x) "ecology deals with the ways in which organisms are moulded by their surroundings, how they may use their surroundings and how an area is altered or affected by the presence of an organisms."

3.1.1 Scope

As seen from its definition, the goal of ecology is to find out how everything in the biosphere is among organisms, populations, communities, ecosystems, and the biosphere.

Meaning of different terms used above are given hereunder:

(i) An *organism* is any form of life.

(ii) All organisms of the same kind constitute a *species.*

(iii) Each species is composed of smaller units, known as *populations.* For example, white oak trees in a forest, people in a country or world constitute a particular population. *A population* can also be defined as a group of individual organisms of the same species in a given area.

(iv) Each organisms and population has a *habitat,* the place or type of place, where it naturally thrives.

(v) A *community* is made up of all the populations of plant and animal species living and interacting in a given habitat or area at a particular time. Thus, a *community* is a group of populations of different species in a given area.

(vi) *Ecosystem* (or Ecological System) can be defined as a functioning, interacting system composed of one or more living organisms and their effective environment, both physical and biological. Thus, *ecosystem* is the whole biotic community in a given area plus its abiotic environment.

(vii) All of the earth's ecosystems together make up the *biosphere.*

Ecology plays an important role in agriculture (crop rotation, need control etc.), management of grasslands (range management), forestry, biological surveys, pest control, fishery biology, and in the conservation of soil, wildlife forest, water supplies etc.

The concept of ecology has now been extended from plant and animal ecology to other ecologies of the biosphere, like forest ecology, grassland ecology, lake ecology, river ecology, agricultural ecology, industrial ecology, urban ecology, population ecology and social ecology etc.

3.1.2 Some other terms related to Ecology

1. *Species:* A species is a natural biological unit tied together by the sharing of a common gene pool. Species can also be defined as a uniform interbreeding population spread over time and space.
2. *Vegetation:* Vegetation is the totality of plant growth including large or small populations of each species inter-mixed in a region. In other words, we may say that vegetation is the sum total of plant population covering a region.

3. *Flora:* It is the vegetation species content of the region irrespective of the numerical strength of each species.
4. *Habitat:* The place, where an organism lives, is known as the habitat of that organism. The habitat of an organism represents a particular set of environmental conditions suitable for its successful growth.
5. *Biome:* Biome is a major ecological community of organisms maintained under a climate zone.
6. *Biomass:* Biomass is the standing crop expressed in terms of weight (*i.e.* organisms mass) of the living matter present.
7. *Conservation:* Conservation means to ensure the preservation of a quality environment for the organisms in terms of their nutrition, recreation etc., and to ensure at the same time through planned manipulations, a continuous yield of useful plants.
8. *Pollution:* Pollution is an undesirable change in the physical, chemical or biological characteristics of air, land and water, that may or will harmfully effect human life directly or indirectly.

3.1.3 Levels of Organisations in Ecology

The main levels of organisations of ecology are :

1. **Individual** : Organism is an individual living being that has the ability to act or function independently. It may be plant, animal, bacteria, fungi etc.
2. **Population** : Population is a group of organisms usually of the same species, occupying a defined area during a specified time. The main factors that make population increase are birth and immigration, while the main factors for decrease in population are death and emigration.
3. **Community** : A community is a local association of several populations of different species. A community can also be defined as a group of mutually adjusted plants and animals inhabiting a natural area.

   In order to survive, individuals of any one species depend on individuals of different species with which they actively interact in several ways. For example, animals require plants for food and trees for shelter. Plants require animals for pollination, seed dispersal, and soil microorganism to facilitate nutrient supply.
4. **Ecosystem** : An ecosystem is defined as a structural and functional unit of biosphere consisting of community of living beings and physical environment, both interacting and exchanging materials between them. It includes plants, trees, animals, fish, birds, microorganisms, water, soil and people.
5. **Biome** : The terrestrial part of the biosphere is divisible into enormous regions called biomes, which are characterized by climate, vegetation, animal life and general soil type. Biome is a major ecological community of organisms maintained under a climate zone. No two biomes are alike. The climate determines the boundaries of a biome and abundance of plants and animals found in each one of them. The most important climate factors are temperature and precipitation.
6. **Biosphere** : Biosphere is the part of the earth where life can exist. Biosphere represents a highly integrated and interacting zone comprising of atmosphere (air), hydrosphere (water) and lithosphere (land).

3.2 BASIC CONCEPTS AND PRINCIPLES OF ECOLOGY

Ecology has certain basic fundamental principles which govern the various aspects of organisms *e.g.*, evolution and distribution of plants and animals, extinction of species in different components of biological communities, life-cycle of organisms, interactions and inter-relationships among the organisms and physical environment etc. These concepts and principles may be analysed on structural and functional basis separately.

Fundamental concepts and principles of ecology are:

1. All living organisms and their environment are mutually reactive affecting each other. Animal population, flora and vegetation are inter-dependent through the environment and are mutually reactive.
2. Environment is a complex of several inter-related factors and is dynamic *i.e.* varies with time and space.
3. The species make effort to maintain their uniformity in structure, function, reproduction, growth and development by preservation of its genetic pool. The species may increase their capacity of tolerance towards changing environment.
4. Organisms too modify their environment as a result of their growth, dispersal, reproduction, death, decay etc. The dynamic environment and organisms make ways for the development of different kinds of organisms through a process known as *succession.*
5. Under similar climatic conditions, there may simultaneously develop more than one communities, some reacting to climax stage, other under different stages of succession. This complex of several communities in any area, represented by an assemblage of different kinds of plants, animals etc., sharing a common climate is called a *biome.*
6. Ecosystem brings physical environment, plants and animals together in a single framework which facilitates the study of interactions between biotic and abiotic components.
7. At the largest, the whole biosphere becomes an ecosystem, in which biotic and abiotic components are intimately related to a series of large-scale cyclic mechanisms. These cycles help in the transfer of energy, water, sediments, chemical elements etc., in various components of biosphere.
8. Sustained life on the earth is a characteristic of ecosystem.
9. Nothing actually disappears when we throw it away, because all the materials are rearranged and cycled and recycled through a series of cyclic pathways in the natural environment.
10. We live on a planet earth whose resources are finite.
11. Nature has spent literally millions of years defining a stable ecosystem.
12. The same physical (right from the origin of the planet earth and its atmosphere) and biological (since the origin of first organism) processes which operate today, operated in the past not necessarily with constant magnitude and frequency with time and will operate in future at rates that will vary as the environment is influenced by human activity.

13. Nature hazards affect adversely the biological communities in general and man in particular. For example, after floods, diseases spread at faster rate.
14. Man, being an active agent of environmental change, modifies the ecosystem through the exploitation of natural resources. He reduces the ecological diversity and complexity of the ecosystem by removing a host of biotic communities for different purposes.
15. There is inbuilt self-regulating mechanism in a natural ecosystem through which any change caused by external factors in the ecosystem is counter-balanced by the responses of the system to the change in such a way that ultimately ecosystem or ecological stability is restored. The ecological diversity and complexity enhance the ecosystem stability. Here, ecosystem stability means balance between production and consumption of each element in the ecosystem.
16. There is marked positive correlation between primary productivity of ecosystem and solar radiation. The other factors, besides solar radiation, which control the ecosystem productivity, include abiotic factors (rainfall depth of water, climate and chemical factors—nutrients) and biotic factors.
17. The inorganic and organic substances are circulated among the various components of the biosphere through a series of closed system of cycles collectively known as *'biogeochemical cycles'* in such a way that total mass of these substances remains almost the same and are always available to biotic communities. These inorganic and organic substances are required by the organisms as nutrients for the sustenance of their lives.
18. All living organisms and physical environment are naturally reactive.
19. Ecosystem functions through the input of energy mainly solar radiation which is trapped by green plants and is used to prepare food through the process of photosynthesis.

    The first and second laws of thermodynamics are completely applicable in this process.

First Law of Thermodynamics: In any system of constant mass, energy is neither created nor destroyed but it can be transformed from one type to another type. The energy inflow in a system is balanced by the energy outflow.

Second Law of Thermodynamics: When work is done, energy is dissipated and the work is done when one form of energy is transformed into another form.

3.3 ECOLOGICAL SUCCESSION

The communities of living biological species are not permanently stable over time. They do change, involving replacement of one community with the other. This replacement of one community with the other is called the *ecological succession*. This is a continuous process, atleast until a final stage, called *climax,* is reached. Even after a climax is reached, it is doubtful as to whether the community will always keep its composition unchanged. Communities are never stable, but are dynamic, changing more or less regularly over time and space.

Environment is always kept on changing over a period of time due to:

(i) Variations in climatic and physiographic factors, and
(ii) The activities of the species of the communities themselves.

These influences bring about marked changes in the dominants of the existing community, which is, thus, sooner or later replaced by another community at the same place. This process continues and successive communities develop one after another over the same area, until the terminal final community again becomes more or less stable for a period of time. This occurrence of relatively definite sequence of communities over a period of time in the same area is known as *ecological succession.*

In simple words, **succession** may be defined as a natural change in the structure and species composition of a community. Thus, the succession refers to change in a community, following either disturbance or the collonisation of newly exposed substrata.

The process of *'ecological succession'*, or sometimes terms of following three parameters:

1. It is an orderly process of community development that involves changes in species structure and community processes with time, it is reasonably directional, and hence, predictable.
2. It results from modification of the physical environment by the community.
3. It culminates in the establishment of stable ecosystem as it is biologically possible on the given site.

Causes of Succession

(a) *Initial or Initiating causes:* These are climatic as well as biotic. The former includes factors, such as erosion and deposits, wind, fire etc., and caused by lighting or volcanic activities; and the latter includes various activities of organisms. These causes produce the bare areas or destroy the existing population in an area.

(b) *Continuing causes:* These are the processes as migration, aggregation, competition, reaction etc. which cause successive waves of population as a result of changes.

(c) *Stabilising causes:* These cause the stabilisation of the community. Climate of the area is the chief cause of stabilization.

Basic Types of Succession

1. *Primary Succession:* In primary succession, the living organisms establish themselves for the first time in that habitat. This first community of living organisms that develop in a habitat is called the pioneer community. The pioneer community will modify the sub-stratum, and the new community, called the *secondary community*, will come upto replace the pioneers. Each community will exist for sometime, and then it will be replaced.
2. *The Secondary Succession:* It differs from the primary succession in the sense that in this case, the habitat or sub-stratum would have been previously occupied by a living community, but due to *catastrophic event,* like flood, fire or volcanic eruption etc., the previous community disappeared in the past, and now new organisms start inhabiting that habitat, in present times.
3. *Autogenic Succession:* After the succession has begin, in most cases, it is the community itself which, as a result of its reactions with the environment, modifies its own environment and this causing its own replacement by new communities. This course of succession is known as *autogenous succession*.

4. *Allogenic succession:* In some other cases, however, the replacement of the existing community is caused largely by any other external condition and not by the existing organisms. Such a course is referred to as *allogenic succession*.

3.4 POPULATION ECOLOGY

The term *population* is used in many ways and can have different meanings. In ecology, ***population*** is a group or assemblage of organisms of the same species that live at a given time in a particular area. The *population ecology* is concerned with inter-relation between individuals of the same population as well as those of the other species. We call these as intra-specific and inter-specific relationships. When the inter-relations are of beneficial nature to the participants, then it is called as co-operation, but when these are harmful, then we call it as disoperation.

Population ecology can be defined as the study of the individuals of the same species where the processes as aggregation, interdependencies between individuals etc., and the various factors governing such processes are emphasized. Population ecology is a new subject, the foundation of which was laid by animal ecologists in the first half of the twentieth century. The increasing population of the world has attracted the attention of not only the ecologists but also of behavioural scientists and has led to the development of such new facets in ecology. Population ecology, thus, has to play a vital role in number of socio-economic problems, both at national as well as international levels.

Under favourable conditions, the group of individuals increase in numbers. An increase in this number results in other interdependent species growing in the area. Such an increase brings about harmful as well as beneficial effects (interactions) on the species. Harmful effects are mainly due to increased competition for nutrition, space etc. The beneficial effects include the protection, influences on reproduction etc.

The subject of population ecology can be discussed in following three parts:

1. Population characteristics (properties).
2. Population dynamics.
3. Regulation of population.

3.4.1 Population Characteristics

Various characteristics which are unique to the group as a whole are described hereunder:

(i) *Population Size and Density:* Total size is generally expressed as the number of individuals in a population, population is also reflected by its *density*, which is the number of individuals per unit area or volume. This is known as numerical density. We can also measure biomass per unit area or volume and this is known as biomass density. The unit of area or volume in measuring density varies with the size of species. For example, this may be expressed as number of micro-organisms per litre of water. For larger animals or plants, it may be acre or hectare, 1 million cells of *phytoplanktons* (as *algae*) per cubic meter of water, or 100 kg of fish per hectare of water surface.

By measuring the size or density of a given population from time to time, we can obtain important information about its rate of increase and can also predict future changes in its size. Since, population is a changing entity, we are interested not only in size and density but also in nature of its change. As varying from place to place, population density also varies in time. Population may remain constant or they may fluctuate *i.e.*, they may steadily increase or decrease. Such changes are the main focus of population ecology. In such a situation, there may be four factors: *Natality* (production) and *immigration* (individuals entering a population from elsewhere) lead to increase in density; while *mortality* (death of individuals) and *emigration* (moving out of population) lead to decrease in density. A *rate* of change in population is obtained by dividing the change by period of time elapsed during the change.

(ii) *Spacing:* Spacing is an important feature of every population. There are three possible ways in which the individuals in a population may be spaced relative to one another. Accordingly, spacing may be *uniform*, *random*, or *clumped*.

(iii) *Age Structure:* In most populations, individuals are of different ages. The proportions of individuals in each age group is called age structure of that population. Age distribution is important, as it influences both, natality and mortality of the population.

Age ratio is the ratio of immature individuals to adults present in a population. A low age ratio than the usual or normal indicates the poor reproduction. Thus, the ratio of the various age groups in a population determines the current reproductive status of the population, thus, anticipating its future.

Breeding age is a characteristic of a species and is represented by the age at which the animals attain sexual maturity and affects reproductive capacity and growth of a population.

(iv) *Natality (Birth-rate):* The new individuals of any organism are born, hatched, germinated, arise by division etc. *Natality rate* is the number of offsprings produced per female per unit time.

(v) *Mortality (death-rate):* It refers to death of individuals in the population.

A birth-death ratio *i.e.*, $\frac{\text{birth}}{\text{death}} \times 100$ is called *vital index*.

(vi) *Life Tables:* There are the tables in which information on natality and mortality in different ages and sexes are compiled. With the help of these tables, it is possible to estimate the growth or decline of a population. Each life table has columns for age of individuals, number surviving to each age, the number dying in each age group, fertility rate, the number of young born by each age group etc. The information obtained from these tables provide the net reproductive rate of the population. These tables also help in providing mortality. These details are then used to calculate the rate of population growth.

(vii) *Adaptation:* This is the most important characteristic. Each population is fully adapted to its environment and is capable of using the available resources. *Adaptation* is a consequence of evolution and depends on the genetic variability in a population.

3.4.2 Population Dynamics

Populations are not static but are of dynamic entities, which are always in a state of flux. Thus, population dynamics means changes in population sizes. The study of population dynamics is called 'demography'. The main aim of any demographic study is to quantify the changes in a population by finding out the number of births, deaths, immigrants and emigrants.

Approaches to study of population dynamics are:

(i) Mathematical models,

(ii) Laboratory studies,

(iii) Field studies.

3.4.3 Regulation of Population

This deals with the factors which govern population density. The logistic model and its derivatives assume that (*i*) a population will level off of its carrying capacity, (*ii*) that there is an upper limit to population density set by the environment, and (*iii*) that the population is regulated at or around that level.

As already discussed, population density can only be increased by natality or immigration, and decreased by mortality or emigration. The population density can be influenced by (*a*) density dependent, (*b*) density independent factors. Density-dependent factors include *competition* and *predation* disease etc. Density independent factors do not vary systematically in their effects as density changes. Abiotic environmental factors are often density independent but not always. Floods, fires, extreme cold or volcano may damage all population regardless of their densities.

The size of a population is determined by the interactions either among its members or with other species of competitions, predators or parasites. The role of *competition* in regulating population is directly effective by causing the mortalities, nest destruction and less food supplies. It also results in lowered rates of reproduction, increased predation, dispersal into other regions and decreased health.

The *predator-prey* relationship provides a typical example of density dependent regulation. The explosive increase of some species, when their predators were removed, shows that they had been regulated by their predators. The removal of mountain lions resulted in a great increase in population of deer at some places.

In general, we find that density-independent (abiotic) factors set the ultimate limit or carrying capacity which cannot be permanently exceeded. The density dependent factors, on the other hand, generally stabilize population at or somewhat below the ultimate level. The population of any species may be said to be stabilized or regulated when population fluctuates in restricted manner about a mean. When population encounters very favourable environmental conditions, the fluctuation may go well above the average size of the population. Soon, however, there will be a tendency to revert back to average level.

3.5 COMMUNITY ECOLOGY

As already discussed, population ecology is concerned with the individual population (a group of individuals of single spacies) in relation to environment. However, community ecology deals with the groups of different kinds of population in the area. In nature, different kinds of organisms grow in association with each other. A group of several species (plants and/or animals) living together with mutual tolerance (adjustment) and beneficial interactions in a natural area is known as *community*. In a community, organisms share the same habitat growing in an uniform environment.

Ecological community can be defined as a group of population of different species living in the same local area and interacting with one another. The group or association of plants growing together in a particular habitat is called *plant community*.

A community can also be defined as:

- a group of mutually adjusted plants and animals inhabiting a natural area; or
- a local association of several populations of different species, or
- an assemblage of the population of living organisms in a prescribed area or habitat.

A self-sustaining and self-regulating ecological unit is known as major community. It is made up of a large number of smaller minor communities that are not altogether self-sustaining.

A community ecology is concerned with all kinds of interactions among populations. It studies ranges of various populations. The habitats occupied by populations, the food chains and patterns of energy flow, ecological niches, etc. It also considers some other aspects of community, such as community structure, species diversity, community stability etc.

General Characteristics

All communities have certain general characteristics in common. Like other living entities, a community grows, develops, passes through a relatively stable mature phase, reproduces and ultimately dies. Such community life-cycles, taking hundreds and thousands of years, result from an interaction between organisms and their environment.

Communities do not have exact limits but tend to overlap each other. Animals frequently shift from one community to another because of seasonal or other variations. Some of the major characteristics of the community are as follows:

1. *Species Diversity:* Each community is made up of much different organisms, such as plants, animals microbes. The number of species and population abundance in community vary greatly.
2. *Growth Form:* Community is described in terms of major growth forms as trees, shrubs, herbs, mosses etc. In each growth form as in trees, there may be different kinds of plants as broad-leaved trees, evergreen trees etc.
3. *Dominance:* In each community all the species are not equally important. There are relatively only few of these, which determine the nature of the community. Such species are known as *dominants*.
4. *Succession:* Each community has its own developmental history. Succession have been discussed separately.
5. *Trophic Structure (Self-sufficiency):* The feeding relations of the species in the community determine the flow of energy and materials from plants to herbivores to carnivores.

6. *Stratification:* It involves vertical changes within the community. In this community, possesses a number of vertical layers of species. In grassland communities, there is *subterranean floor*, containing basal portions of the vegetation, and *herbaceous substratum* consisting of upper parts of grasses and herbs with a characteristic fauna.

   Stratification in a forest community is more complicated where following subdivisions are recognised: (*a*) subterranean subdivision, (rich in humus with roots of plants, soil fungi, bacteria, protozoans etc.), (*b*) herbaceous vegetation, (*c*) forest floor (have dead organic matter, litter and fungi, bacteria, animals active in its decay), (*d*) shrubs, and (*e*) trees.
7. *Ecotone:* Intermediate transitional zone between two distinct communities is known as *Ecotone* or a tensional zone. An ecotone can be defined as a zone of transition between adjacent ecological systems having a set of characteristics uniquely defined by space and time scales and by the strength of interactions between adjacent ecological systems.
8. *Turnover:* Individuals of the various populations emigrate or die out and are replaced by others. Since this flux is automatically self-adjusting, the community remains internally balanced and exhibits a numerical steady-state.

3.6 INTER-RELATIONSHIP (ECOLOGICAL RELATIONSHIPS)

Organisms live together influencing each other's life directly or indirectly. Such vital processes, such as growth, nutrition and reproduction depend upon the interactions between the individuals of same species (intraspecific interaction), or between those of different species (interspecific interaction). Moreover, interdependency may exist between species which are taxonomically widely different, such as between trees and bacteria, or between elephants and ants. The relationship may be beneficial to both, harmful to both, or beneficial or harmful to one and neutral for the other.

Interspecific Relationship

The members of two or more species may establish inter-relations through their co-operation (beneficial coactions) or disoperations (harmful coactions). These inter-relations may be of following categories:

1. Symbiosis

Symbiosis means "living together". It is a beneficial coaction between two or more different species in which one or more species are benefitted and neither species is harmed. These include:

(i) Mutualism,

(ii) Commensalism,

(iii) Protoco-operation.

These interactions, also known as "*positive interactions*", benefit each other in respect of food, shelter or transport and the two partners may be in close contact or one of them may live within a specific area of the other or attached its surface.

2. Antagonism

It is a harmful coaction between two different species in which atleast one of the species is harmed. It consists of the following:

(iv) Competition,

(v) Parasitism,

(vi) Predation,

(vii) Antibiosis.

In these interactions, also known as *"negative interactions"*, members of one population may eat members of the other population, compete for foods, excrete harmful wastes or otherwise interfere with the other population:

1. **Mutualism.** It is a relationship between two or more individuals of different in which all are benefitted by one another. Mutualism may be *facultative*, when the individuals involved in the association are capable of existence independently. It is *obligatory* when the relationship is imperative to the existence of one or both species.
2. **Commensalism.** It is an association or relationship between the two different the organisms in which one is always benefitted while the other is neither benefitted nor harmed. This relationship may be permanent or temporary and the benefit derived from the other organism may be in the form of protection, transportation, living space and food. A commensal may live permanently attached to its host, or may live within its body.
3. **Protoco-operation.** It is a relationship between two species, which is favourable but not obligatory to both *i.e.*, not essential for the survival of either population.
4. **Competition.** Competition refers to the interaction in which two individuals or two species struggle for limit amount of common resource (food, water, nesting space, light etc.). In such competition, the species that is more efficient in using the resource will stay, while the other will be eliminated. The interspecific competition plays a significant role in population regulation and in setting distributional limits. Thus, it directly affects the structure of biological communities.

   Competition is usually keenest between individuals of the same species (intraspecific competition, because they have identical requirements for food, shelter, mate etc.
5. **Parasitism.** Parasitism is the type of relation in which parasite derives benefit at the expense of the host. A parasite generally does not kill its host, but cause tolerable harm to its host. The parasite may be external or internal.

   Thus, we can say that a parasite is the organism living on or in the body of another organism and deriving its food more or less permanently from its tissues. Those organisms which derive their nourishment only partly and remain in contact with their hosts only for a short period of their lifecycles are not true parasites. Majority of parasites are micro-organisms, of which fungi, bacteria, mycoplasmas and viruses parasite plants as well as animals.

Parasitism is mainly food coaction, but the parasite derives shelter and protection from the host as well. The balance between parasite and host is upset if the host produces antibodies or other substances, which are harmful for the normal development of the parasite.

Most also, generally, try to avoid parasites by some means or the other. The common method is the secretion of certain juices, which do not allow the parasites to come out of their cysts as in protozoan parasites. Sometimes, body fluids may develop such qualities that may not allow the parasites to fourish. Some host species may develop immunity to parasites.

Parasites are major contribution to mortality in numerous species. In humans, it is generally agreed that parasitic infections are a major factor in infant mortality in developing countries.

6. **Predation.** *Prey-predator relationship* is a relationship in which one animal (prey) is removed from the ecosystem by the other (predator). A lion feeding on a deer or a frog feeding on an insect are the examples of this relationship. A single species may have many predators and a single predator may feed upon several prey species.

The word *predation* is used in a general sense to describe the eating of one species by another. In contrast with a parasite which derives nourishment from its host without killing, a predator is free living which catches and kills another species for food. Predators feed upon the adults, or larvae or even eggs of their prey.

Factors affecting predation: In a natural population, the act of predation is affected by a number of factors including certain characteristics of the predator (body size, hunger level, efficiency of search, pursuit and capture), the prey (the body size, visibility, evasive ability, ante predator defenses) and of the environment (visibility, availability of alternate prey).

Examples of predatory forms: Some of the common examples of predation are:

(*i*) *Browsing and grazing:* Herbivores kill the plants and use unharvested herbs, shrubs or even trees as their food. Sometimes, these predators pore much problems of management of natural and artificial vegetations. Many insects and ruminants browse lightly over the vegetation. Cattle, camels, goats etc. frequently brows the tender shoots of shrubs and trees and sheep graze the grasses. Grazing in shrubby vegetation generally increases the number and sizes of the shrubs by removing the competitive grasses.

(*ii*) *Seeds and seedling destruction:* Animals, such as *insects*, squirrels, mice, rodents etc. consume large quantities of seeds as food. Moreover, they browse seedlings of shrubs and trees, and damage most of them by trampling.

(*iii*) *Plants as food:* Aquatic plants are frequently eaten by animals, like ducks, fish, muskats etc.

(*iv*) *Carnivore plants:* A number of plants consume insects and other small animals for their food. These are adapted to attracted, catch and digest their victims.

Role of Predation: Predation is helpful in the regulation of population size and community structure in all types of habitats. Predators with a wide range of prey species are better fitted for survival than those which feed upon only one or two prey species.

7. **Antibiosis.** In term 'antibiosis' generally refers to the production of "chemical poisons", "antibiotics" or "allopathic agents" that are specifically harmful (antagonistic) to other species. Such antibiotic substances are known to be produced by many kinds of bacteria, fungi and actinomycetes. Antagonistic substances formed by algae have been shown to affect other algae adversely.

   The term "antibiosis" also refers to the complete or partial inhibitation or death of one organism by another through the production of some harmful substance or environmental condition. Here, none of them derives any benefit. The phenomenon of antibiosis is common among microbial world.

3.7 HUMAN ECOLOGY

Human ecology is the study of the mutual relationship of environment, community and human actions. Human are one of the interacting entities in ecosystems. They are contributors, participants and receivers in ecological processes, just as other species functioning in the flow of energy and cycling of nutrients and whose population is subject to growth as well as regulation. At the sametime, humans are the most powerful species in effecting alteration in ecosystems and ecological processes.

The environment serves not only human kind but other species as well. In this sense, human ecology is not different in kinds from any other kinds of ecology. Therefore, humans have no more prerogative to have domination over the land than any other species. Rather, because of their capacity for cognition, reflection, and prediction, humans have responsibility for the environment for surpassing the responsibility of other species. That responsibility is one of being a steward and guardian. In contrast to almost all other species, humans can voluntarily control, or atleast regulate, their environment and population—as well as their behaviour, genetic heritage and evolutionary future. We shall explore in more detail, the nature of human ecology and in subsequent chapters shall discuss various aspects of the human population; pollution of air, water and land etc.

History of Human Ecology

The formal history of human ecology dates atleast to 1921 as a sub-discipline of the social sciences. Although a less tightly defined history can be traced back to the classical Greek philosophers. The Greek philosophers were the first to note that human ecology's roots lie in the intuitive and empirical knowledge of ancient man who lived, perform in intimate relations with the natural world.

Humans and its Relationship with Environment

Two processes go on simultaneously in human life. On one hand, man establishes harmony with his geographical, cultural environment, other individuals and animals. On the other hand, he also tries to control his environment and changes it according to his requirements. He learns to live in hills, valleys, plains and on seacoasts and adapts himself according to nature. He controls the nature through his intelligence, skill and technology.

In the 'primitive social system', man used to adapt himself according to the nature, as his capacity of control over nature was limited. For hunting, fishing and feeding his cattles, he depended on jungles, rivers, seas and grazing fields. Gradually, tools were developed and his dependence on nature decreased a little. With the development of tools, rearing of animals and clearing of the forests, the system of agriculture began. Gradually, permanent villages developed. Means of transport and village markets developed in order to exchange surplus production.

Technology and man's control over nature are correlated. In the primitive societies, man's technological capacity was not so developed, compared to this peasant communities have a better technology and could control nature to some extent. In the agrarian system, the human energy was replaced by the animal energy for agricultural activities. Man used wells, ponds or canals to irrigate his fields during the period of shortage of rainfall. Crafts and cottage industry help in making the instruments of production and, therefore, the technology of the pre-industrial cities was more developed than that of agrarian villages. The modern industrial cities are developed by the modern technology and the effort to control environment is seen everywhere.

3.8 HUMAN SETTLEMENTS

Human population lives in settlements of different shapes and sizes. Clusters of houses make a *settlement* which may consist of hamlets of 6 to 12 huts or compact villages accommodating a few hundred houses. Bigger settlements are called towns or cities. Large cities are known as *metropolis*. The basic unit of residence is a house which may be a small hut or big mansion or an apartment.

The term *'settlement'*, therefore, refers to "a group of houses or huts with a certain layout plan and includes the buildings meant for residential or other purposes, such as pens for animals or stores for equipment or machinery, as well as the streets or the roads which connect them together.

Two types of settlements are normally recognised. These are rural, and urban settlements. These differ from each other depending upon their occupation, size and population.

1. Rural Settlements

These consists of houses, streets and the surrounding farmland. These may be either clustered or isolated. The people in rural settlements are mainly engaged in agriculture and other primary activities. These settlements are of small size, having small population, may be upto few hundred people. Density of population in rural resettlement is less and modern facilities are often absent. The types of rural settlements are determined by: (*i*) Physical factors, like altitude, climate, drainage availability of water; (*ii*) Ethnic or cultural factors, such as tribal cast or communal identity, and (*iii*) Historical factors, like safety. Most rural houses are made of locally available material.

2. Urban Settlements

In India, urban settlements are those places which satisfy the following criteria:

(a) a minimum population of 5000,

(b) atleast 75% of male working population engaged in non-agricultural jobs, and

(c) a density of population of atleast 400 persons per sq. km.

Therefore, in urban settlements, people are mainly engaged in industry, trade, transportation and services. These are bigger in size and are often equipped with modern facilities.

Patterns of Settlement

As discussed above, patterns of settlement differ from place to place and is the result of the interplay of cultural or physical factors. Differences in settlement are also due to a great extent to the economy or the way of making a living adopted by a particular group of people. A house gives an idea about the life that he carries on. The knowledge of settlement throws light on the differences in the mode of living in different parts of the country.

The mode of settlement may be mobile or fixed.

Mobile Settlements (Nomadism): Some people move from place to place in search of their living without having a permanent residence. These people are called *nomads*. Their houses are temporary and are easily movable from one place to another. Most of the nomads of the world inhabit desert margins, arctic tundras or tropical forests. Many tribal groups belong to the mobile population of the world.

Fixed or Sedentary Populations: These are tied to specific place on the surface of the earth. Such populations have more permanent patterns and forms.

Patterns of settlement are generally determined by the occupation of the human group. For instance, the hunting, fishing favours nomadism because mobility is essential to the human group that follows this economy. Agriculture which makes permanent attachment to land favours scattered or clustery patterns settlement. Manufacturing and demands concentration of people in certain small area, materials are easily available.

3.9 ORIGIN AND EVOLUTION OF LIFE (ANIMAL)

Evolution of life means origin, growth and development of life through time. Thus, evolution is the indicator of historical development of animal species. History of life on the earth is preserved in fossils. Fossils are the hardened remains of plants and animals once lived on the earth.

The universe began to form over 15,000 million years ago. The earth is much younger and is about 4600 million years old. This age of the earth has been calculated by the scientists with the help of radioactivity and fossil records.

The only sources of the evidences of the origin of the oldest animals on the earth's surface are fossil preserved in the sedimentary rocks of different geological periods. These fossils provide only the rough estimates about the origin of animals because the fossils have also undergone large-scale modification and transformations.

The earth has existed for about 3500 million years before the first sign of life appeared. As the fiery planet earth cooled, its atmosphere formed. How did life on the earth began, is not for sure. Perhaps minerals from rocks formed and reformed new chemical combination innumerable times – thus, an unusual chemical combination appeared: a living cell which could use energy to feed itself and reproduce copies of itself. However, man appeared on the earth about a million years ago, out of which only about 500 years of existence is known.

Following is the chronological sequence of the origin and evolution of animals:

1. Pre-Cambrian Period (more than 600 million years ago)

According to H. Clemmay (1976), the fossils of oldest animals have been found in Zambia (Africa), which indicate that earliest animals might have evolved in the said area some 1000 million years ago. The primitive earliest animals belonged to the category of *'detrivores'* which derived their food from the detritus because no autotrophic plants could be evolved by that time. The fossils of softbodied invertebrate animals, like jellyfish, segmented worms, and soft corals have been found from the Ediacara Hills of South Australia. These organisms were evolved in marine environment which dateback to 680 to 580 million years ago. The fossils of similar organisms have been found in modern Australia, Europe, Africa and North America. All these contents might have been united together during that period (pre-Cambrian period).

2. Cambrian Period (600 million to 500 million years ago)

The climate during this period became warmer, while it was cooler during pre-Cambrian period. Consequently, algae, fungi and bacteria as plants and several varieties of microscopic organisms and invertebrate animals (such as unicellular organisms of the Phylum Protozoa, sponges, sea urchins, see cucumbers, starfishes etc.).

3. Ordovician Period (500 million to 440 million years ago)

First appearance of vertebrate animals in fresh water environment was noticed in this period. These vertebrates were characterized by jawless bodies but by bony armour around their bodies so as to protect them from other organisms and from the severity of the environment. Snails, corals and sponges developed in this environment.

4. Silurian Period (440 to 400 million years ago)

Global climate again changed from warmer to cooler type, and geological and tectonic activities were dominated by large scale mountain building. The primitive inland seas were dominated by marine algae, whereas first land plant in the form of clubmoss was developed in the surroundings of the inland seas. Because of the richness of land and water flora a large variety of animals were evolved and there was wide expansion of already evolved invertebrates.

5. Devonian Period (400 to 350 million years ago)

Because of the increased temperature again, and broad distribution of uniform climates maximum development in the life-form of primitive animals took place. The primitive fishes developed bony gills which were later on transformed into jaws. In this period fresh water fishes developed habit and capability of involving and attacking other organisms and ability to compete with other organisms to get food. The first amphibians and spiders were evolved and wide expansion and dispersal of primitive fishes took place. Many land animals developed burrowing habits. Because of the development of first forests resulted into phenomenal increase in the numbers and type of species of animals due to availability of abundant food. Several varieties of insects and flies were originated.

6. Carboniferous Period (350 to 270 million years ago)

In the first part of this period, several types of plants and animals were evolved; increase in number of insects; and development in amphibians further and a few of them were evolved into reptiles. In later part of carboniferous period, the climate of Gondwanaland (including South America, Africa, India, Australia and Antarctica as well, these were united together) changed into very cool climate, due to which major parts of Gondwanaland were extensively glaciated. Great Ice Age of this period resulted into widespread extinction of several animal species, but the animals which became successful in withstanding the severe climatic conditions and ultimately could survive, became very strong and powerful. Amphibians still remained dominant animal but reptiles also increased their numbers.

7. Permian Period (270 to 225 million years ago)

This period also remained covered with ice-sheets as in of carboniferous period. All land-masses were still united together as *'Pangaea'*. This period was dominated by many events like formation of mountains, their erosion and resultant degradation, origin of deserts, advancement and retreat of ice-sheets etc. Amphibians and reptiles still dominated. Reptiles acquired the properties of brisk walking, running and swimming.

8. Triassic Period (225 to 180 million years ago)

'Pangaea' was broken into two bid landmasses (Laurasia and Gondwanaland) which were separated by Tethys sea, but were still linked by a few land bridges. There was development of world-wide tropical and sub-tropical climates and several deserts and phenomenal increase in the number of invertebrate animals, such as spiders, scorpions etc. First aquatic reptiles were evolved. A few reptiles developed the characteristics of primitive flight mechanisms.

9. Jurassic Period (180 to 135 million years before present)

This period saw a world-wide warm climates and beginning of the disruption of 'Pangaea' and continental drift though at a very small scale. This period was characterised by the abundance of first birds, appearance of primitive mammals, evolution of dianosaurs which could live in marine habitats, evolution of primitive birds from such reptiles which had the primitive flight mechanisms, evolution of first lizards and crocodiles etc. A few plants developed flower, like structure which helped in developing insects.

10. Cretaceous Period (135 to 70 million years before present)

In this period, there was evolution and development of flowering plants. Among animals, there was phenomenal increase in the number of dianosaurs. Different species of bony fishes were evolved.

11. Tertiary Epoch Period (70 to 1 million years before present)

This period is sub-divided into (*i*) Eocene period (70 to 40 million years B.P.), (*ii*) Obligocene period (40 to 25 million years B.P.), (*iii*) Miocene period (25 to 11 million years B.P.), and (iv) Phiocene period (11 million to 1 million years B.P.).

This epoch was characterized by the origin of the youngest and tallest Alpine folded mountains (*e.g.*, Rockies, Andes, Alps, Himalayas etc.), large scale displacements and drift of continents and sea-floor spreading and the attainment of the present position of the continents and ocean basins. This process is still continuing.

This period was characterized by the modernization and extension of flowering plants, evolutionary explosion of mammals, maximum diversity in the insectivores and herbivores, evolution of carnivorous mammals; evolution of hoofed animals like horse, rabbits and rodents, evolution of even toed herbivores like deer and cattle, and later on odd toed animals like rhinoceros and hippopotamus, evolution of ancestors of whales in the sea and ancestors of modern bats in the air, and elephants.

In the later part of this epoch high mountainous areas began to become cool. The modern mammals were evolved in this later part. Development of grasslands, phenomenal increase in number and species of mammals, migration elephant-like animals took place in this period.

12. Quaternary Epoch (from 1 million years before present to present time)

During this epoch, there were four glacial periods characterized by advancement of ice sheets and four interglacial periods, when there was retreat of ice-sheets. These glacial periods have affected and modified the plants and animals on a large scale. There was further increase in herbaceous vegetation but decrease in trees and forest because of covering of ground surface by ice-sheets. Due to the four glacial and four warm interglacial periods, a few large animals faced mass extinction, like dinosaurs because they were slow-moving animals due to heavy weight of their bodies and large tails. Brisk-walking and fast-running animals could survive because they rapidly migrated from the areas of extreme environmental conditions to relatively favourable areas.

Though the modern man was evolved about 10,000 years ago but the initial stage of the evolution of man began from "tertiary epoch". Man-like animals were evolved from animals living on trees branches, such as different species of Monkeys, Langur, Gorillas and Chimpanzee. The Cental-east Africa is considered as the original place where the early man was evolved some 40,000 years ago.

Dispersal of Animals

The spreading of animals from the places of their origin to other areas is called, "animal dispersal" which may be gradual, rapid, seasonal or forced. Dispersal of animals is also caused due to continental drift and climatic changes. The important factors which affect dispersal and migration of animals are generally grouped into two categories *i.e.*: (*i*) Environmental or physical factors, and (*ii*) Biotic factors. These factors are discussed separately in detail.

3.10 EVOLUTION OF PLANTS AND SPECIATION

Plant evolution means the origin, growth and development of plants through time. Thus, the evolution is the indicator of historical development of plants. The process which is responsible for the origin of species is called *speciation*. Following are the important processes and mechanisms which determine the nature of evolution of plants:

1. ***Variation:*** Variation is common feature of all plant species because there is a wide range of variation and diversity in the plants. Variation is found even in different populations

of a single species in terms of their height, size, shape of size of leaves, number, colour and structures of flowers etc. Variation is found even among the offsprings of the same parents. Inherited qualities get modified by the environmental conditions during the growth and development of plants.

2. *Natural selection:* In the process of natural selection, the heritable variations in the populations form the basis of evolution of species. The process of natural selection means that a few of the advantageous qualities inherited from the parents of a species by a few populations of that species are such that these qualities enable a few individual members of that species to survive in their environment and to become adapted to the environmental conditions. While some other individual members of the species lack in the advantageous qualities because these could not be inherited by them from their parents. The plants which do not possess the advantageous qualities are eliminated.

Charles Darwin assumes the principles and mechanisms of the evolution of species by the process of 'natural selection' and 'survival of the fittest' (known as *Darwin theory of evolution*).

Following are the main principles of "natural selection" based on the Darwin Theory of the origin of species:

(i) There are heritable variations in the individual species. Though each heritable variation is small but it is very significant because it forms the very basis of evolution of species.

(ii) Heritable variations provide certain characteristics which become advantageous to some individuals to become more efficient and better able to survive in their environment for the struggle of their lives. These advantageous characteristics when passed on the succeeding generations, they become superior and further efficient in the struggle for existence and survival.

(iii) The individuals of species which lack the advantageous characteristics are most likely to be eliminated. During the struggle for existence, the winner leads to evolution of species and the laser leads to extinction.

(iv) This process of natural selection and adaption to environmental conditions lead to gradual modification and diversification of species over long period of time.

3. *Artificial Selection:* Artificial selection occurs through human activities, when he creates new environment. For example, man creates new seeds to increase the crop-productivity through artificial measures. The process of artificial selection carried out by man creates new hybrid and high-yielding seeds on one hand and eliminates several naturally occurring seeds.

4. *Isolation:* The process of isolation stops the exchange of genes between different members of species: This reproductive isolation is caused due to a variety of barriers, grouped into two categories. (*i*) *External barriers* operate from outside the plants and prevent cross-pollination between different populations of species of plants. These barriers include Geographical isolation, ecological isolation, seasonal isolation and mechanical isolation. (*ii*) *Internal barriers* prevent hybridization even after cross-pollination has taken place.

5. *Gradual speciation:* As already discussed, the process which is responsible for the origin of species is called *speciation*. The speciations are of two forms, namely gradual speciation and abrust speciation. In gradual speciation, initial breeding and reproducing plants are evolved in any region having homogeneous environment. The number of such plants gradually increases and their habitat also expands, intruding in the adjoining area of different environment. Thus, these intruders or newcomer plants develop their own sub-populations or sub-groups and already developed plants of the adjoining areas. Subspecies are evolved through the selective influences of these newly developed sub-populations.

6. ***Abrupt speciation:*** Abrupt speciation is the process of sudden evolution of new species. This happens when there is sudden change in the number of chromosomes in the plants.

7. ***Mutation:*** Mutation is another process of speciation, wherein the spontaneous evolutionary change gives birth to new species.

Dispersal of Plants

The present day distribution of plants over the globe is the result of 'migration' of plants over time and ecologically suitable areas. The spread or migration of plants takes place mainly through: (*i*) vegetative propagation, and (*ii*) seed dispersal. Since plants do not have mobility like animals, hence, plants spread or dispersal through vegetative propagation is extremely slow process. Thus, the dispersal of plants become effective through external factors mainly through the dispersal of seeds. The main agents of seed transportation are wind, ocean currents and animals including man who transport seeds of various sorts in varying amounts in different ways. Birds also carry such seeds which stick to different parts of their bodies.

History of Evolution

As already discussed, the interpretation of evolutionary changes in both types of organisms (*i.e.*, plants and animals) on the basis of available records of fossils of plants and animals preserved in the sedimentary rocks are grouped into two schools of thoughts:

(i) School of thoughts based on *'Gradual (or Geographical) Speciation'* states that the evolution of all branches of the whole (plant and animal) kingdoms takes place slowly and side by side over long period of time. Sometimes a few groups evolve more rapidly than the other groups.

(ii) School of thoughts based on *'Abrupt speciation'* believes that the organic world in the state of equilibrium and there is gradual evolution of species but sometimes the equilibrium is disturbed or interrupted, and thus, there are short periods between two periods of stability or species equilibrium. These periods are characterized by sudden happenings and catastrophes when there is abrupt and sudden evolution of species.

Most scientists believe one group of species suddenly advances at the cost of other group of species.

The historical evolution of the plants can be traced through the geological periods as given below:

1. As already discussed earlier in the section on 'evolution of animals', evolution of organisms on the earth's surface begins from the *'Pre-Cambrian Period'* (more than 600 million years ago) when probably the first life in the form of single-celled or unicellular microscopic organisms resembling the present-day cocoid or spheroid bacteria, was evolved in aquatic environment. These early organisms were an aerobic heterotrophs (these organisms did not use oxygen because at that time there was no oxygen in the environment).

2. It is generally believed that in the later period, the initial heterotrophic organisms might have given birth to autotrophic plants which could have become able to manufacture their food themselves. The initial autotrophic organisms included purple bacteria, blue-green algae and green algae. These initial autotrophic plants were also originated in water areas. These aquatic plants are supposed to have formed oxygen through biological processes.
3. It is further believed that the origin and colonization of plants on the land surface began in late '*Silurian*' (440-400 million years ago) and early '*Devonian*' periods (400–350 million years ago) when the first terrestrial green plants were evolved from the aquatic red and brown algae. During the beginning of 'Devonian' period, several new species of flora were developed. The first tree-like plants were evolved during middle and late '*Devonian*' period. By the end of this '*Devonian*' period, the flora attained uniform character because of the development of vegetation in uniform climatic conditions and the vegetation communities were transformed into forests.
4. All the land masses of the earth continued to be united upto '*Carboniferous*' period (350 to 270 million years ago). The disruption of this united landmass and drifting of its broken landmasses had a great influence on the development of world flora. There is no unanimity about the time of disruption. Some of the scientists believe that this united landmass was disrupted during '*Jurassic*' period (180 to 135 million years ago).

   There were two united landmasses *e.g.*, *Laurasia* (or *Angaraland*) consisting of Asia, Europe and North America; and *Gondawanaland* including South America, Africa, India, Australia and Antarctica.

   The northern segment of united landmass *i.e.*, *Laurasia* was characterized by the spread of dense luxuriant forests of tall arborescent plants during '*carboniferous*' period. These plants had some plants having the height of more than 30 meters. These vegetations were buried under sediments and compressed and ultimately were converted into peats and coals.

   During carboniferous period, there was a global climate change, when most of Gondwanaland was covered with ice-sheet and this has drastically changed the earlier vegetation of Gondwanaland which was similar to the vegetation of Laurasia. The deglaciation of ice-sheets resulted into the origin of new sets of climatic conditions over Gondwanaland *e.g.*, cold and humid temperate climates.
5. There was a large scale change in the climate, plant and animal life during '*Mesozoic Era*' (225–70 million years ago). During this period, there was large-scale drifting of continents and ocean basins. There were disrupted into many parts which drifted in different directions to acquire their present position.

   Jurassic (180 to 135 million years ago) and *Cretaceous periods* (135 to 70 million years ago) provided ideal climatic conditions for the development of a variety of plants and animals. Sub-tropical humid climate developed over major areas became most favourable for evergreen forests.

   During *Cretaceous period*, the world flora underwent a tremendous transformation because of the emergence of flowering plants. The evolution of flowering plants was so rapid and abrupt that they soon changed into dense forests. This sudden spread of dense forests of flowering plants resulted into marked decrease in the number of earlier species and thick population. By the end of *'Cretaceous period'*, the vegetation community was well-organised *e.g.*, top layer of tall flowering plants, second layer of shrubs, third layer of herbaceous plants and fourth or ground layer of mosses and lichens.

6. The western part of Gondwanaland is generally considered as the birthplace of flowering plants. From this part of Gondwanaland, flowering plants were dispersed to other places. During *'Tertiary Period'* (70 to 1 million years ago) origin of Alpine mountains (like to Rockies, the Andes, the Alps, the Himalayas etc.) and related climatic changes, glaciation of major parts of North America and Euracia, and during *'Quaternary period'* (1 million years ago to present time) frequent changes in climate and sea-levels because of sub-glacial and interglacial periods etc. have largely affected the life-form and pattern of vegetation at global and local scales.

QUESTIONS

1. Define the term 'Ecology'. Discuss its scope.
2. Discuss basic concepts and principles of Ecology.
3. Explain the terms 'population' and communities in Ecology.
4. What is 'ecological succession'? Discuss in detail.
5. (a) Enumerate various causes of succession.
   (b) Discuss various types of succession.
6. What is 'Population Ecology'? Discuss various population characteristics.
7. Write short notes on:
   (a) Population dynamics (b) Regulation of population
8. Discuss the term 'community ecology' in detail. Enumerate general characteristics of communities.
9. Write an essay on 'Ecological Relationship'.
10. Write short notes on:
    (a) Mutualism (b) Competition
    (c) Parasitism (d) Predation
11. Define the term 'predation' and give an account of predation prey relationship.
12. What is human ecology? Discuss the history of human ecology in brief.
13. Write a note on "humans and its relationship with environment".
14. Discuss IIuman-Settlement with reference to ecology.
15. Write a detailed note on 'evolution of animal life'.
16. Discuss the evolution of plants.
17. What do you understand by 'speciation'? Discuss in detail.
18. Write short notes on:
    (a) Dispersal of plants (b) History of evolution of organisms
19. Explain the following in brief:
    (a) Species (b) Vegetation (c) Flora
    (d) Habitat (e) Biome (f) Biomass

4
Chapter

ECOSYSTEMS

4.1 CONCEPTS

Plants, animals and other organisms together with the physical environment with which they interact, constitute the *Ecological System* or *Ecosystem.*

Ecosystem is a system which comprises the physical environment and the organisms which live therein. Thus, an ecosystem consists of both, living or biotic and non-living or abiotic components. The abiotic components of the physical environment influence the types of organisms living in an era, different from those in sea-water. The non-living components of the ecosystem consist of chemical substances found in the soil water and atmosphere.

An ecosystem can also be defined as:

(i) An ecosystem can be defined as a functioning, interacting system composed of one or more living organisms and their effective environment, both physical and biological.

—*F.R. Fosberg*

(ii) An ecosystem is a particular category of physical systems consisting of organisms and inorganic components in a relatively stable equilibrium, open and of various sizes and kinds. —*A.G. Tansley*

(iii) An ecological system or ecosystem is a unit that includes all of the organisms in a given area interacting with the physical environment so that a flow of energy lead to clearly defined trophic structure, biotic diversity and material cycle within the system. — *E.P. Odum*

An ecosystem can be very small or very large such as a tropical rain forest, an ocean, a lake, a desert, a grassland, or even a domestic aquarium as long as it consists of a self-regulating community of plants and animals interacting with one another and with their non-living environment. All of the earth's ecosystems together make up the biosphere. The biosphere is a giant ecosystem.

The differences among various ecosystems found in different parts of the world are caused by differences in temperature, precipitation, and availability of life-sustaining chemicals or nutrients from the soil, water and air.

4.2 PROPERTIES OF ECOSYSTEMS

1. The ecosystem is what we choose it to be. It can be the biosphere, the earth, forests, rivers, ponds, islands, wheat fields, an aquarium, a desert. Ecosystem is, therefore, a handy operational term that can be applied at any scale.
2. There is an ecosystem within an ecosystem and yet another ecosystem within it.
3. The ecosystem is an intricate network consisting of biotic and abiotic factors, connects through complex, interaction that arise between the biotic and abiotic components.
4. The ecosystem is a unit of the nature that is in itself an independent functional entity.
5. Ecosystems vary in complexity.
6. Despite the variations in different ecosystems, the underlying basic ecological principles and processes are similar.
7. Ecosystem of any given space-time-unit represents the sum of all living organisms and physical environment.
8. Ecosystem is composed of three basic components, namely biotic, abiotic (habitat) and energy components.
9. It occupies certain well-defined area.
10. It is an open system which is characterised by continuous input and output of matter and energy.
11. Ecosystem tends to be in relatively stable equilibrium unless there is disturbance in one or more controlling factors.
12. An ecosystem is powered by energy of various sorts but the solar energy is most significant.
13. An ecosystem is a structured and well-organised system.
14. Ecosystems are resource systems.
15. Ecosystem concept is monistic in that environment (abiotic component), man, animals, plants and micro-organisms put together in a single framework.
16. Ecosystem is a functional unit wherein the biotic components and abiotic components (including energy component) are intimately related to each other through a series of large scale cyclic mechanisms *viz.* energy flow, water (hydrological) cycle, bio-geo-chemical cycle, mineral cycle, rediment cycle etc.

4.3 FUNCTIONS OF ECOSYSTEMS

In order to know the functioning of an ecosystem, it is essential to understand as to how plants and animals are related with each other through the *food-chain*; how *energy* passes from one organism to the other; and how the minerals circulate between the biotic and abiotic components of an ecosystem. Therefore, for understanding the functions of any ecosystem, we should study the following processes:

1. Food-chain and Food-web.

2. Flow of energy
3. Minerals cycles (Bio-geochemical cycles)

For better understanding, both structure and functions must be studied together. From the operational view point, the living and non-living components of ecosystem are so inter-woven into the fabric of nature that their separation from each other becomes practically very much difficult.

Food-chain and Food-web, flow of energy and Biogeochemical cycles of ecosystems have been discussed in the following paragraphs alongwith "structure of ecosystem:"

4.4 STRUCTURE OF ECOSYSTEM

Ecosystems have following major components (or elements):
1. Abiotic (non-living) or Physical components
2. Biotic (living) components

1. Abiotic or Physical Components

Abiotic or non-living components comprises land and soils, water, air and sunlight. Abiotic components of the ecosystem consist of chemical substances found in the soil, water and atmosphere. These chemicals may be organic substances, like carbohydrates, fats, proteins and vitamins, and biogenic substances, like carbon, nitrogen, hydrogen, phosphorous, calcium and potassium in large quantity and iron, manganese, magnesium, zinc, cobalt etc. in small quantities required to sustain life of different biotic communities.

Abiotic components of the physical environment influence the types of organisms living in an area. For example, the organisms living on the landmasses are different from those in sea-water. Climatic conditions bring about variations in the types of plants and animals found in different landmasses. Other abiotic elements of the climate are temperature, rainfall, duration of sunlight, winds, nature of soil, slope of the land, composition of water bodies etc.

If the whole biosphere is considered as an ecosystem at global scale, it includes lithosphere, hydrosphere and atmosphere.

The *Lithosphere* refers to the solid layers of rock material on the earth's surface, both on the continents and ocean floors. The average thickness of lithosphere is about 100 km and consists of minerals, rocks, soils, mountains, sand dunes etc.

The *hydrosphere* refers to the layer of water on the surface of the earth in the form of oceans, lakes, rivers and other water bodies. Water occurs on the land in the form of ice-sheets in the polar regions and on high mountains water also occurs below surface of the land in the form of underground water. Water also occurs in the form of water vapour in the lower layers of the atmosphere. The heating by the sun is responsible for the circulation of water in the hydrosphere through evaporation, condensation and precipitation. In this process of circulation, water is consumed by plants and animals in the biosphere.

The air envelope that surrounds the earth is called the *atmosphere*. The atmosphere consists of a mixture of gases having a relatively uniform composition in the lower layers. An average sample of pure dry air consists of 78% nitrogen, 21% oxygen and 0.9% argon. Other gases, such as carbon dioxide, hydrogen, helium and ozone are present in minute quantities. The lower layers of the atmosphere also contain water vapour in variable quantities.

The density of the atmosphere decreases with height. In the lowest layer, known as *troposphere,* temperature of air decreases with height. Above the troposphere, another layer is called the *stratosphere* (about 40-45 kilometer thick), in this layer, the temperature remains constant for some height and then increases with height. The ozone in this layer absorbs the harmful ultraviolet radiation from the sun. The other layers are known as mesosphere and thermosphere.

2. Biotic (living) Components

The **biotic component of the environment** consists of plants (flora) and animals (fauna) including man as an important factor. All the organisms work to form their social groups and organisations at several levels. They also work to derive matter from physical environment for their sustenance and development. This process generates economic environment. Man as an organism has three aspects, namely physical, social and economical. *'Physical man'* requires habitat (space), water, food and like other biological populations, releases wastes into the ecosystem. *'Social man'* establishes social institutions, formulates laws, principles and policies to safeguard his existence, interest and social welfare. *'Economic man'* derives and utilises resources from the physical and biotic environment with his skills and technologies. When exploitation of this economic function exceeds the critical limit, the equilibrium of the environment/ecosystem is disturbed and several environmental and ecological problems crop up which become detrimental not only to man himself but also to the whole population of species in a given ecosystem.

Table 4.1: Components (Elements) of Ecosystem.

Abiotic	*Biotic*
Energy	Green plants
Temperature	Non-green plants
Water	Decomposers
Atmospheric gases and wind	Parasites
Fire	Animals
Gravity	Man
Soil	
Topography	
Radiation	
Materials	

Producers, Consumers and Decomposers

Biotic components are broadly divided into two major groups, on the basis of their nutritional relationship:

(a) Producers (or autotrophic organisms)

(b) Consumers (or heterotrophic organisms)

- *Producers:* The green plants are *producers,* as they have the capacity to manufacture their own food from existing supplies of water and carbon dioxide, utilising sunlight as the source of energy. This process is known as photosynthesis, as organic materials are synthesised using sunlight from minerals, and water in the soil and carbon dioxide from the atmosphere. Oxygen, the much needed gas for survival, is released as a by-product of photosynthesis. The presence of green pigment called chlorophyll in the leaves of plants makes photosynthesis possible. In ocean waters, phytoplankton are producers, as they produce their own food using solar energy.
- *Consumers:* Organisms, mostly animals, that feed either directly or indirectly on producers are called consumers energy from the sun as plants do. To obtain energy and nutrition, they eat plants directly or feed on plants indirectly by other organisms. Consumers may belong to one of the following four classes on the basis of their food sources:
  - (i) *Herbivores* (plant eaters) are primary consumers that eats only plants.
  - (ii) *Carnivores* (animal eaters) are secondary and higher-level consumers, which feed on plant-eating animals.
  - (iii) *Omnivores* (plant and meat eaters) are the consumers that eats, both plants and animals.

    The rabbit that feeds on green plants is a primary consumer (herbivores), and the fox that eats the rabbit is a secondary consumer (carnivores). The humans are omnivores as they eat both plants and animals.

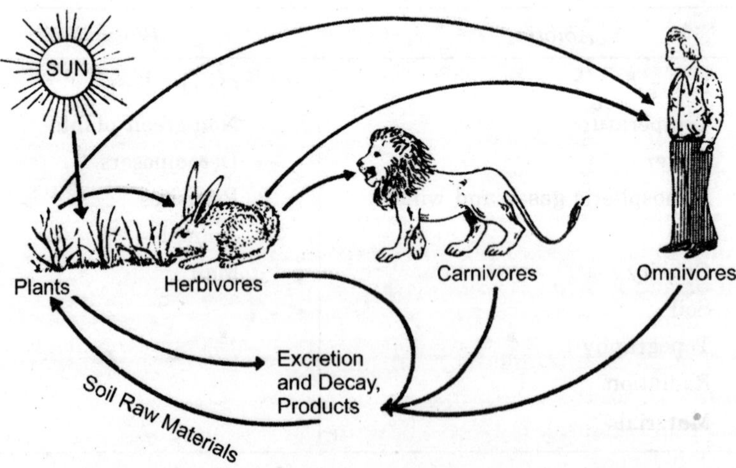

Fig. 4.1: Nutritional relationship among biotic components of an ecosystem.

  - (iv) *Detritus Feeders (or decomposers):* This category of consumers feed on dead or decomposed tissues of plants and animals, and obtain both energy and nutrients from dead organic matter helping in re-circulation of the materials in the environment. Numerous kinds of bacteria, fungi, earthworms, termites etc. are the examples of decomposers. Deadwood and leaves-undergo decay, rot or decomposition in which its complex molecules are broken into simpler chemicals, these are returned to the soil and water for reuse by producers *i.e.* green plants. Thus, it helps in completing the cycle of soil nutrients.

Subsystems of Biotic Components

Biotic components of the biosphere/ecosystem/environment consists of following three subsystems:

1. Plant system
2. Animal system
3. Micro-organisms

Out of these three subsystems, plants are the most important, as they alone produce organic matter which is used by themselves and by animals and micro-organisms either directly or indirectly. Plants also make the cycling and re-cycling of organic matter and nutrients for different components of the biospheric ecosystem:

1. ***Plant System:*** Plants are found on any land in different forms *e.g.*, woodland, forest, grasslands, marshlands etc. These different forms of plants are collectively known as vegetation. Out of the abiotic or physical factors, climate and soil mostly affect the species of plants, their structure and growth form. In turn plants also affect and control the properties of soils and climatic conditions of their physical habitat. Thus, plant community not only affects but also determines the productivity of the land of their habitat.
2. ***Animals System:*** As discussed earlier, animals depend on autotrophic green plants. Main functions of animals are: (*i*) to use organic matter made available by the green plants, (*ii*) to re-arrange the organic matters, and (*iii*) to decompose organic elements. Organic matters are available to animals in the three forms, *viz*: (*i*) from living plants and animals, (*ii*) from partially decomposed plants and animals, and (*iii*) from organic compounds in solution forms.
3. ***Micro-organisms:*** These are also known as decomposers, as these decompose the dead plants and animals and other organic matter in different forms. During the process of decomposition of organic matter micro-organisms obtain their food. A large number of micro-bacteria and fungi are included in the category of micro-organisms.

The number of animals and micro-organisms is not easy to be identified and named. The distribution of animals, their evolution, dispersal, extinction, relationships between animals in the ecosystem, relationships between plants and animals and their mutual interaction are the subjects of interest.

4.5 ECOLOGICAL BALANCE

When a biosphere is considered as an ecosystem, all types of plants and animals—starting from the minutest bacteria, fungi etc., to huge trees, elephants, lions, whales etc.

become part of its biotic component. The relative number of these different types of organisms in the biosphere is such that there is no scarcity of food for any organism.

In such a balanced ecological system, the smaller organisms are much more in number, and their growth and reproduction is also much faster than those of larger organisms, which are fewer in number and reproduce slowly. This ensures the availability of sufficient food for larger animals.

In the natural environment in the biosphere there exists a perfect balance or equilibrium between the various organisms, and this is known as the **ecological balance**. In this equilibrium state, the relative numbers of different organisms in a particular environment remain constant.

The ecological balance of nature may get disturbed, when changes take place in the natural environment (the changes may be in physical factors, such as temperature, precipitation, evaporation, light intensity, salinity of sea water etc.), which may consequently change the relative numbers of different organisms in the biosphere.

This affects the dearth of food for certain particular organisms, which may ultimately lead to large scale mortality of those particular organisms. In such circumstances, a new equilibrium is finally re-established under these changed conditions.

After the arrival and reproduction of man on the Earth, a large scale impact has been caused on the biosphere, due to his unchecked actions, such as large scale deforestation, hunting of animals. The man has also developed new types of plants and domesticated animals. His habit of using pesticides, insecticides and chemical fertilizers in the agricultural farms has affected the relative proportion of various organisms in the biosphere. Such continuing imbalance in the biosphere, if not checked, will prove disastrous to the very existence of man himself.

4.6 CAUSES FOR ECO-IMBALANCES

4.6.1 Human Activities

Human activities and technological developments leading to luxurious life have caused severe irreversible unbalances and disturbances in the natural environment. The main activities that have caused these alterations in the ecosystems and the pollution of the environment in general, are categorized as under:

1. Agriculture
2. Industrialisation
3. Urbanisation
4. Mining
5. Transportation

1. Agriculture

With the development and increase in human population, the agriculture on more and more land, has been practised to grow crops. Presently humans are not only growing food crops to eat and quench their hunger, but also growing several kinds of cash crops. They have brought a large part of the forest land into agriculture and also started using chemical fertilizers and pesticides to increase crop production.

The present system of agriculture is highly mechanised as this has become imperative, as the human population is increasing rapidly. Mechanised agriculture is consuming tremendous amount of energy, particularly fossil fuel like diesel. These products (fertilizers, pesticides and fossil fuels) are seriously polluting our environment and destabilizing it.

The pesticides used for killing pests do not kill the pests alone, but also kill some useful organisms, like the honey bee, birds, beetles etc. They also kill the useful organisms and micro-organisms present in the soil. As a matter of fact, the pesticides poison the whole terrestrial food chain. The pesticide poison has even spoiled our drinking water as well as the entire food, leaving all of us to undergo slow poisoning deaths.

As human population is increasing, the developing countries are forced to convert many forest areas, wet lands, and other productive areas to agricultural land. During last few decades, man has converted major portions of the forest land to the agricultural land and damaged the ecosystem and the environment in general. The continued destruction of forests may have serious impacts on climate, hydrological cycle, and mineral balance on earth.

2. Industrialisation

Agriculture and industrial revolution together, supported with rapid growth of human population, has increased our demand of natural resources, such as air, water, minerals, forests etc. All of these resources are exhaustible. The industries cause maximum environmental instability by polluting air, water and land.

This unsustainable utilisation of our natural resources is going to prove deadly for our future generations. The rapid uncontrolled growth of the industries, as at present, is throwing their waste-waters into our precious water sources, like rivers and lakes and are contaminating and spoiling them. This will ultimately make clean drinking water totally unavailable for us.

The industries are also releasing several poisonous gases and substances in our atmosphere. This is polluting the air, which is very essential for sustaining life. Therefore, there is need to think over these fundamental issues, and do something to stop unsustainable developmental activities, and initiate to conserve our natural resources.

3. Urbanisation

In the process of urbanisation, mass migration of people from rural areas to industrialised urban areas take place, due to increased job opportunities.

In view of fast urbanisation taking place, a large amount of land has been consumed to build such giant urban centres. Mostly precious farming lands in river basins had been lost in urbanisation. The urbanisation changes the green, wet and earthen land to the concrete and asphalt paved land. This increases the rain run-off by many folds.

Secondly, the urbanised areas produce enormous quantities of sewage and domestic waste waters, which on being discharged into the rivers, either in treated, untreated or partially treated condition, pollute the rivers and adversely affect the aquatic ecosystems.

Thirdly, the air pollution and land pollution, adversely affects the urban areas to great extent, thereby affecting the nearby ecosystems and thereby creating local, regional, and even global environmental problems.

Fourthly, construction works taking place in urban areas requires (*i*) large amount of sand, which is removed from the river beds, resulting into unstabilization of riverine ecosystem, and (*ii*) large amount of boulders and stone aggregates, taken from hills, thereby destroying the hills by large scale mining in the area.

Fifthly, wood is another major requirement of urbanisation, which is required in large quantities for furniture, doors, windows, almirahs, cup-boards, etc. Forests are destroyed to meet this demand, causing serious consequences for forest eco-systems. As we know that forests regulate climate, prevent soil erosion and modulate water flow very effectively.

Further, large cities have large demands of available resources, such as potable and usable water; energy including electricity, gas, petrol and diesel; building materials; and minerals. All these required demands produce hazardous wastes that pollute our environment.

4. Mining

Mining is the extraction of minerals and ores from the Earth's crust. Mining provides us coal for energy requirements; ores for making chemicals, fertilisers, minerals and metals, clays for making crockery, rocks and stones for making buildings, bridges, roads etc.

In response to the increased demand due to industrialisation, urbanisation and modernisation, the exploitation of mineral resources has increased many fold, resulting into large scale mining.

The mining, whether it is at the surface or underground, changes the whole landscape and the ecosystem existing in the mining areas. The past mining activities have given rise to vast areas of wasteland, which are now useless from the point of view of agriculture or agro-forestry.

5. Transportation

The excessive development of road transport, using diesel and petrol, have created a number of adverse impacts on the environment. The **adverse impacts of transportation** can be grouped into the following categories:

1. **Energy Consumption:** The road transportation is responsible for increasing the production and use of fossil fuel, which is not only harmful to our environment, but is past depleting the availability of this limited resource.
2. **Land Consumption:** Huge area of land is cleared and locked upto provide transportation corridors and parking areas.
3. **Loss of Agricultural Land:** Increasing road traffic forcing the residents to move to countryside, resulting conversion of more and more farmland to urbanland, thereby reducing the available agriculture land. Thus, the productive fertile farmland is being continuously lost to the concrete, barren median strips and suburban lawns etc.
4. **Loss of Wild Life Habitat:** Meadows, wetlands, and forests are lost alongwith farmland, greatly reducing wildlife habitat of several species.
5. **Air Pollution:** Vehicle emissions are the biggest source of air pollution in all the large and medium cities. Cars and trucks emit a wide variety of harmful gases, such as carbon monoxide (CO), nitrous oxides (NO_x) and other products of incomplete combustion.

6. **Water and Land Pollution:** Chemical gases and particles emitted by cars and trucks do eventually fall out of the air on the ground surface, either by gravitational force or with rain drops. The cars and trucks also often leak oil, gas, brake fluid, engine coolant, etc., thereby causing land pollution and also water pollution due to flow of these pollutants into sewers and storm water drains.
7. **Noise Pollution:** Noise levels increasing with the increase of traffic, adding to the stress of modern city life.

4.6.2 Natural Hazardous Events

Following are important natural processes which may unstabilise the environment:

1. Volcanic eruptions
2. Earthquakes
3. Tsunamis
4. Forest fires
5. Cyclones and Tornadoes
6. Floods
7. Landslides and Avalanches
8. Droughts

It has been established that the environmental changes brought about by the natural events are usually temporary in character and are reversible. The ecosystems altered by the natural disastrous events, do have the capacity to recover, though the time taken may be somewhat more, in case the damage caused is tremendous.

Above listed natural hazardous events and their environmental impacts have been discussed at relevant places in the book.

4.7 FOOD CHAINS

All organisms including human beings need food which provides energy for growth, maintenance and reproduction. A part of the energy provided by food is used for biological processes and the rest is dissipated to the environment as heat energy by the process of respiration. Undigested food is excreted and enters the decomposition path.

In any ecosystem, various living organisms (plants, animals) are arranged in a definite sequence according to their food habits. Plants are producers which are eaten by herbivores, which in turn are eaten by carnivores.

This transfer of food energy from the source in plants through a series of organisms (arranged in a linear manner) with repeated eating and being eaten is known as a *food chain* in an ecosystem.

As we have discussed earlier, green plants manufacture their own food which becomes source of food energy for all types of organisms in the biosphere. Plants manufacture their food with the help of carbon dioxide (taken from atmosphere through their leaves during day time), inorganic salts like phosphorous and nitrates and water from soils through their roots by the process of *root osmosis* and sunlight. Plants convert water and carbon dioxide into starch and sugar with the help of sunlight. This entire process is called as *photosynthesis*.

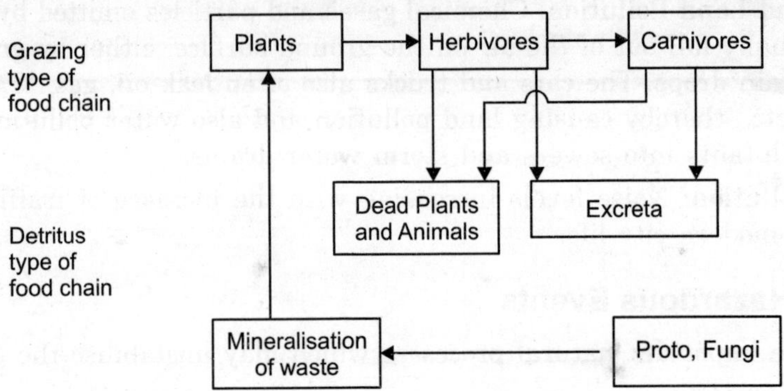

Fig. 4.2: A simple grazing and detritus type of food chains and their interrelationship.

A food chain is a flow chart that follows the course of food energy through an ecosystem, and the energy flow though ecosystem is unidirectional from lower trophic level to higher trophic level. Thus, green plants occupy, in any food chain, the first trophic (nutritional) level—are called the *primary producers.* The energy, as stored in food matter manufactured by green plants, is then utilised by the plant eaters — the herbivores, which constitute the second trophic level and are called the *primary consumers.* Carnivores that eat herbivores are at the third trophic level, and are called the *secondary consumers.* Fourth trophic level animals are those which take their food either directly or indirectly from all the three lower trophic levels. Such animals (as man) are called omnivores.

Thus, a *food chain* is a sequence of energy transfer from the lower trophic levels to the higher trophic levels.

An example of food chain indicating its trophic levels and kinds of organism are shown hereunder:

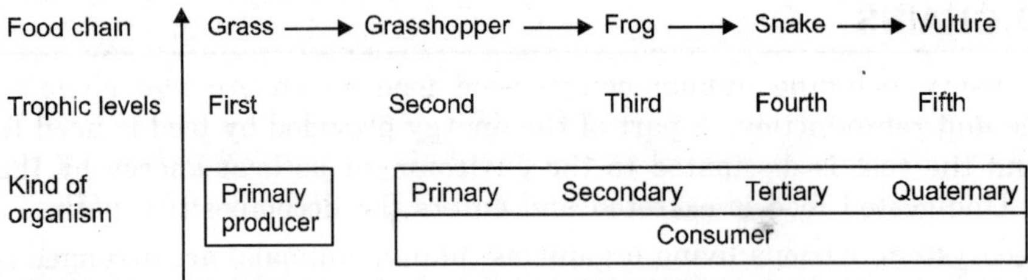

4.8 FOOD WEB

Under natural conditions in the same ecosystem, depending upon the variety of organisms as based on their food habits, a number of linear food chains operate at a time. These chains are inter-linked with each other at several points. This inter-locking pattern of a number of food chains forms a web-like arrangement known as a *food web.*

When the feeding relationships in a natural ecosystem become more complicated, the food chain does not remain simple and linear rather it is complicated by several inter-connected overlapping food chains. Such complicated food chains are food web.

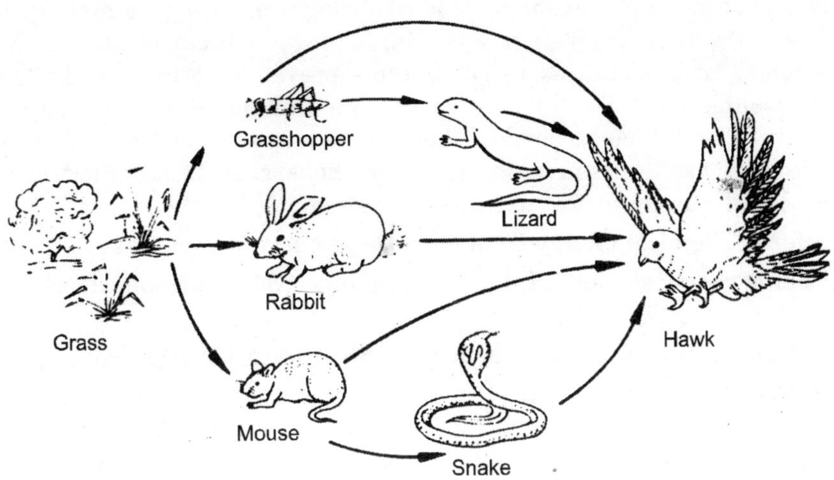

Fig. 4.3: A diagramatic representation of a food web in a grassland ecosystem.

4.9 FLOW OF ENERGY

Energy flow is the passage of energy through the trophic levels of a food chain. The behaviour of energy in ecosystem can be termed energy flow due to unidirectional flow of energy. As we know that solar radiation is the basic input of energy which enters the ecosystem. This solar energy passes through the hierarchy of trophic levels in a food chain and ultimately becomes output from the ecosystem as energy is lost through the respiration from each trophic level.

As we have already discussed, the radiant solar energy is trapped by green plants (primary producers) and is used to prepare food (chemical organic matter) through the process of photosynthesis. Thus, autotrophic green plants transform a part of solar energy into food or chemical energy which is used by the green plants *i.e.* primary producers at trophic level 1, to develop their tissues and thus, it is stored in the primary producers at trophic level 1. The chemical energy stores in trophic level 1 becomes the source of energy either directly or indirectly to all the animals of different trophic levels in food chain in a natural ecosystem. Some portion of energy is lost through respiration from trophic level 1 and same portion of energy is transferred to plant eating animals (herbivores) at trophic level 2. Some portion of plants which fall down, on the ground and ultimately consumed by decomposers, and thus, transferring some energy from trophic level 1 to the decomposers living in the soils.

From the above, we see that when a cow grazes, chemical energy stored in grasses is transferred to the cow. Thus, the chemical energy consumed by herbivores animals helps in the building of their own tissues at trophic level 2 and thus, the energy is stored in them. This stored energy is in the bodies of herbivores now becomes the source of energy for carnivores animals at trophic level 3. A substantial portion of chemical energy is lost through respiration from herbivores at trophic level 2 because they have to consume energy for their movement etc.

Similarly, a substantial portion of potential chemical energy stored in the bodies of carnivores is lost through respiration from trophic level 3 because the carnivores animals have to run for greater distances to catch their preys. A portion of chemical energy is transferred to trophic level 4 or by omnivores. The animals at this level also take energy from trophic level 1 and 2. Some portion of energy is also released through respiration from this trophic level 4 *i.e.* by omnivores. Thus, we have seen that following are the main pathways of flow of energy in the natural ecosystem:

(i) Transfer of chemical energy from each trophic level to the next higher trophic level,

(ii) Transfer of chemical energy from dead organisms of each trophic level to decomposers, and

(iii) Loss of energy in the form of heat through respiration from each trophic level and from decomposers.

The heat energy released through respiration is lost to the atmosphere and is not again available to the organisms for reuse. Thus, it is evident that the energy flow in the ecosystem is unidirectional and non-cyclic.

The percentage of energy transferred from one trophic level to another is called *ecological efficiency*. The efficiency of energy transfer from one trophic level to another varies from 5% to 20% depending on the types of organisms and environmental conditions. On an average, only 10% of energy is transferred from one trophic level to another. Such low efficiency is due to the fact that not all the organisms present at one level become easily available as food for the consumers at higher level. Since ecological efficiency is low, the capacity of ecosystem to support organisms at the higher trophic levels is limited.

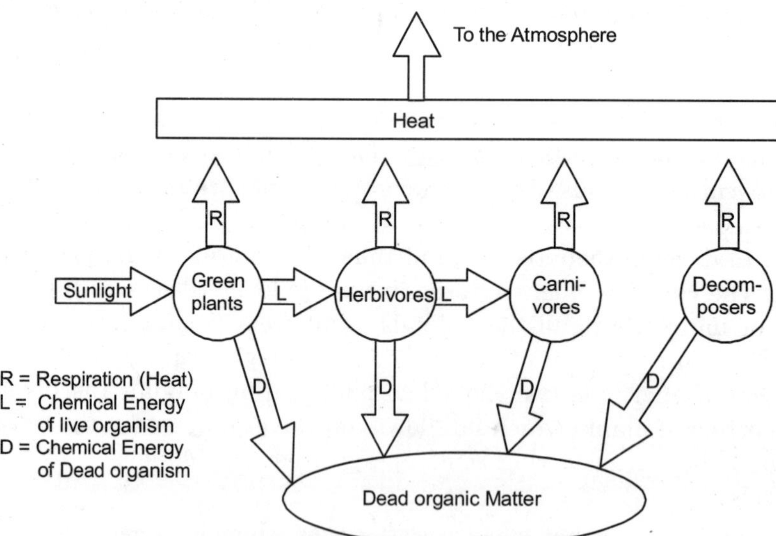

Fig. 4.4: Energy flow in the ecosystem.

Thus, we can say that after fourth or fifth link in food chain, no enough energy is available to support another trophic level. Extending this concept further, in case we want to support more human population on the earth, this could be possible by providing them with vegetarian diet rather than non-vegetarian food.

4.10 BIOGEOCHEMICAL CYCLES (BIOSPHERE CYCLE)

The living organisms do require various chemical elements and their salts for supporting life. The six major essentially required elements are: Carbon, Hydrogen, Oxygen, Nitrogen, Phosphorus and Sulphur. Other important elements required are: Calcium, Magnesium, and Potassium. Some other minor elements required by the organisms are: Iron, Manganese, Copper, Zinc, Chlorines and Iodine. Although most of these elements, which are required for supporting life on the earth, are found in the Earth's crust, yet certain elements like hydrogen, oxygen, nitrogen and carbon dioxide are largely available in the atmosphere. The combination of hydrogen and oxygen also largely occurs as water (H_2O) in all the three forms.

These chemical elements lend to circulate in the *biosphere* in characteristic paths from the physical environment to the organisms and back to the environment. These more or circular paths are known as *bio-geochemical cycles.* Such a cycle is so named, because these involves living organisms (represented by the word *bio*), soils and the earth's crust (represented by the word *geo*), and the chemical element in circulation (represented by the word *chemical*).

Following are the main bio-geochemical cycles:

(i) Hydrological cycle. (ii) Nitrogen cycle. (iii) Phosphorus cycle.

(iv) Carbon cycle. (v) Oxygen cycle.

The producers (green plants) with the help of radiant energy and minerals (C, H, O, N, P, Ca, Mg, Zn, Fe, etc.) taken from their soil and aerial environment build up complex organic matter, like carbohydrates, fats, proteins etc. Some ecologist viewpoint is that since green plants produce carbohydrates and not energy and since they convert radiant energy into chemical form, these must be called as converters. However, the term 'producer' is so widely used that it is preferred to be retained as such.

These are the **cycles of circulation of elements and minerals between the biotic and the abiotic components of an ecosystem.** These cycles are discussed hereunder:

1. **Hydrological Cycle** : Already discussed in the chapter on "The Earth and the Biosphere."
2. **Nitrogen Cycle** : In this cycle, the nitrogen gas in the air serves as a reservoir pool, and this nitrogen is fixed by certain plants with the help of micro-organisms and natural electrification process, forming plant proteins. As an output, the growth of the plants and vegetation occur, which leads to the growth of herbivores. The excreta of the animal life as well as the dead bodies of the living organisms (plants as well as animals) are finally decomposed by bacteria to produce ammonia, then nitrites, which by demineralisation goes back to the atmosphere as nitrogen gas.
3. **Phosphorous Cycle** : In the phosphorous cycle, the phosphorous is sourced from the Earth's crust, making it a sedimentary type of a bio-geochemical cycle. All the elements, like Phosphorous, Calcium, Potassium, Magnesium etc., which originate from the sediments of the earth's crust and follow the sedimentary cycles through soil, water and living organisms.
4. **Carbon/Carbon dioxide Cycle** : In this cycle, the carbon present in the atmospheric air as carbon dioxide gas enters the biological world through the process of

photosynthesis. During photosynthesis, green plants use solar energy to combine carbon dioxide and water, producing energy rich carbohydrates (sugar etc.) and in the process releasing oxygen to the atmosphere.

4.11 ECOLOGICAL PYRAMIDS

We find some sort of relationships between the *numbers, biomass* and *energy* contents of the producer ecosystem. These relationship, when represented in diagrammatic ways, are referred to as *ecological pyramids.* The relationships represent certain common characteristics *e.g.*:

(i) The number of species tends to decline successively from the base of the trophic level (trophic level 1) to the top of the trophic level, and this graphic representation is called *pyramid of numbers.*

(ii) The total biomass tends to decrease progressively from the base of the trophic level through higher trophic levels to the top and is represented graphically by *biomass pyramid.*

(iii) The energy availability at each successive higher trophic level decreases. Its graphic representation is called as *energy pyramid.*

1. Pyramid of Numbers

These include only the number of species and not their sizes. The animals (organisms) at the base of a food chain are relatively abundant, whilst those of the end are relatively few in number, and there progressive decrease in between the two extremes.

On an average, the primary producers or the auto-trophic green plants are small in size but very large in number at the base of the pyramid (representing trophic level 1) because these have to provide food to the animals herbivores trophic level 2 which are smaller in number, then the plants at the trophic level 1, but larger in size. Similarly, the number of animals (carnivores) at the trophic level 3 is smaller than the number of animals at the trophic level 2, but the size of animals becomes further larger. This chain continues.

In the forest ecosystem, the situation is reversed because the primary producers (trees) are very large in size at the base of trophic level. In such cases, a single large tree is capable of supporting a large number of primary consumer animals (herbivores birds feeding upon the tree fruits) at trophic level 2, and the number of parasite living and feeding upon the birds body is still greater. The shape of this pyramid is inverted.

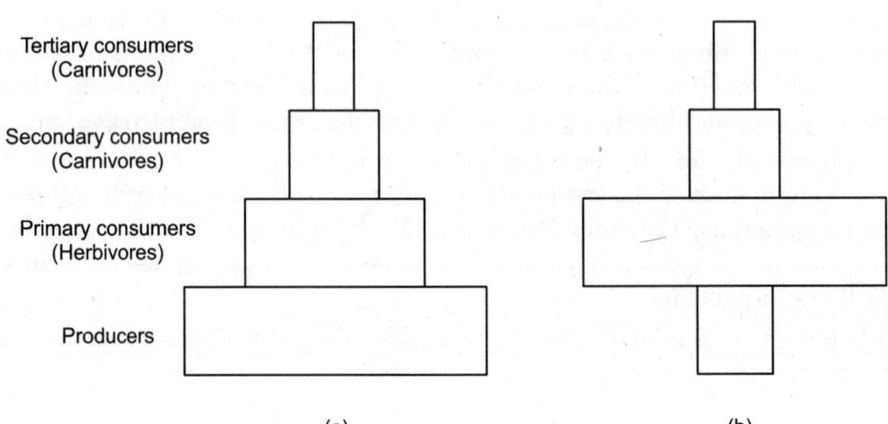

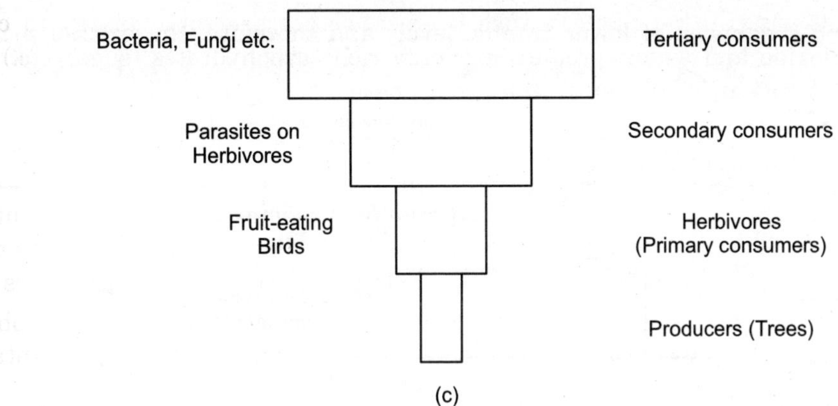

Fig. 4.5: Pyramids of numbers (Individuals per unit area)
(a) Grassland/pond ecosystem, (b) Forest ecosystem, (c) Parasitic food chain (Inverted pyramid)

2. Pyramid of Biomass

Pyramids of biomass represent the total weight of organic matter, of each trophic level instead of number of species, as is done in previous case. Figure represents the two types of ecosystems. First case, where the pyramid of biomass is upright, represents the biomass of a single tree, which naturally is very high than the biomass of a number of birds feeding upon the tree. Similarly, the biomass of even a very large number of bird parasites is lesser than that of the birds. Therefore, in this case of upright pyramid, the biomass decreases with successive tophic levels.

The pyramid of biomass becomes inverted in a pond or lake ecosystem, where the biomass of diatoms and other phytoplankton is quite negligible as compared with that of the small herbivores fish that on small fishes is still greater. Similar condition exist in most aquatic ecosystems.

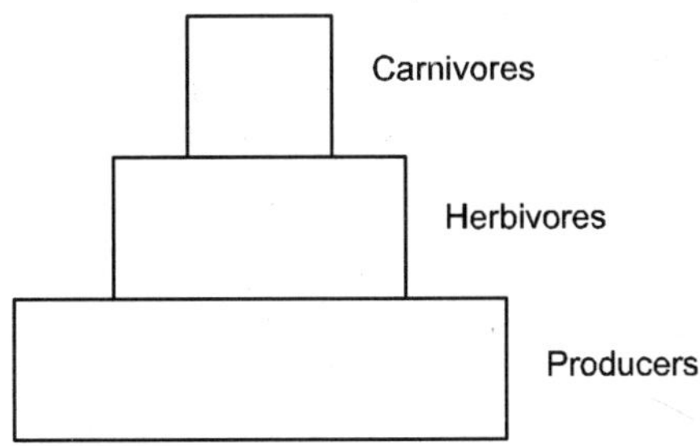

Fig. 4.6: Pyramid of biomass (Dry weight per unit area).

3. Pyramid of Energy

Unlike the pyramids of numbers and biomass, the shape of pyramid of energy is always upright. The pyramid of energy represents the total amount of energy utilised by different trophic level organisms of an ecosystem in unit area over a set period of time (usually in kcal per square metre per year or per day). The quantity of energy trapped by green plants is highest among organisms of lower trophic levels and therefore, the base of pyramid is broad.

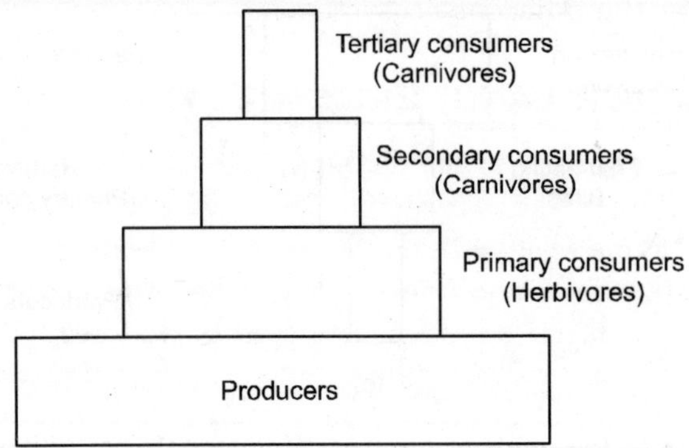

Fig. 4.7: Pyramid of energy (k.cal per unit area within unit time) in any ecosystem.

Similarly, in an aquatic ecosystem, the large numbers of phytoplankton quickly complete their life-cycles and sets of new populations. Thus, the cumulative energy contents that these generation after generation of phytoplankton trap in course of a year is certainly much more than trapped by only a few generations of herbivores fishes in the corresponding time and space. The energy content trapped by the carnivores living on the herbivores fishes is the least.

Therefore, if primary producers or organisms of any trophic level are destroyed, the organisms of the next higher trophic level will automatically perish for want of food or source of energy.

4.12 TYPES OF ECOSYSTEMS

Based on the particular type of habitat, they are classified as:

1. ***Terrestrial Ecosystems or Biomes.*** They are also known as 'land ecosystems'. These can be further classified as:

(a) Tundra

(b) Forest Ecosystem

(c) Grassland Ecosystem

(d) Desert Ecosystem

(e) Cropland Ecosystem

2. ***Aquatic Ecosystem.*** These can be further classified as:

(a) *Freshwater Ecosystems.* These may be:

(i) Running water as rivers, spring or stream.

(ii) Standing water as lake, pond, pools.

(b) *Marine Ecosystem.* These may be:

(i) Open Ocean Ecosystem

(ii) Coastal Estuarine Ecosystem

Ecosystems normally do not have distinct boundaries. Each ecosystem blends into adjacent ones through a transition zone that contains many of the plants and animals and other characteristics found in the adjacent ecosystems.

4.13 TERRESTRIAL (OR LAND) ECOSYSTEMS

These are of greater concern to us, as our needs for food and other materials are met from terrestrial ecosystems. The surface of the land has a cover of vegetation of great diversity depending on climatic conditions. Moisture and temperature are the two important limiting factors of terrestrial ecosystems. Water is essential for plants growth because nutrients needed for growth are supplied in a dissolved state from the roots to the leaves through the medium of water. Every plant has optimum temperature requirements for germination, growth and reproduction and also a range of temperature beyond which it cannot survive. Plants growth ceases when temperature falls below 6°C.

The inter-relations between organism and environment on the land constitute "Terrestrial Ecology". Due to variation in the topographic features of valleys, mountains, and slopes, certain differences occur. These differences are reflected in both the material and biotic diversities. Altitudinal and latitudinal variations cause shifts and differences in the climate patterns. Due to this variation in climate the plant and animal life vary in different terrestrial ecosystems.

4.13.1 Tundra

Tundra is a "barren land" found where environmental conditions are very severe. There are two types of tundra:

(a) Arctic Tundra

(b) Alpine Tundra Extends

Arctic Tundra extends as a continuous belt below the polar ice cap and above the tree line in the northern hemisphere. It occupies the northern fringe of Canada, Alaska, European Russia, Siberia and island group of Arctic Ocean. On the south pole tundra is very small since most of it is covered by ocean. Animals of tundra are reindeer, musk ox, arctic hare, caribous, lemmings and squirrel.

Alpine Tundra occurs at high mountains above the level of Arctic mountains which are found at all latitudes. The alpin tundra therfore shows day and night temperature variations.

4.13.2 Forest Ecosystems

Forests are classified into evergreen and deciduous forests. Forests occur on the areas of humid climatic with abundant water supply:

(a) Tropical Evergreen Forest

These are also known as 'tropical rain forests'. These occur in equatorial regions and Tropical coastal regions having heavy rainfall, where the temperatures are warm throughout the year and precipitation occurs almost daily. The hot and humid climate supports broad-leaved evergreen plants. A fully developed tropical rainforest has atleast three layers of vegetation. The tallest trees form an open canopy but the lower crown levels block most of the available light. Animal life is abundant and diverse including ground-dwellers and tree-dwellers. This covers about 8 per cent of the earth's surface but contains more than half of its flora and fauna. There is no mass shedding of leaves during any season of the year. Shedding of old leaves and growth of new ones takes place throughout the year.

Since the temperature is high, the organic matter is decomposed rapidly by bacteria fungi, detritus-feeding ants and termites. Roots of the plants quickly absorb the nutrient minerals. They are very productive, excellent in species richness and variety. The animals include monkeys, snakes, ant eaters, tropical birds, bats, large carnivores animals and a variety of fish in the rivers. Of all the species of insects known, about 70 to 80% occur in these forests.

(b) Deciduous Forest

Seasonality (hot summers and cold winters) is the characteristics of the temperate deciduous forest. The annual precipitation ranges between 75 and 150 cm. The soil of these forests consist of a topsoil rich in organic materials and a deep, clay rich lower layers. With the onset of autumn most of the trees and shrubs become leafless. These forests comprises broad-leaved hardwood trees, such as oak, hickory, which form a dense canopy that overlies saplings and shrubs. The fauna found in these forests include frogs, turtles, snakes, lizards, squirrels, rabbits, deers, bears, foxes and birds.

The deciduous forests were among the first biomes to be converted to agricultural use. These alongwith overgrazing and deforestation have contributed to the degradation of lands.

Components of a Forest Ecosystem

(a) *Abiotic Component.* These are the inorganic as well as organic substances present in the soil and atmosphere. In addition to the minerals, some dead organic matter is also found in forests.

(b) *Biotic Component.* The living organisms present in the food chain is found in the following order:

(i) *Producers.* It includes trees, climbers, shrubs and ground vegetation.

(ii) *Consumers.* In forest ecosystems *primary consumers* are herbivores, such as animals feeding on tree leaves like ants, flies, beetles, leaf hoppers, bucks, spider, alongwith large grazing animals, like deer, *neelgai,* elephants, rabbit, monkeys, moles, squirrels, flying foxes, fruit bats etc. *Secondary consumers* in forest ecosystems are the carnivores, like snakes, birds, lizards, fox etc. feeding on the herbivores. *Jertiary Consumers* in the forest systems are the carnivores, like lion, tiger etc. that eat carnivores of secondary level consumers.

(iii) *Decomposers.* Forest ecosystems have wide variety of micro-organisms including fungi, algae, bacteria etc.

4.13.3 Grassland Ecosystems

Grassland ecosystems support more species of animals than any other terrestrial habitat. A common feature of all grasslands is intermittent, erratic rainfall, amounting to about 25 to 100 cm annually. The irregularity of rain, porosity and drainage of soil prevent a continuous or ample supply of water to plant roots. Pre-dominant vegetation of this ecosystem is grass species, but sometimes also allows the growth of a few trees and shrubs. Summers are hot and winters are cold in temperate grasslands.

Grassland soil has considerable organic material because the grasses above soil surface die off each winter and contribute to the organic contents of the soil.

These grasslands have been used for cultivation of our food crops or are used as pastures for our domestic animals.

(A) *Biotic Components*: Different components of this ecosystem are as follows:

1. *Producers.* Grassland ecosystems include grasses, few herbs, shrubs and ground vegetation.
2. *Consumers.*
   (a) *Primary Consumers.* The grass provide food and allow the growth of large population of herbivorous animals, like cows, buffaloes, deer, sheep, goat, rabbit, mouse, insects, termites.
   (b) *Secondary Consumers.* These include carnivores, like frogs, snakes, lizards, fox, wolf, jackals, birds etc. feeding on the herbivores.
   (c) *Tertiary Consumers.* These are the hawks and top carnivores, like, lion, tiger etc. that feed on the secondary consumers.
3. *Decomposers.* These decompose the dead organic matter and help to bring minerals back to the soil and make them available to the producers. These include wide variety of micro-organisms including fungi, algae, bacteria etc.

(B) *Abiotic Component*: These are the inorganic well as organic substances present in the soil and atmosphere. In addition to the minerals some dead organic materials are also present.

4.13.4 Desert Ecosystems

Deserts are arid areas with annual rainfall lower than 25 cm. These are the regions with high evaporation rate and hence, available water also becomes saline quickly. Climate of deserts is quite different than that of other biomes. Difference between daily maximum and minimum temperatures is great. The skies over the deserts are sunny and days are very hot (40°C and above) atleast during summer. Nights are generally cold, even in summer.

A region is arid when it is characterised by a severe lack of available water to the extent of preventing the growth and development of animal life. **Semi-arid** climate is the climate of region that receives precipitation below potential evaporation but not as low as desert climate.

Large deserts are in the interiors of continents, where moist sear air or moisture bearing winds do not reach. There are also local deserts, which are rain-shadow areas on the downward side of mountain ranges. Deserts are mainly located in western India and Deccan Plateau, where the climate is extremely dry. Whereas the **cold deserts** such as in Laddakh are located in the high plateau of Himalayas.

The most typical desert landscape seen in Rajasthan is a part of **Thar desert**. This desert has sand dunes, and also areas covered with sparse grasses and few shrubs, which grow if it rains in the area.

Winds in deserts are more or less continuous, and rainfall is irregular. Lack of vegetation and presence of wind and stormy rains cause tremendous soil erosions most plants in the desert are small but grow rapidly, bloom and produce seeds all within few days after rain. Such plants live relatively for a small period. Most desert plants have small leaves, or none at all, or their leaf surfaces are often reduced to spines and thorns, minimising water loss by evaporation. Some have very long roots, reaching deeply buried water, while others like cacti absorb water rapidly after a rain and store the same in spongy internal tissues.

As a result of sparse vegetation, desert soil is low in organic material but is often high in mineral content, particularly the salts.

Desert animals tend to be small. During daylight, they remain in their shelter; and during morning, evening and in night, they become active and come out for hunt feed themselves.

Desert animals can stay dormant during the driest periods. Frogs, reptiles, rodents, jack rabbits, foxes and owls are the examples of the desert animals. Areas of scanty vegetation with semi-arid scrubland are used for camel and goat and sheep grazing.

Areas having a little moisture such as along the water courses are used for growing crops like *jowar* and *bajra* (millet).

Different components of this ecosystem are as follows:

(A) ***Abiotic Component in Desert.*** Due to high temperature and very low rainfall, organic substances are poorly present in the soil.

(B) ***Biotic Components in Desert.***

(i) *Producers.* These are shrubs, especially bushes, some grasses and a few small trees. Desert plants tend to have reduced leaves or no leaves, an adaptation that conserves water. In cacti, the photosynthesis is carried out by the stem, which also expands to store water, the leaves are modified into spines, which discourage herbivores.

(ii) *Consumers.* Desert animals are also adapted to scarcity of water and extremes of temperature. In most deserts, the large mammals and birds are comparatively rare or are absent altogether. Some animals, like the camel are adapted to extreme desert conditions. Reptiles, especially lizards and snakes are found desert. Desert animals include many burrowing forms, which may avoid the direct heat of the sun. In addition to desert adapted (which remain in their shelter during day time) insects, there are few amphibians (frogs and toads). Desert animals include rodents. Carnivores, such as fox and some birds of prey, like owls feed on the rodents and jack, rabbits. During the driest months of the year, many desert insects, amphibians, reptiles and mammals tunnel underground, where they remain inactive.

(iii) *Decomposers.* Due to poor vegetation, the amount of dead organic matter is very less, hence the decomposes are poor in desert ecosystems with only few species of thermophilic fungi and bacteria.

Threats to Desert Ecosystem

Growth in human population and different types of developmental activities are adversely affecting the natural ecosystem of the desert. The extensive irrigation systems for converting these lands has changed the several natural characteristics of the desert regions. Rapid evaporation of the canal water brings the salts to the surface thereby converting the region to become highly unproductive as it becomes saline. The special species that evolved in this unique ecosystem over millions of years are likely to become extinct soon.

Conservation of Desert Ecosystems

People residing in desert area have traditionally protected their meagre water resources. The Bishnoi tribe in Rajasthan is known to have protected their *khejdi* trees and the black-buck antelope for several generations. The national parks and wildlife sanctuaries in desert and semi-arid areas are required to protect residual patches of this ecosystem.

The Indira Gandhi canal in deserts of Rajasthan is destroying this natural desert ecosystem, as it will convert the region into intensive agricultural land.

Development projects in the deserts will alter its landscape and it also reduces the habitat available for various species, which are also at the verge of extinction.

Thar Desert

The Thar desert also known as the **Great** Indian Desert is spread over Punjab, Haryana, Rajasthan and Gujarat and covers an area of about four lakh sq km. Though smaller than the Sahara in Africa, it is the most populated desert in the world. The only river in the region is Ghaggar, which enters Rajasthan from Punjab and dries in the desert.

The Thar region has shrubs and trees like Khejri, Babul and Rohida, and fruits such as Ber and Pilu.

Sheep, goats and camels are the common animals used by the people in this area. The black buck, common India hare red lynx, jackal and wild dog are available in the thar desert. Lizards and snakes of different variety are available in the region. Local communities like the Bishnois are traditionally protecting the trees and the animals.

4.14 AQUATIC ECOSYSTEMS

Since water bodies cover nearly three quarters of the earth's surfaces, such ecosystems are vastly in existence on our planet. Major portion of water is contained in oceans, while only a tiny part is contained in lakes and rivers. Due to large scale differences in the quality of waters in various types of water bodies, they offer different types of environment, leading to development of different types of biological life. In all aquatic ecosystems, water is the medium in which biotic and abiotic components coexist. Water is the medium through which organic, inorganic and sediment material is distributed. The heat properties of water control circulation patterns within the ecosystem. The amount of dissolved salts, oxygen, sedimentation and amount of light absorbed by a water-body also determine the kind of community that can develop.

Marine water contain more than 3.5% of salt, while fresh water of rivers, lakes etc. do contain less than 0.02%. In some areas, we may find some water bodies, which are even more saline than the oceans, such water bodies may be called *hyper-saline* and may usually be found in arid areas, where rivers flow into a lake that has no outlet.

Depending upon the quality and nature of water, aquatic ecosystems can be classified into following three broad categories:

1. Freshwater ecosystems
2. Marine ecosystems, and
3. Estuarine ecosystems.

4.14.1 Fresh Water Ecosystems

Although these occupy a small portion of the earth's surface, they are of great importance to man because they provide drinking water as well as water for domestic and industrial needs.

Important fresh water ecosystems are:

1. Flowing water ecosystems (Rivers and Streams) and
2. Standing water ecosystems (Lakes, Pond and Marshes)

1. Flowing (Moving) Water Ecosystem

These are rivers and streams. The rate of flow of water and velocity are quite variable over the distance and also over the turbulence. The flowing water ecosystem depends upon the type of soil through which the river is passing (such as beds of rock, gravel, sand, silt or mud etc.) and also upon the velocity of flow, which depends upon the reaction and bed gradient. The flow velocity and erodibility of the bed and catchment soil may determine the amount and type of *sediment load.* When the river carries a lot of sediment, its water will be muddy. The turbidity of the flow will not allow height to penetrate through much depth below the surface.

The nature and composition of flora and fauna depend on the source of river and the land environment. The perennial rivers originating in springs or from melting glaciers, have cool water with rapid currents in their higher reaches. Floating population of plankton are generally, absent in such waters. In the middle reaches, the water current become slower and the river-body becomes wider. Sediments are deposited on the river-bed. The surface gets more of sunlight raising the temperature of water, and therefore, rate of photosynthesis and biological activities are also high. In the lower reaches, the current diminishes still further, the water is often muddy, cutting down light at the river-beds chemical characteristics of water depends upon the terrain through which it is moving.

The kinds of organisms vary greatly depending upon the strength of current, light penetration, sediment load, type of rudiment material etc.

2. Standing Water Ecosystems

Ponds, lakes and marshes come under this category. The sediments carried into a lake or a pond eventually fill up the water body and turn it into a marsh and then a forest. These do not have constant flow of water, however, wind may produce wave-action in large lakes. In deeper lakes, animals living at the bottom, have considerable pressure due to the weight of the water column and the atmospheric pressure. Like streams and rivers, penetration of light depends upon the sediments brought in regularly.

Aquatic Organism

The aquatic organisms are classified on the basis of their zone of occurrence and their ability to cross these zone. The organisms (both flora and fauna) in the aquatic ecosystem are unevenly distributed.

(i) **Neuston** : These are unattached organisms which live at the air-water interface such as floating plants etc.

(ii) **Periphyton** : These organisms remain attached to stems and leaves of rooted plants or substances emerging above the bottom mud.

(iii) **Plankton** : These include both microscopic plants like algae.

(iv) **Nekton** : These contain animals which are swimmers. These are relatively large and powerful as they have to overcome the water currents.

(v) **Benthos** : These are found living in the bottom of the water mass.

Factors Limiting the Productivity of Aquatic Habitats

Sunlight and oxygen are most important limiting factors of the aquatic ecosystems whereas moisture and temperature are the main limiting factors of terrestrial ecosystem.

The depth to which light penetrates a lake determines the extent of plant distribution.

Wetland Ecosystems

Wetlands are areas intermediate in character between deep water and terrestrial habitats. These habitats experience periodic flooding from adjacent deep water habitats and therefore supports plants and animals specifically adapted to such shallow flooding or waterlogging.

The wetlands may be either inland wetland or coastal wetland, may it be natural or man-made. In these areas the water level remains near or above the surface of the ground for most of the year.

Marshes and swamps known as freshwater wetlands are usually covered by shallow water for a part of the year and contains water-tolerant vegetation. The soil is water-logged and accumulates organic materials. Wetland plants (mostly grass-like plants) are highly productive and provide enough food for a variety of organisms. They are valued as wildlife habitat for water bird species, beaver, musk, offers rats and game fishes.

The wetlands are mainly of following types :

1. **Bogs** : These are the wet spongy areas of land, and formed from deep lakes that are filled with dead compacted plant material.
2. **Marshes** : These are flat treeless land areas covered with water. These areas are permanently wet and flooded land area.
3. **Swamps** : These are the areas of muddy watery land covered by trees and bushes. The swamps may be: deep water swamps, shallow water swamps of mangrove swamps.

(i) *Deep water swamps are* found near the large slow moving rivers that flood regularly. Their floods spread water over the adjoining low land turning such land into permanently flooded swamps.

(ii) *Shallow water swamps* are usually found in areas where soil stays moist or water covered for only a part of the year.

Both the above swamps are the fresh water swamps, and provide home to insects, frogs and fishes. These small creatures provide food for long-legged bird.

(iii) *Mangrove swamps* have salt water and lie along tropical sea-coasts, and are named for the mangrove shrubs or mangrove trees.

Functions of Wetlands

- Habitat to aquatic flora and fauna including migratory birds.
- Filtration of sediments and nutrients from surface water.
- Water purification and ground recharging.
- Maintenance of stream flow.
- Provide water, fish, fodder, fuel etc.
- Control rate of run off in urban areas.
- Source of livelihood to local people.
- Supporting specific diversity.

These support biodiversity, millions of people, protect from storms, flood control, improve water quality, supply food, fiber and raw materials.

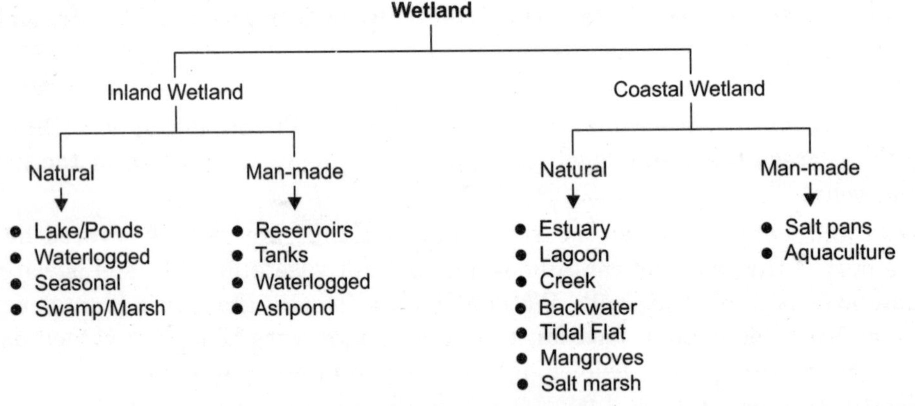

India's Wetlands

- India has a total of about 27000 wetlands of which about 23000 are inland wetlands and about 4000 coastal wetland.
- These occupy about 18% of the country's area of which 70% are under paddy cultivation.
- Natural wetlands in India range from high altitude wetlands in Himalayas, flood plains of the major river systems, saline wetlands of the arid and semi-arid regions, coastal wetlands such as lagoons, backwaters, estuaries, swamps etc.

Components of Freshwater Ecosystems

The freshwater ecosystems have following components:

(a) ***Abiotic Components.*** The major abiotic components in freshwater ecosystems are heat, light, pH value of water, and the basis of inorganic and organic compounds, such as water itself, carbon dioxide, oxygen, calcium, state, but most of them are present as stored in particulate matter as well as in living organisms.

(b) ***Biotic Components.***

(i) *Producers.* These are green plants and photosynthetic bacteria. The producers fix radiant energy and with the help of minerals derived from the water and mud, they manufacture complex organic substances as carbohydrates, proteins lipids etc. Producers may be free floating macrophytes, and minute floating or suspended phytoplankton.

(ii) *Consumers.* Herbivores are the *primary consumers* who are feeding directly on living plants or plant remains. Some mammals, such as cows, buffaloes etc. also visit the ponds casually and feed on marginal roots macrophytes. Some birds also regularly visit the pond feeding on some hydrophytes. *Secondary consumers* are the carnivores which feed on the herbivores. These are mainly insects and fish. *Tertiary consumers* are some large fish as game fish that feed on the smaller fish.

(iii) *Decomposers.* The main decomposers are bacteria, fungi and actinomycetes. These are also known as *micro-consumers,* since they absorb only fraction of the decomposed organic matter (dead plants and animals).

Global Estimates of Water contained in Biosphere

S. No.	*Segment*	*% of total water*
1.	Oceans	97.3
2.	Fresh water	2.7
	Total	100.0

Fresh Water from different Sources

S. No.	*Source*	*% of total fresh water*
1.	Polar ice and glaciers	75.2
2.	Ground water	
	(a) Upto 800 m deep	10.0
	(b) 800-4000 m deep	12.6
3.	Lakes and rivers	0.3
4.	Other sources	1.9
	(moisture and atmospheric vapour)	
	Total fresh water	100.0

Average Precipitation and Runoff in World

S. No.	*Item*	*Quantity (approx.)*
1.	Total land area	1,45,000 B. m^2
2.	Precipitation	
	(a) volume	1,09,000 B. m^3
	(b) depth	752 mm
	(c) percentage evaporation	57%
3.	Runoff	
	(a) volume	47,000 B. m^3
	(b) depth	324 m.m
	(c) percentage of precipitation	43%

4.14.2 Marine Ecosystems

Major oceans of the world, Atlantic, Pacific, Indian, Arctic and Antartic cover approximately 70% of the earth's surface. Each ocean represents a very large and stable ecosystem. Marine waters are saline with an average salinity of about 3.5%. Salinity is least near the poles and river mouths. Ocean currents—both at or near surface and at the bottom help mixing of water between various levels. Warm water from equator moves to polar regions while cold water moves from poles to equator. Another important feature is that all the major marine waters throughout the world are inter-connected.

Most important physical nature of marine ecosystem is the *tides*. The level of sea-water in the oceans rise and falls, usually twice a day. The height of the tides varies with the season, local topography and phases of the moon. The shore-lines of Oceans possess different substrata at different places. It may be bare rocky or sandy. Each type of shore has its own community of plants and animals. Sea-floor is covered with a variety of rediments having minerals and organic matter. Depth of sea is variable. Light does not reach to more than few metres (say 30–200 metres) from the surface. Temperature at surface varies from freezing point (–1.9°C) at the polar regions to about 25–30° in the tropics. In temperate seas, there is considerable variation during different seasons. Temperature also varies with depth.

Gases, like oxygen and carbon dioxide are also dissolved in sea-water. Dissolved oxygen is essential for marine animals except those that breathe air. Ecology of shallow and deep waters as well as estuarine part is different from each other.

The components of marine ecosystem are:

Abiotic Components

High sodium, calcium, magnesium and potassium salt concentrations, variable dissolved oxygen content, light and temperature make a unique physico-chemical condition in marine water. The concentration of dissolved nutrients remains low, which constitute an important limiting factor to determine the size of marine population. All the above mentioned minerals alongwith dissolved gases, like oxygen and carbon dioxide do help in supporting diverse biological life in the sea. Nitrates, nitrites and phosphates are very essential for *phytoplankton*. Silicates are essentially required by sea animals and also by the plants. The concentration of oxygen is good in shallow seas, where there is a considerable amount of photosynthesis and wave action, splashing and turbulence.

Biotic Components

1. *Producers.* It includes phytoplankton, large seaweeds, mangrove vegetations, and some microscopic algae. These organisms show a distinct zonation at different depths of water in the sea.
2. *Consumers.* The consumers are dependent for their nutrition on the primary producers. Herbivores are the primary consumers, who feed directly on producers. These are mainly crustaceans, mollusks, fish etc. Secondary consumers are the carnivores fishes, such as *Hering Shad, Mackerel etc.* feeding on the herbivores. Tertiary consumers are the top carnivores fishes, like Cod, Haddock, Halibut etc. that feed on other carnivores of the secondary consumers level.
3. *Decomposers.* The microbes active in the decay of dead organic matter of producers and macro consumers are mainly bacteria and fungi.

Problems Associated with Oceans

Earlier, traditional view was that the ocean is an inexhaustible wealth and it can accommodate all users and all kinds of uses and that any amount of waste will be absorbed by the marine system. Now, we have come to know and experienced that the ocean is a finite resource and that human activities are already caused an extremely adverse impact on the ocean. Because of pollution and overfishing, many fish species have disappeared forever.

Since deep sea has minerals, biological resources and petroleum, many countries are now developing techniques to tap all this sea wealth. Oil exploration and very heavy traffic in shipping create consequent problems like large oil spills from giant tankers. The dumping of hazardous waste including nuclear material into ocean is raising pollution levels. Near the shore heavy discharges from industry and sewage systems are also adding the pollution.

In the past there had been no control over the exploitation of the wealth of the ocean. Now, "UN Convention on the Law of the Sea" (UNCLOS), which came into force in November 1994 has made a landmark agreement. This convention has set the territorial limit as 12 nautical miles from the shoreline (one nautical mile is equal to 1.85 km). This convention establishes for every country an Exclusive Economic Zone (EEZ) upto a distance of 200 nautical miles from the shoreline. Every country has the exclusive right to exploit the resources of its EEZ. Beyond the EEZ is the International Area.

According to the Convention, the ocean has to be used only for peaceful purposes and should be managed in the interests of all including future generations. It has also laid down the mechanism to control the ocean resources and to resolve the disputes.

India has a coastline of over 7,000 km and an EEZ of two million sq. km, which is two-thirds of our land area. The Sealed Authority has granted exclusive rights to India to explore 1,50,000 sq.km in the central Indian ocean.

Case of Antartika and Southern Ocean

Antartika is an icy continents, surrounded by Southern Ocean. The Antartika ice cap contains 70% of the world's fresh water. A thin layer on the ocean supports very small shell fish (known as krill) which is consumed by other fish, seals, penguins and whales.

The environmental changes in the recent past has shown that mean temperature of the Antartika Peninsula has gone up by 2.5°C in summer and 5.6°C in winter. With the increase in temperature huge pieces of the ice are breaking off. In future to come, this process if continued could raise global sea levels. The reason for this is global warming due to burning of fossil fuels.

Due to hunting in Antartica, the result is extinction of seal, elephant seal, and the blue whale.

Chilika Lake–Revival of Lake

Chilika lake situated in coastal Odisha and has an area of about 1000 sq. km, is the largest brackish water lagoon in a sea. The lake is rich in fauna having home to a variety of residential and migratory birds, many fish species and the dolphin.

In the later part of the 20th century the lake had been facing a number of ecological and social problems. Heavy siltation resulting from massive deforestation in the catchment area reduced the lake area and was continuing at the rate of about 2 sq. km every year. Its opening to the sea was blocked which reduced the salinity level and many fish species were affected. The poor drainage of water led to waterlogging in the agricultural lands around the lake.

Chilika was recognised as a biodiversity hotspot in 1982. The number of fish species dropped from 126 in 1930 to only 69 in 1995. About 800,000 people in the catchment area depend on the lake for their livelihood and the increase in population strained the resources further. The average annual fish catch dropped from 8,800 tons to 1,600 during the decade of 1990s. Illegal prawn culture aggravated the problem. By the end of 1990s, the lake appeared to be dying. At that time, this site was given a status of 'Ramsar site' requiring priority conservation attention.

4.14.3 Estuarine Ecosystems

An estuary is a passage where the tidal water from the sea comes in contact with river current. It is a place of transition between river and sea. The estuaries are semi-enclosed coastal waterways with free connection to the open sea, and within which sea-water is diluted with fresh water from the river. River water being higher than the sea-water, 'flows' over the sea-water. Because of the tidal action, however, it mixes with fresh water, creating a zone of water with intermediate salinity.

The overall characteristics of the estuary depends on several factors, such as: the amount of freshwater, the nature of tidal action, the season of the year, the salinity of sea-water, nutrient transport from surrounding land etc.

Abiotic Components

In this transitional outfall reach of the river, the river water flows towards the sea, while sea-water comes back into this river zone alongwith its wave action. The movement of water masses over each other, however, creates turbulence and formation of eddies. These eddies, which often contain nutrients carried by the river, are called the *nutrient traps.* Since the river flow largely increases during rainy season, the salinity of the estuary will be very low in rainy season.

Biotic Components

The biological life found in estuarine ecosystem include plants and animals which can withstand high variability of salinity. Estuaries are highly productive for biological life, since there is rich inflow of nutrients from the river as well as from the sea, with abundant solar energy. Marsh grasses, seaweeds, seagrasses, benthic algae and phytoplankton are the main producers of estuaries. *Vascular plants* which are usually absent in marine ecosystems, are found in estuaries. *Prawns* and fishes, crabs, and fishes of several types and of large economic value are obtained from such ecosystems.

4.15 MAN-MADE OR ARTIFICIAL ECOSYSTEMS

So far we have discussed natural, self-regulating ecosystems without much direct interference and manipulations by man. In addition, there are artificial, man-engineered ecosystems, where, in order to obtain more food, clothes, timber and other useful products man becomes responsible for the replacement of natural systems. In order to have maximum production, man makes much planned manipulations in the physico-chemical environment. These include addition of fertilisers to soil, use of chemicals for disease control, irrigation practices. Man-made manipulations include villages and cities, orchards and plantations, gardens and parks with their plants and animals, large dams and reservoirs, lakes, canals and small fishery tanks.

A *cropland-ecosystem* or *agro-ecosystem* is, therefore, an artificial system primarily to grow species like wheat, maize, paddy, sugarcane, pulses, oil-seads, vegetable cotton etc. The entire human civilisation is supported by agriculture. Essentially, agriculture is the process of cultivating soil in order to raise economic plants. Humans have converted very large areas of forest and grasslands into croplands.

Abiotic Components

Abiotic components of this ecosystem are similar to natural ecosystems including climatic conditions of the region, organic and mineral contents in the soil (particularly C, H, O, N, P, K) and atmosphere. In cropland, man generally uses a number of chemical fertilisers. The physical condition of the farm depends upon latitude, overall climatic and edaphic factors. The physical deficiencies overcome by artificial irrigation and use of fertilisers and manures.

Biotic Components

The type of crop depends upon the climate, season and choice of the farmer. Usually, one crop is raised in a cultivated field and there is a strong selective pressure to reduce the diversity. The cropland ecosystems, thus, lack the diversity of natural ecosystems.

Most important man-made modifications in the biotic community is the use of domestication of animals. Animal population of cultivated lands comprises earthworms, nematodes, pollinators and pests, insects, rodents, many birds, rats, rabbit and other smaller reptiles etc.

Carnivores, like frogs and some birds that eat insect are the secondary consumers. Tertiary consumers are carnivores like snakes and hawks which feed on the secondary consumers.

The decomposition of dead organic matter of plants and animals make the minerals available again. Decomposers are the microbes present in soil as well as air. These are mainly bacteria, actimonomycetes and fungi.

4.16 DEGRADATION OF ECOSYSTEMS

We have so far seen that the life is based on ecosystems and the ecosystems provide a variety of products which are very essential for the human civilisation. However, ecosystems are frequently disrupted by human activities, leading to the extinction of species of the plants and animals that can live only in a particular natural ecosystems.

Some species (known as 'keystone' species), if eliminated, seriously affect the ecosystem. Extinction of species occurs due to changes in land use. Forests are deforested for timber, wetlands are drained to create agricultural land, and semi-arid grasslands used as pastures are converted to irrigated fields. The pollution from industries and the waste from urban settings also lead to the poisoning and extinction of several species.

The depletion of natural resources is because of two reasons, namely:

(i) Our rapidly exploding population that require increasing resources to sustain itself, and

(ii) Changing life-style, the growth of affluent societies that consume and waste a very large proportion of resources and energy.

The increasing extraction of resources is at the cost of our natural ecosystems, leading to its degradation. It is, therefore, necessary that we should think before wasting resources by conserving, reusing and recycling them.

Degradation aspect of ecosystems have been discussed in detail in the chapter on "Environmental Pollution".

4.17 ECOSYSTEM BENEFITS (SERVICES)

Benefits provided by ecosystem are as under :

Nature provides following four types of services to us free of cost :

1. **Provisioning Services** : The products obtained from ecosystem include food, wood, fresh water, breeding stocks.
2. **Regulating Services** : Regulation of ecosystem processes provide benefits such as : regulation of climate, water purification, flood control, air purification and control of human diseases, recovery from unpredictable events.

3. **Cultural Services** : These are the non-material benefits obtained from ecosystems such as recreation, education from nature, feeling the beauty of nature, ornamental plants.
4. **Supporting Services** : These are the services which are necessary for the production of all other ecosystem services, for example : biomass production, production of atmospheric oxygen and soil formation.

QUESTIONS

1. Write short notes on:
   (a) Tundra
   (b) Wetland ecosystem
   (c) Spoiling of Oceans.
2. Define 'Ecosystem'. Discuss the concept of ecosystem.
3. Explain the components of an ecosystem.
4. Distinguish between:
   (a) Producers and Consumers
   (b) Food Chain and Food Web
   (c) Biotic and Abiotic Components of an Ecosystem
5. (a) What is meant by bio-geochemical cycles?
   (b) Explain the concept of food-chain and food-web.
   (c) Explain photosynthesis and transfer of energy from one consumer to another.
6. (a) Discuss nutrient cycles.
   (b) Explain the term 'stability of an ecosystem.'
   (c) Explain the significance of 'photosynthesis.'
   (d) Discuss the movements of energy and mineral matter in an ecosystem.
7. Define ecosystem. Give an account of the structure and functions of an ecosystem.
8. Explain the following:
   (a) Ecological pyramids
   (b) Food web
   (c) Productivity.
9. What is food chain? Give comparative account of grazing and detritus food chains.
10. What are the factors that affect net primary production implants?
11. Write short notes on the following:
    (i) Pyramid of biomass (ii) Pyramid of numbers
    (iii) Pyramid of energy (iv) Ecological efficiencies
12. What do you understand by ecological pyramid? Describe the relationships between the numbers, biomass and energy contents of the producers and consumers of different order in an ecosystem.

13. (a) State the two laws of thermodynamics.
    (b) Explain ecological adaptation.
    (c) Explain the meaning of the term 'wetlands'.
14. Explain the different types of forest ecosystem.
15. (a) What are the characteristics of desert plants?
    (b) What are planktons?
16. Distinguish between:
    (a) Terrestrial and aquatic ecosystems
    (b) Marine and freshwater ecosystems
    (c) Species richness and species diversity
17. Describe freshwater environment and mention various plants and animals that live there.
18. Give salient features of each of the terrestrial ecosystems.
19. Write the names of major biomes. How one biome differs from another? Give an account of the characteristics of the tropical rainforest biome.
20. Give an account of habitat conditions and the biotic communities in a fresh-water body. How the habitat conditions differ from that of a forest.
21. What are the major ecological differences between a desert and a forest and grassland?
22. Give an account of ecosystem in estuaries.
23. Distinguish between 'evergreen' and 'deciduous' forests.
24. (a) Define and explain an 'aquatic ecosystem.'
    (b) Differentiate and explain the two important categories of aquatic biomes.
25. Explain and distinguish between the following:
    (a) Fresh water and Marine ecosystems.
    (b) Estuarian and Freshwater ecosystems.
    (c) Slow moving and Fast moving riverine ecosystems.
    (d) River and Lake ecosystems.
26. Discuss various properties of ecosystems.
27. Discuss subsystems of 'biotic components'.
28. Discuss 'man-made' ecosystems.
29. Write a detailed note on ecological balance, and causes for ecological imbalances.
30. (a) Explain benefits of ecosystems.
    (b) Explain degradation of ecosystems.
    (c) Revival of Chilka Lake.
31. Write a detailed note on "Desert Ecosystems".

5

Chapter

BIODIVERSITY AND ITS CONSERVATON

5.1 INTRODUCTION

The term *"Biodiversity"* is a concise form of *"Biological Diversity"*. The term 'Biodiversity' means the variety available on the Earth amongst the different living organisms, including plants as well as animals. *Biological diversity*, or Biodiversity as it is often called, is, hence, defined as *"the variety and variability among living organisms and the ecological complexes in which they occur."* Biodiversity involves different species of plants and animals. Diversity in simplest words can be defined as 'the number of species found in a community.

A measure of diversity is known as *species* richness. The richness or poorness of biodiversity of a region depends upon the number of their species *i.e.*, larger the number of species, richer the biodiversity and vice-versa. The ecological balance of a region is directly related to greater (richer) biodiversity. Thus, 'species richness' can be expressed as the number of species in a sample or a community.

5.2 LEVELS OF BIODIVERSITY

Before discussing the levels of biodiversity, we must understand the term 'species'. *'Species'* is defined, "as a group of similar organisms that interbreed or share a common lineage or generation."

Following are the three types or levels of biodiversity:

1. Genetic Biodiversity

All finer levels of organisation, biodiversity is the genetic variation within same species, both, among geographically separated populations and among individuals within single population. The genetic biodiversity occurs amongst the same species due to enormous number of combinations, which can be formed by the genes found in that species. When the genes within the various members of the same species show different versions or varieties due to different combinations, a lot of variability between the individuals of the same species do occur.

Genetic diversity covers distinct population of the same species, such as thousands of traditional rice varieties in India. This shows variations at the genetic level, and differ in colour, size, shape, smell, taste and nutrient contents.

2. Species Biodiversity

Species biodiversity or *organisational biodiversity* is the variability among the different species of organisms of a community. Biodiversity at its most basic level includes the full range of species on the earth, from micro-organisms, such as viruses, bacteria that protects through the multi-cellular kingdoms of plants, animals and fungi.

Species diversity broadly represents the species richness and their abundance in a community. The term 'species diversity' considers the number of species present in a habitat, alongwith considering the number of individuals of each species. This term reflects the relative abundance of populations of various species present in a habitat.

Thus, species diversity refers to the variety of species within a region. It can be measured in following two ways:

(i) The number of species in a region—(species richness).

(ii) The relationship of species to each other—(Taxonomic diversity).

For better understand the Biodiversity, the terms *population* and *communities*, must be known. They are defined as under:

Population: It denotes groups individuals of any one kind of organisms. Therefore, one or more individuals of the same species, co-occurring in time and space, may be termed as the population of that species.

Community: It includes all the populations of organisms with similar life habitat, occupying a given area.

3. Ecosystem Diversity

Ecosystem (or ecological) diversity shows the variations in ecological homes, food chains and food webs, nutrient cycling etc. provided by different ecosystems. Wide variations, occur among various ecosystems, due to their having different physical conditions, like climate, rainfall, temperature, altitude, types of soils etc. Tremendous diversity is, therefore, found within the ecosystems.

In view of the above, there may occur considerable diversity within the various types of ecosystems, such as forest ecosystems, desert ecosystems, grassland ecosystems etc.

The ecosystem diversity is of great value, as it gives rise to tremendous diversity in biological life. Therefore, the ecosystem diversity must be preserved in order to maintain biodiversity.

Measurement of Biodiversity

Biodiversity is measused in following two ways :

1. **Species Richness** : It is the measure of number of species found in a community. This method is used for :

(a) Measuring the diversity within a particular area or ecosystem.

(b) Comparing diversity between ecosystem.

(c) Measuring overall diversity for the different ecosystems within a region.

2. **Species Evenness** : It measures the properties of species at a given site. Low evenness indicates that only a few species dominate the side.

5.3 BIOGEOGRAPHICAL CLASSIFICATION OF INDIA

India has a varied landscape rich in natural resources. India is shielded by the world's highest mountains, the Himalayas in the North. The southern part of India takes the shape of a peninsula and divides the Indian Ocean into the Bay of Bengal to the south-east and the Arabian Sea to the south-west. India, being a vast country, shows a great diversity in climate, topography and geology, and hence, the country is very rich in terms of biological diversity.

All types of climates except the polar climate are generally available in the country, producing enormous variety in its flora and fauna. We do have a long coastline and different marine ecosystem, besides several huge fresh water rivers. India has large scale differences in soil conditions, amounts of rainfall, maximum and minimum temperatures. It also has plains and hilly areas. Therefore, India's biological diversity is one of the most significant in the world, since India with 2% of the total landmass of the world, contains about 7% of the world's known wildlife.

The biodiversity of India is divided as follows :

1. **Realms of India** : Realm is a continent or sub-continent sized area with unifying features of geography, fauna and flora. The Indian region is composed of following two realms :

   (a) The Himalayan region represented by **Palearctic Realm,** and

   (b) The rest of the sub-continent represented by **Malayan Realm.**

2. **Biomes of India** : Biomes are the main groups of plants and animals living in areas of certain climate patterns. A biome includes the way in which animals, vegetation and soil interact together. India has following 5 biomes, namely:

   (a) Tropical Humid Forests. (b) Tropical Dry or Deciduous Forests.

   (c) Warm Deserts and Semi-deserts. (d) Coniferous Forests.

   (e) Alpine Meadows.

3. **Bio-geographic Zones**

Depending upon the climate, rainfall, geography, vegetation and communities, India can be divided into 10 zones or regions. Each of these regions contain a variety of ecosystems, such as forests, grasslands, lakes, rivers, wetlands mountains and hills. These 10 biogeographical zones (regions) of India are:

Biogeographical Zone	*Area Covered*
1. Trans-Himalayan zone (Cold Desert zone)	Cold snow covered mountains in the region of Leh and Laddakh.

2. Himalayan zone	The Himalayan ranges and valleys of Kashmir, H.P., Uttarakhand, Assam and North-eastern States.
3. North-Eastern India	Brahmputra valley and North-eastern hills.
4. Indo-Gangetic plain	Stretching from Eastern Rajasthan through U.P. to Bihar and West Bengal.
5. Desert	Thar desert and Rann of Katchh.
6. Semi-arid plains	Punjab plains, Gujarat and Rajputana.
7. Deccan Peninsula	Indian States—south of Vindhyas including A.P., Odisha and M.P. on Eastern Plateau, Tamil Nadu on Southern Plateau; Karnataka, Maharashtra and Gujarat on Western Plateau.
8. Western Ghats	Malabar coast, Western Ghats.
9. Coastal land	West coast and East coast.
10. Indian Islands	Andman and Nicobar Groups of Islands, and Lakshadweep Islands.

A brief description of all these biogeographical zones is given here under:

1. Trans-Himalayan Zone

This region lies on the rain shadow (no rain) zone of the Himalayan ranges located at 4500 to 6000 m height above the mean sea-level. These areas consist of bare rocks and glaciers and do find alpine vegetation. Some common trees of this region are: *fir, kail, populus, juniper* etc. Some of the scently and medicinal plant species are also found here. Some prominent wild animal species found in this region are: *wild yak, snow sheep, goat, snow leopard, marbled cat, tibbetan wild ass.* Some of the important bird species of this region are: *snow goose, kaikal, snow petridge, black kite, tibbetan petridge and goshowk etc.*

2. Himalayan Zone (Region)

The *Eastern Himalayas,* one of the wettest regions of the world, offers the greatest biodiversity. These Himalayan ranges occur at altitude of about 1800 to 3500 m. This region is a very active centre of organic evolution and is the cradle of flowering plants. Animal wealth of this region is also very diverse including large number of species of *birds,* animals of *goat family* and certain species *antelope family,* the *musk deer,* the *snow leopard,* the *brown bear,* the *black bear* and the *red panda.*

The *Western Himalayas,* on the other hand, are much winder and colder with a drier climate. This region has *tropical moist deciduous forests* and has *chir, blue pine deodar,* fir etc. Animals found in this region include : *Yak, bear, goral, serow. Snow leopard* and *brown bear* inhabit the alpine region. *Sambar, muntjac* etc. are seen in the lower altitude areas.

The major centres of biodiversity of this region are : *Valley of flowers, Pithoragarh, Gori valley, and Karakorum.*

3. North-Eastern India

This region includes north-eastern states and sometimes treated as a part of Eastern Himalayas. Due to the presence of Garo hills, Khasi-Jaintia hills, Naga hills, Mizo hills etc. in this region, very heavy monsoon rainfall occurs in this area. Therefore, this zone is very rich in biodiversity.

Most parts of this region, therefore, falls in the category of tropical moist evergreen forests.

63% of mammalian fauna are found in Assam. The area is rich in smaller carnivores. However, country's highest population of elephants (herbivores) is also found in this region.

4. The Indo-Gangetic Plain

This region comprises of vast land area falling between Himalayas and Vindhya mountains. The soil of this region is formed by the alluvial deposits of the 'Ganges' and its tributaries. Large portions of the land of this region is under cultivation, while certain portions, particularly in the states of Jharkhand and Madhya Pradesh are under forests, which are of tropical dry deciduous type. The south-eastern and portion of the Indo-Gangetic Plain merges with the littoral and mangrove regions of *Sunderbans*. This Sunderban area provides an excellent wet land ecosystem, providing fish and fire wood on a large scale.

The fauna of this region includes *elephants, black buck, rhinoceros, crocodile, turtles and gazelle* (a kind of deer); and birds like *Bengal floricans, dense water fowl* community.

5. Desert (Hot)

The hot desert is characterised by low and fluctuating rainfall, high solar heat, low atmospheric humidity, strong winds with sandy plains and dunes.

Natural vegetation consists of tropical thorn forests and tropical, dry deciduous forests, sandy deserts with seasonal salt marshes and mangroves are found. *Phog* shrubs are grown on sand dunes. Flora wealth of the desert include 682 species of flowering plants. Species of fruit plants, vegetables, gums and medicinal plants are also found in this area.

The Thar desert possesses several species of *insects, reptiles and birds. Black buck, gazelle Nilgai, wild ass* (confined to Rann of Kutchh). *Desert fox, great Indian bustard, Chinkar, and desert cat* are the other animal species found in the area.

6. Semi-Arid Plains

The semi-arid region in the western India includes the areas of Rajasthan, extending to the Gulf of Kutchh and Cambay and the whole Kathiawar peninsula.

The natural vegetation consists of tropical thorn forests and tropical dry deciduous forests, moisture forests (in extreme north *i.e.*, in Punjab). The sandy plains have a few scattered trees of *acacia* and *prosopis*. The rocky habitats are covered by bushes of *euphorbia*, while species of *salvadora* and tamarix occur mainly near saline depressions.

The *lion of Gir forest* in Gujarat is the endemic animal species of this zone.

7. Deccan Peninsula

This is a large area of raised land (varies from 900 m in the west to 300 m in the east) covering about 43% of India's total land surface. The area has four major rivers, *i.e.*, Narmada, Godavari, Krishna and Cauveri that support the wetlands of this region which have fertile black and red soils.

Large parts of this area are covered by tropical forests. Tropical dry deciduous forests are in the northern, central and southern part of the plateau. The eastern part of the plateau in A.P., Chhattisgarh and Odisha has moist deciduous forests.

Animal species, like *tiger, wild bear, sloth bear, gaur, sambar and chital* are common in the entire alongwith small population of *wild buffaloes* and *elephants and barasingha.*

8. Western Ghats

The Western ghats represent the prominent mountain range, stretching more than 1600 km length and running parallel to the west coast of India. These ghats receive considerable rainfall on the western side and lesser rainfall on the eastern side of the ghats. Therefore, the western side have *tropical moist* evergreen forests (tropical rain forests), while the eastern side have tropical moist deciduous forests.

The biodiversity of the region is very rich and contains large variety of species of *flowering plants, shrubs, orchids, palms, grasses and ornamental plants.* Out of the total plant species of this region, about 40% are endemic (*i.e.*, not found else where). The area has many economically important plants, such as banana, rice, black paper, ginger etc.

These ghats provide very rich diversity of fauna (animal life) also *Nilgiri Langur, Lion, Tailed macaque, Nilgiri Tahr, Malabar grey hornbill*, and most *amphibian* species and *Lizards* are endemic to this area. Other animals found here include the *tigers, leopards and elephants.*

9. Coastal Region

India has a long coastline along the Arabian sea and Bay of Bengal. The coastal region is highly productive and has rich ecosystems in flora and fauna. The coastal inter-tidal zones provide wetlands, which are rich in vegetation and other animal organisms.

A large number of species of animals are found to be living on and near to the various sea coasts. They include *dolphins, crocodiles, dugong* and *avifauna,* besides various species of fresh water *turtles* and tortoises, and marine turtles. Several *fishes crabs* are also found along coasts and beaches. The oceans adjoining the coasts also have a very rich biodiversity.

10. The Indian Islands

India has a group of 325 islands in *Andman and Nicobar Group of Islands* in Bay of Bengal, out of which only 21 islands are inhabited. Another group of 36 major islands exist in the Arabian sea in the name of *Lakshadweep group of islands*.

The Andman and Nicobar Islands have various types of forests, such as the *tropical evergreens* as well as *moist deciduous forests.* These islands also have the inter-tidal zones giving rise to mangrove forests.

These have large varieties of *flowering plants, avifauna, giant robber crab,* besides *turtles, wild bear,* Andman water snakes.

Animal life found on Lakshadweep islands mainly consists of *turtles, typical coral organisms,* ornamental fishes and dugongs.

Bio-geographic Provinces

The above mentioned 10 bio-geographic zones have been further divided 25 bio-geographic provinces. This classification is based on various factors such as altitide, moisture, topography, rainfall etc. The bio-geographical princes contained in each of the bio-geographic princes are given below :

S.No.	*Bio-geographic Zones*	*Bio-geographic Provinces*
1.	Trans Himalayan zone	1. Himalayan Ladakh Mountains 2. Tibetan Plateau 3. Sikkim
2.	Himalayan zone	1. North West Himalaya 2. West Himalaya 3. Central Himalaya 4. East Himalaya
3.	The Indian Desert	1. Thar Desert 2. Kutch Desert
4.	The Semi-Arid Plains	1. Punjab Plains 2. Gujarat Rajputana
5.	The Gangetic Plains	1. Upper Gangetic Plains 2. Lower Gangetic Plains
6.	The Western Ghats	1. Malabar Plains 2. Western Ghat Mountains
7.	The Deccan Peninsula	1. Central Highlands 2. Chota Nagpur 3. Eastern Highlands 4. Central Plateau 5. Deccan South
8.	The Coastal Lands	1. West Coast 2. East Coast
9.	North Eastern India	1. Brahmaputra Valley 2. North East Hills
10.	Indian Islands	1. Andamans 2. Nicobars 3. Lakshdweep

5.4 IMPORTANCE AND VALUES OF BIODIVERSITY

The world is inhabited by great diversity of life forms: animals and plants, living in diverse habitats and possessing diverse qualities. This diversity is vital to human survival, as they provide food and materials for livelihood. Since the beginning of human civilization, man has depended on biodiversity around himself. The biotic resources support human-livelihoods and aspirations to meet the changing needs. Indeed, the continuing loss of biodiversity will result into the imbalance between human needs and wants and nature's capacity.

Biodiversity, besides basic needs of maintaining an ecological balance, provides us enormous goods and services having huge economic value. In addition, biodiversity has a social, scientific, educational, cultural, recreational and aesthetic values.

The multiple uses of biodiversity *i.e.*, biodiversity value may be classified as follows:

1. Economic Value

A. Consumptive Use Value

These are the direct use values where the biodiversity products are harvested and consumed directly, such as food, drugs and medicines, fibre, fuel woods etc:

(i) *Food:* A large number of plants are being consumed by human beings and their domesticated animals (cattles) as food. For food, man gather or harvest the plants, or do hunting. To meet the food requirements of cattles, he converts the forest to grazing land.

(ii) *Drugs and Medicines:* Plants from the wild are being used for producing medicines. A large population of the world depends upon plants or plant extracts for medicines. The *Penicillin* uses as an antibiotic is derived from a fungus called *Penicillium*. Similarly, *Tetracycline* is received from a *bacterium*. *Quinine,* the medicine to cure from malaria is obtained from the bark of Cinchona tree. A large number of marine animals are supposed to possess anti-cancer properties which are yet to be explored systematically.

(iii) *Fuel:* Fuel wood is being used by people living in the forests or nearby areas. The fossil fuels, like coal, petroleum and natural gas are the products of fossilized biodiversity.

B. Productive Use Value

Productive use value of biodiversity is the value assigned to the products that are commercially harvested for sale in formal markets. Such saleable products include: fish, timber, fuel wood, silk from silk worms, wool from sheep, fur from many animals, lac from lac insects, fodders, fruits, cereals, medicinal plants. Musk from musk dear. Many industries like, paper and pulp industry, plywood industry, pearl industry, rubber industry, railway sleeper industry, silk industry, textile industry, fishery industry, leather industry, pharmaceutical industry etc. are almost totally dependent upon the productive use value of biodiversity.

A considerable trade relying upon biodiversity also includes meat, pets including aquarium fishes, reptiles and birds etc.

Agriculture, forestry and fisheries play an important role in the economy of the country. Illegal trade through smuggling takes place for sandlewood, tusks of elephants, musk from musk dear, skins of animals, like tigers, leopards, cats, snakes etc.

2. Social and Cultural Value

These are the values associated with the social life, customs and religion of the people. Many plants are considered holy and sacred in our country, like *Peepal, Tulsi, Lotus flowers, Beetle, Coconut* etc. Many animals and birds, like langurs, monkies, cows, bulls, snakes, peacocks, crows etc. do have an important place in our customs.

3. Ethical and Religious Value

The ethical and religious value of biodiversity is rooted in the understanding that humidity is part of the nature and that we are just one species among others. All species have an inherent right to exist. Future generations also have an inherent right to know them and also to use them. It involves ethical issues, like, *"all like must be preserved"* and *"live and let live"*.

Religions also have a significant impact on our attitude towards natural resources. These values mean that we may or may not use a species, but their existence in nature gives us a pleasure. We all feel sorry when we learn that "passenger pigeon" or "dodo" is no more on this earth.

4. Aesthetic Value

The beauty of nature, its plants, animals, birds etc. is a great thing to watch and enjoy. The aesthetic value of biodiversity has been expressed in many forms, through art, poetry, songs, literature etc. people from far and wide, spend a lot of time and money on visiting wild areas and wildlife, and this type of tourism is known as *"eco-tourism"*. Wild life sanctuaries and zoos attract lakhs and lakhs of tourists every year.

5. Scientific and Educative Value

The study of plants and animals of forests and oceans provide enough research opportunities to the scientists, who are busy in finding cure for diseases, like cancer and AIDS. People preparing films, showing the actions of animals and birds inside the deep oceans and forests, are a highly educative programme and helping in creating awareness among the people towards the value of nature and the natural environment.

6. Option Value or Potential Use Value

Option value is the value of knowing that there are biological resources existing on this biosphere that may one day prove to be an effective option for something important in future. Thus, the option value of biodiversity suggests that any species may prove to be a miracle species some day.

Option value includes the future potential of biodiversity that are presently unknown, and need to be explored. Therefore, each time we loose a species, we lose its potential use. There is a possibility that we may have some potential cure of cancer or AIDS existing in marine ecosystem or in tropical rain forests.

Option value includes the values, in terms of the option to visit areas where a variety of flora or fauna, or specifically some rare or endangered species, or species confined to that particular area alone (known as endemic species) exist. For example, India has about 60% amphibians and 50% lizards as endemic, therefore, if one wants to see such a species, he has only one option *i.e.*, to visit India and our ecosystem.

7. Ecosystem Service Value

It refers to the services provided by the ecosystems, like prevention of soil erosion and floods; maintenance of soil fertility, cycling of nutrients; cycling of water, their roles for absorbing carbon dioxide, reducing global warming etc.

The Indian landmass is bound by Himalayas in the north, the Bay of Bengal in east, the Arabian sea in the west and the Indian Ocean in the south. The wide variety physical features and climatic conditions have resulted in a rich biodiversity. These vary from the humid tropical Western Ghats to the hot deserts of Rajasthan, from the cold deserts of Leh and Laddakh and icy mountains of Himalayas to the warm coasts of peninsular India. India is, therefore, very rich in biodiversity with a sizable percentage of endemic flora and fauna.

8. Medicinal Value

Hundreds of plants are used in the traditional medicines. It has been estimated that more than 50% of the world's population use plants as medicines. Many people have a vast and unique knowledge of the local plants and their medicinal values. Modern pharmaceutical companies depend heavily on the medicinal plants in their search for drugs.

5.5 INDIA AS A MEGA-DIVERSITY NATION

India is one among the 12 mega-biodiversity countries in the world and is divided into 10 biogeographical regions as discussed earlier. The wide variety in physical features and climatic conditions have resulted in a diversity of ecological habitats, like forests, grasslands, wetlands, coastal and marine ecosystems and deserts, which harbour and sustain immense biodiversity. India is considered to be the homeland of 167 important plant species of cereals, millets, fruits, vegetables, pulses, fibre crops, and oil seeds. India also has 114 breed domesticated animals. About 5150 species of flowering plants (30% of the world's endemic flora) are endemic to the country. It is estimated that 62% of the known amphibian species are endemic to India.

38 species of mammals, 69 species of endemic birds, and 214 species of reptiles are endemic to India or Indian subcontinent. Moreover, nearly 60% of the world's tigers, 80% of the world's one-horned rhinoceros, 100% of the Asiatic lion, 65% of the Asian elephant and 80% of the world's gharials are found in India.

Mega-diversity Nations of the World	
Australia	Madagascar
Brazil	Malaysia

Cameroon	Mexico
China	Myanmar
Columbia	Peru
Costa Rica	Philippines
Ecuador	South Africa
Ethiopia	Venezuela
India	Zaire
Indonesia	

5.6 GLOBAL BIODIVERSITY

It is estimated that there are about 8 to 10 million biological species are existing on the earth. According to Hindu mythology, the total species are 84 lakhs (8.4 million). However, so for only about 1.8 million (18 lakhs) *i.e.*, about 21.4% of the total number of species are known and have been documented by the scientists in the world, which shows that majority of species are still to be discovered.

The distribution and magnitude of the biodiversity existing over the entire earth is a product of about 3 to 5 billion years of its evolution including speciation, migration, extinction, and now human influence. In recent times, due to increased human interference, a significant number of species are being lost.

Out of all the terrestrial ecosystems, the tropical rain forests provide the maximum biodiversity. These tropical rain forests are inhabited by millions of species of plants, animals, birds, insects and micro-organisms etc. About 80% of the global biodiversity live in tropical forests which is only 7% of the total land surface. The marine biodiversity is even much higher than the terrestrial biodiversity. Estuaries, coastal water and oceans exhibit tremendous biological life and efforts are being made to identify and classify them.

Most of the bio-rich nations are near the equator. China, Brazil, Malaysia and Indonesia are also very rich in biodiversity. The world is now realising the biological resources as the 'global resources'.

5.7 INDIAN BIODIVERSITY

India occupies only 2.4% of the world's land area, but its documented contribution to the world's biodiversity is around 7%. The total number of species found in India is around 1,28,000; out of which 47,000 are plant species and 81,000 are animal species. Therefore, there is a large proportion of the Indian biodiversity is still unexplored.

In addition, India's contribution to agro-biodiversity is also impressive. India stands seventh in the world as far as the number of species contributed to *agriculture* and *animal husbandary* is concerned. India has contributed such successful animal species as *water buffalo* and *camel*, and such crop species as *rice* and *sugarcane*. India has also been a secondary centre of domestication of such animal species as *horse* and *goat* and such plant species as potato and maize.

As a result of several thousand years of our civilization, we have produced more than 60,000 varieties of cereals (including rice, wheat etc.), more than 25000 varieties of pulses, and more than 20,000 varieties of oil seeds in the Indian sub-continent.

In terms of **plant diversity**, India ranks tenth in the world and fourth in Asia. India represents nearly 11% of the world's known floral diversity. Important floral groups found in India are : Algae, Fungi, Bacteria, Lichens, Bryophytes, Petridophytes, Gymnosperms, Angiosperms.

Agriculture remains one of the dominat drivers and mainstay of economic growth in India. The large number of distinct agro-ecosystems, characterized by variations in adaphic, climatic and geographic features has contributed to diverse cropping patterns and systems across the country.

India has vast resources of livestock and poultry, which play a vital role in rural livelihood security. In terms of livestock population, India ranks first in buffaloes, second in cattle and goats, third in sheep, fourth in ducks, fifth in chicken and sixth in camels in the world.

Common Fauna in India

Like flora our fauna (animal life) is also equally rich and varied. There are about 75000 known species. There are about 2000 species of fish in fresh and marine water. There are about 1200 species of birds and in addition there are 140 species of amphibians, 420 species of reptiles, 340 species of mammals, 50,000 species of small insects and worms. Among the mammals we have elephant found in hot wet dense forests of Assam, Kerala and Karnataka. Camel and wild asses are found in hot and arid deserts. One horned rhinoceros lives in swampy and marshy lands of Assam and North Bengal.

Indian bison, Indian buffalo, the nilgai, Chausingha (four-horned antelope), black buck, gazel and deer (this include swamp deer, spotted deer, musk deer and mouse deer) are the unique Indian animals.

Indian lion is found in Gir forests of Saurashtra in Gujarat. Bengal tiger is found in Sunderbans. Other animals of cat family are: leopards, clouded leopards, snow leopards (confined to upper reaches of Himalayas).

Several species of monkey including langur and lion tailed mecaqune having hair around the face. Among the birds, geese, ducks, pigeons, cranes, hornbills and sunbirds are common in India.

Tiger is the national animal and peacock is the national bird of India.

Common Flora in India

Main flora available in India, regionwise are as under:

In *Himalaya foothills* including *tarai* formations and *shivalik* ranges mainly have sal, silk-cotton trees, giant bamboos, tall grassy meadow. High altitude Himalayan region has coniferous-pine, dwarf hill bamboo and alpine. Eastern Himalayas have mainly oaks, fir, yew and junipers, lichens, orchids alongwith western Himalayan fauna.

Peninsular India has *sal* in north and east, while *teak* in southern plateau has dominant trees. Desert of Rajasthan has scattered and thorny scrub species with reduced leaves, cacti as main plants.

Tropical rainforests comprising north-eastern India and the Western Ghats in South including Malabar Coast. These areas have mosses, ferns, orchids, lianas and vines, palms, cardamom and ginger.

5.8 HOTSPOTS OF BIODIVERSITY

Areas which exhibit high species richness as well as high species endemism are termed as hotspots or biodiversity. Therefore, biodiversity hotspots (also termed as biological hotspots) are the areas that are unusually rich in biological species, most of which are endemic (*i.e.*, not found elsewhere), and are under a constant threat of being over-exploited. An area is designated as a hotspot when it contains at least 0.5% of the total global plant species as endemic, and also the region must have lost more than 70% of its original habitat.

The term hotspots was introduced by Myers in 1988. Initially, he identified 12 hotspots on global level, which later has been revised by him in the year 2000 to 25. These 25 hotspots cover less than about 1.5% of the world's land area. About 44% of all the terrestrial plant species and 35% of all the terrestrial invertebrate species of the world are endemic and found in these 25 hotspots.

The hotspots are the areas that are severely threatened by human activities. The hotspot concept targets regions where the threat is greatest to the greatest number of species and allows conservations to focus cost-effective efforts there. Two of these 25 hotspots are existing in India extending into neighbouring countries. These are: Eastern Himalayas, and Western Ghats. These Indian hotspots are not only rich in floral wealth and endemic plant species, but are also rich in reptiles, amphibians, swallow tailed butterflies, and some mammals.

It would also be pertinent to mention here that the evolution of new species take enormous number of years, and that too, in isolated and untouched environment. Today, the fast growing human populations in the hotspots contribute to their deterioration by the introduction of exotic species, the illegal trade in endangered species, industrial logging, slash and burn agricultural practices, developing tourism infrastructure, mining and the construction of highways, dams, oil wells etc.

5.8.1 Biodiversity Coldspots

The areas that have relatively low biological diversity but are also experiencing a high rate of habitat loss are called biodiversity coldspots. Although a biodiversity coldspot is low in species richness, it can also be important to conserve, as it may be only location where a rare species is found. Extreme physical environments (low or high temperatures or pressures) inhabited by just one or two specially adopted species are coldspots that warrant conservation.

5.9 THREATS TO BIODIVERSITY

The components of biodiversity are so inter-dependent that any change in the system leads to a major imbalance and threatens to the normal ecological cycle. Extinction of biological species is the natural process of evolution. The species have, thus, evolved, died out and substituted by newer ones, during their evolution process which is a continuous natural process. In pre-historic periods, the process of loss of old species and evolution of new species has occurred at a very slow rate. The rate of extinction was as low as one species per 10,000 years. However, between 1600 to 1950 AD, this rate was increased to one species per 10 years; and currently, it is 1000 species per year.

The loss of diversity is also very severe in agricultural ecosystems. Thousands of wild crop varieties were replaced with a few hybrid species. This resulted in the disappearance of our genetic resources of crop plants, especially in wheat and rice. With the disappearance of those several varieties of crop plants, the associated micro-organisms and fauna were also lost. Indiscriminate use of fertilizers and chemicals have reduced the microbial species diversity. Livestock populations have also been homogenised, and their diversity is already extinct.

Marine life and habitats are similarly under serious threat of destruction, leading to large scale threats to marine biological life. Human beings have also destroyed most of the wetland ecosystems, and mangrove forests.

Impact of Biodiversity Loss

The poor people, who are dependent on biodiversity for their daily survival, feel the impact first. The industrialised countries, also soon start experiencing the adverse effects of biodiversity loss. Most of their food crops, medicines, textiles, paper etc. originate from plants in developing countries. The destruction of forests means that the less carbon will be absorbed and natural climate control mechanisms are lost. This is responsible for major impact on the world's climate.

It must be realised that all the other problems are potentially reversible, but the loss of biodiversity is not reversible.

Threatened or Endangered Species

It has been experienced that for a species to survive and flourish, enough numbers must be present in the habitat to make reproduction possible. If density and population size fall below a desirable level, the numbers start going down.

A **threatened or vulnerable species** is one, which though found in reasonable size in its natural habitat, but the numbers are declining. Unless **conservation measures** are taken, it is likely to move into the endangered category.

A species is declared as **endangered** when the number of survivors is so small that it could soon become extinct over its habitat. Unless it is **protected**, it will move into the critically endangered category, before it goes extinct.

Causes for Biodiversity Loss

Biodiversity loss is caused when either a particular species is destroyed or the habitat (which is essential for survival) is damaged. The extinction of species takes place when they are exploited for economic gain or hunted. Extinction of species is also caused due to environmental reasons like ecological substitutions, biological factors which are caused either by nature or man.

1. **Natural Causes** : Floods, earthquakes, landslides, rivalry among species, and diseases.
2. **Man-made Causes** : Habitat destruction, uncontrolled commercial exploitation, hunting or poaching, converting rich biodviersity site for human settlement or industrial development, extension of agriculture, pollution, filling of wetlands etc.

Extinction of Species

A number of species are becoming extinct every year due to human economic activities *e.g.*, extension in agricultural land, increase in agricultural productivity, construction of dams and reservoirs, accelerated soil erosion, industrial development, urbanisation etc. Similarly, species of marine organisms are being destroyed due to pollution of sea-water (mainly oil slick).

Extinction of species is the most serious aspect of the loss of biodiversity. Communities can be degraded and reduced in area, but as long as the original species survive, the communities still have the potential to recover. However, once a species is eliminated, it is unlikely to be repeated again. Once a species goes extinct, its chances for further evolution are lost. A species is considered as extinct when no member of the species remains alive anywhere in the world.

Depending upon the decreasing threat of their becoming extinct, the *species may be classified as under:*

1. **Extinct Species:** These are the species which have not been seen continuously for the lost 50 years or more.
2. **Endangered Species:** Endangered species is the species whose number has been reduced to a critical level, or whose habitats have been drastically reduced; and if such species are, not protected and conserved, it is in immediate danger of extinction.
3. **Vulnerable Species:** Vulnerable species are those species whose population is facing continuous decline due to over-exploitation or habitat destruction. Such a species is still found in good numbers, but is under a serious threat of becoming endangered, if casual factors are not checked.
4. **Rare Species:** Rare species are those, which are usually found only at some particular place(s), and are thus, endemic to a local or regional area, within a restricted zone. Sometimes, they may be thinly scattered over a more extensive area, as to make their being sighted on rare occasions.

5.10 FACTORS CAUSING LOSS OF BIODIVERSITY

Though the man induced causes resulting in biodiversity degradation are many, some of the main causes are as follows:

1. Loss of Habitat

The biological species mostly reside either in the forests or in the wetlands, which are being destroyed by humans either for extension of urbanisation or agriculture or for developing pasture lands. The biological species cannot remain alive; if their homes are snatched away from them. The maximum loss of biodiversity is caused by the destruction of tropical forests, which harbour maximum biodiversity.

Sometimes, the loss of habitat is in instalments so that the habitat is divided into small and scattered patches, a phenomenon known as *habitat fragmentation*. Habitat fragmentation is the process where large, continuous area of habitat is both, reduced in area and divided into two or more fragments by roads, fields, towns, canals, power lines etc. Habit at fragmentation may limit the potential of species for dispersal and colonisation. Habitat fragmentation may result in changes in light, temperature, wind etc.

Severe damage has also been caused to the wetlands by the humans, thinking them to be useless wastelands. Many of the *mangrove swamps* have been filled up by dumping malba and other garbage wastes. The mangrove forests along the coasts have also been destroyed to develop hotels and beach houses on the coasts.

Other main factors responsible for habitat loss are:

(a) *Expansion of Agricultural Land:* The large areas with rich biological diversity have been converted to the areas of cultivation with a few species of hybrid crop plants. Similar attempts have also been made in the fields of fisheries and animal husbandary. The biological diversity has, thus, been replaced with biological uniformly and monoculture, in the name of green revolution in agriculture, white revolution in dairying, and blue revolution in fisheries. Of course, the purpose of these revolutions was genuine for ensuring food security to the growing population. The changes in land-use patterns for expansion for agriculture and aquaculture have caused severe losses of biodiversity due to reduction in forest cover and also in grasslands and pastures.

    The modern agricultural practices also require the use of fertilizers, pesticides and other chemicals, which are causing immense hazards to human health, as these are flowing in our water, vegetables, crops, and even in the milk of cows and buffaloes.

(b) *Urbanisation:* Urbanisation has also contribution to the loss of forests and lands. Large chunks of lands have been reclaimed by man for building houses, factories, highways etc. A lot of land has also been used for mining operations, destroying forests leading to the habitat loss of biological life.

(c) *Pollution:* Pollution of air, water and soil and presence of toxic and hazardous pollutants in different components of biosphere have created serious problems. As a result of this, our fresh water resources have suffered a lot and many species of fish and aquatic birds and mammals have been threatened. Therefore, the environmental pollution is considered as one of the immediate causes affecting current levels of biodiversity.

2. Poaching

Poaching is the illegal killing of prohibited wild animals. The persons who carry out this job are known as *poachers*. The poaching is done for earning illegal money through the sale of their hides, tusks, horns, furs, skins etc. The poaching and illegal export (smuggling) of wild products include the live animals and herbal plants and their products. In order to check the activities of poachers, the people's awareness is very important. No one should, therefore, buy furcoat, bag or purse or any other items which are made of crocodile or any such animal skin.

3. Man-wildlife Conflict

Sometimes, the wildlife comes in direct conflict with the humans. Such situations may arise when the wildlife starts causing immense damage and danger to the humanity. In such situations, the affected people may harm or even kill the animals.

Human-wildlife conflict occurs when the needs and behaviour of one party impact negatively on the goals or needs of the other. As human activities continue to intensify in and around protected areas, the human-wildlife conflict escalates. The main causes of the conflict are: loss and damage of forest plantation or agricultural crops; injury of human beings or wild animals, loss of livestock killed by wild animals; competition with livestock for water and for pasture; competition for space (protected areas) with communities; transmission of major disease to livestock; uncontrolled animal movements and migrations.

For example, the shrinking forests covers and non-availability of bamboo leaves (the most preferred food of elephants) within the forests or the sanctuaries, have forced the elephants to come out and damage the crop fields of paddy and sugarcane (as these are also liked by elephants). Several monkeys have similarly been forced to come out into the residential colonies, as they do not find any fruit trees within their forest-habitats, due to the unauthorised felling of trees or their fruits by the people living near the jungles.

Other causes of loss of biodiversity include: (*i*) Introduction of exotic species (generally for horticulture and agriculture) which may kill or eat native species to the point of extinction; and (*ii*) Disease in flora or fauna, animals held in captivity are more prone to higher level of disease.

5.11 ENDANGERED AND ENDEMIC SPECIES

As already discussed, India possesses about 1,28,000 documented species (47,000 plant species and 81,000 animal species). Thus, India's contribution is around 7% of the total number of word's document species of about 18 lakhs. In addition to these, a huge number of species do exist, which are either not known or have not been described or named. Some people have estimated that the country may be in possession of 8 to 10 lakhs species (against this figure for the world are 8 to 10 millions).

As mentioned earlier *endangered species* is that species whose number has been reduced to a critical level, or whose habitats have been drastically reduced; and if such species are not protected and conserved, it is in immediate danger of extinction.

Wild Life (Protection) Act, 1972 (with amendments) has listed various threated and protected species of animals, birds and insects in the country. This contains 79 mammals, 24 amphibians and 47 birds. The *IUCN* (International Union for the Conservation of Nature and Natural Resources) has also provided a list of *mammals* which occur in India and are rated as Critically Endangered (CR) 7 in number, Endangered (EN) 30 in number and Vulnerable (VU) 52 in number, in the *IUCN Red list of threatened* mammals (2003). The *endangered birds* of India as per IUCN list (2003) contains 57 species of breeding birds and 21 species of non-breeding birds.

The IUCN publishes the *Red Data Book* which includes the list of endangered species of plants and animals. According to this list, India contains globally important populations of some of Asia's rare animals, such as the Bengal Fox, Asiatic Cheetah, Marbled Cat, Asiatic Lion, Indian Elephant, Asiatic Wild Ass, Indian Rhinoceros, Gaur, Wild Asiatic Water Buffalo. As many as 3000–4000 higher plants may be under threat in India.

Some **imporant endandered species of India** are:

1. Reptiles : Gharial, green sea turtle, tortoise, python.
2. Birds : Great Indian bustard, Peacock, Pelican, Great Indian Hornbill.
3. Carnivores : Indian wolf, Red fox, Sloth bear, Red Panda, Tiger, Leopard, Striped hyena, Indian lion, Golden cat, Dessert cat.
4. Primates : Hoolock gibbon, Lion-tailed macaque, Nilgiri langur, Capped monkey, Golden monkey.
5. Plants : A large variety of species of orchids, medicinal plants, Rhododendrons, the sandalwood tree etc.

As discussed earlier, *Endemic species* are the species which are native, restricted or peculiar to an area. As already mentioned, India possesses 47,000 plant species, out of which about 7000 are endemic. Most of our endemic plant species are found in Himalayas, Khasi Hills, and Western Ghats. Some of the important endemic plant species include orchids and species, like: *Sapria Himalayana, Uvaria lurida, Nepethes Khasiana, Pedicularis perroter* etc.

A large number out of a total of 81,000 species of animals in our country is endemic. The Western Ghats are particularly rich in amphibians (frogs, toads etc.) and reptiles (lizards, crocodiles etc.). About 62% amphibians and 50% lizards are endemic to Western Ghats.

Endemic species of India has also been discussed earlier under the heading "India as a mega-diversity nation" in this chapter.

5.12 CONSERVATION OF BIODIVERSITY

Since the Rio Conference (1992) , it has been felt throughout the world that the conservation of biological diversity is one of the most important challenges of the present and immediate future. The convention stressed the need of the conservation of the biodiversity for sustainable development and perpetuation of human beings on the earth. It is also being realised that biodiversity possesses such an extremely high economical value as well as ecological value that cannot be ignored to be destroyed by some greedy people.

The preservation and conservation of biodiversity is, therefore, extremely necessary and important, to keep this renewable resources alive and flourishing, for serving humanity for generations to come.

The biodiversity is conserved through following two approaches:

(a) In-situ conservation; and

(b) Ex-situ conservation.

5.12.1 In-situ Conservation

As per Rio convention, in-situ conservation is defined as "the conservation of ecosystems and natural habitats, and the maintenance and recovery of viable populations of species in their natural surroundings, and in the case of domesticated and cultivated species, in the surroundings where they have developed their distinctive properties."

In-situ conservation means the conservation of living resources through their maintenance within the natural ecosystems in which they occur. In-situ conservation is achieved habitats (*i.e.*, ecosystems) by protecting their habitats.

This type of conservation involves conservation of species in its natural habitat *i.e.*, in places where the species normally occurs. In order to ensure the survival of the species we protect the entire natural habitat *i.e.*, ecosystem. When once the natural habitats prevailing in the various types of ecosystems (or biomes) are preserved, the biological animals and plant species living and flourishing in those ecosystems will naturally continue to survive and flourish without any problem factors detrimental to the existence of the species concerned are eliminated and the species is allowed to grow in its natural environment in it has been growing since long.

Protected Areas

In order to maintain the ecosystems intact, and to avoid their being encroached upon or damaged by the human beings, it is necessary to keep aside adequate forest areas, representing different types of ecosystems with their peculiar biological life, as *Protected Areas.*

Identification and notification of a given area as a protected area, aims at promoting, protecting, and improving the already degraded ecosystem, to protect and develop biological species, suitable to that kind of habitat (ecosystem). Such protected areas are geographically protected areas, and are regulated and managed in accordance with the laws and requirements, so as to achieve the specific conservation objectives for the biodiversity specific to each particular area.

In India, we have so far notified 89 national parks, 500 sanctuaries (including the wildlife sanctuaries and bird sanctuaries), and 15 biosphere reserves. Out of the 26 Indian sites in the World Heritage list, following 5 sites come under the natural heritage category. These sites are :

(i) Kaziranga National Park in Assam
(ii) Manas Wildlife Sanctuary in Assam
(iii) Keoladeo Ghana National Park in Bharatpur (Rajasthan)
(iv) Sunderbans National Park in West Bengal
(v) Nanda Devi National Park in Uttarakhand

In addition to the above, there are special projects to protect certain animal species like the tiger and the elephant.

The protected areas in India are classified into the following three categories:

1. Biosphere Reserves

The Biosphere Reserve network programme was launched by the UNESCO in 1971 under its "Man and Biosphere (MAB) programme." The main objectives of the programme are to:

(i) provide long-term in-situ conservation of genetic diversity.
(ii) promote and facilitate basic and applied research and monitoring.
(iii) provide opportunities for education and training.
(iv) provide appropriate sustainable managements of the living resources.

Each biosphere reserve consists of a "core area" and a "buffer zone". The core area is strictly protected in order to maintain its ecological integrity and diversity. However, activities in the buffer zone contribute towards development taking into consideration the traditional culture and evolving landuses etc. A biosphere reserve must represent a sum total of the biological communities interacting within a single life zone (biome) where climate is similar. In Indian subcontinent, 12 such biosphere reserves recognised. These are: Nilgiri (Western Ghats); Nanda Devi (Uttaranchal); Nokrerk (Meghalaya); Manas (Assam); Sunderbans (West Bengal); Gulf of Mannar (Tamil Nadu); Great Nicobar; Similipal (Odisha); Dibru Saikhowa (Assam); Dehang Debang (Arunachal); Pachmarhi (Madhya Pradesh); Kanchanjanga (Sikkim and part of Kanchanjaga hills).

Biosphere Reserves in India

S.No.	*Name of the site*	*Area in sq km.*	*Location (State)*
1.	Amarkantak	1800	M.P., Chhattisgarh
2.	Agasthyamalai	1800	Kerala
3.	Dehang-Debang	5100	Arunachal Pradesh
4.	Great Nicobar	900	Andaman & Nicobar Islands
5.	Dibri-Saikhowa	770	Assam
6.	Gulf of Mannar	10,500	Tamil Nadu
7.	Kanchanjanga	2,600	Sikkim
8.	Manas	2,800	Assam
9.	Nanda Devi	1,900	Uttarakhand
10.	Nilgiri	5,500	Tamil Nadu, Kerala and Karnataka
11.	Nokrerk	800	Meghalaya
12.	Pachmarhi	4,900	Madhya Pradesh
13.	Rann of Kachch	12,500	Gujarat
14.	Similpal	4,400	Odisha
15.	Sunderbans	9,600	West Bengal

2. Sanctuaries

In sanctuaries, some traditional activities like the grazing and collection of firewood etc. are permitted; but ownership of land is not permitted. In 500 sanctuaries, the size of which ranges from 0.61 to 7818 sq. km. However, the most common size range between 100 to 500 sq. km. The sanctuaries are generally species-oriented. Some of the **important wild life sanctuaries in India** are: Ghana Bird Sanctuary (Bharatpur, Rajasthan), Sultanpur Bird Sanctuary (Haryana); Abohar Wildlife Sanctuary (Punjab) for black buck; Hazaribagh Sanctuary (Bihar) for Tiger and Leopard; Wild Ass Sanctuary (Gujarat) for Wild Ass, Wolf, Nilgai and Chinkara; Mudamalai Wildlife Sanctuary (Tamil Nadu) for Tiger, Elephant, Leopard, Jaladapara Wildlife Sanctuary (West Bengal) for Rhinoceros, Elephant and Tiger.

3. National Parks

In these, no habitations or ownership of land are permitted. Even the traditional activities, like the grazing and collection of firewood etc. are prevented in these parks. In India, there are 89 national parks, the size of which varies from 0.04 to 3162 sq. km. However, most common sizes vary between 100 to 500 sq. km. These are hitched to the habitat for particular wild animal species, like lion, tiger, rhino etc.

Some of the **important national parks in India are:** Kaziranga (Assam) for one horned Rhino; Gir (Gujarat) for Indian Lion; Bandipur (Karnataka) for elephant; Periyar (Kerala) for elephant and tiger; Kanha (M.P.) for tiger; Corbett (U.P.) for tiger; Ranthambhore and Sariska (both in Rajasthan) for tiger.

In addition, there are special projects to protect certain animal species, like the tiger and the elephant. For example, *Project Tiger* attempts to save the Indian tiger from the extinction. Crocodiles and snakes are also being protected from extinction by the creation of *crocodile parks* or snake parks.

The protected areas existing in India covers only 5% of our total land areas. This coverage is considered on a very low side, as the IUCN has specified a minimum area of 10%. Therefore, there is a need to identify more reserves. The protected areas must also be integrated with each other, by establishing corridors between adjacent areas wherever feasible, so that the wildlife can move between them.

Difference Between Sanctuaries and National Parks

The Wildlife (Protection) Act of 1972 provides for the declaration of certain areas by the state government as (a) wildlife sanctuaries, and (b) national parks, if the areas are thought to be of adequate ecological geomorphological and natural significance.

The Central Government may, if it is satisfied that an area is of adequate ecological, faunal, floral, geomorphological, natural or zoological significance, for the purpose of protecting, propogating or developing wildlife or its environment, declare it a sanctuary/National Park by notification.

The difference between the two is as under :

1. National Parks enjoy a greater degree of protection than sanctuaries.
2. Certain activities which are regulated in sanctuaries, such as grazing of livestock, are prohibited in National Parks.
3. Wildlife sanctuary can be created for a particular species, whereas the National Park is not primarily focussed on a particular species.

Some Important Sanctuaries/National Parks in India

1. **The Great Himalayan National Park** : This is located in Himachal Pradesh in upper ranges of Himalayas and is the largest sanctuary established to preserve the highlands *icy alpine ecosystem*, and the flora and fauna that flourish in the alpine environment. This National Park is a home for the endangered beautiful *snow leopards*.

2. **The Nanda Devi National Park** : This park is situated in the vicinity of the Nanda Devi (the second highest mountain peak in India–7817 metres) in Uttarakhand and has some of the most unique high altitude flora and fauna. This park is also declared as World Heritage site by UNESCO. In its vicinity are the valley of flowers, Badrinath temple and Hemkund Sahib.
3. **Valley of Flowers National Park** : This park is located in the Garwal Himalayas in Uttarakhand. It is located near Joshimath (a place of pilgrimage for the Hindus).
4. ***The Rajaji National Park*** : This park is situated in the southern foot-hills of Himalayas in Uttarakhand. This park was established in 1983 after amalgamation of 3 wildlife sanctuaries, namely Rajaji, Motichur and Chilla. Today, this park has the highest concentration of wild elephants in the Uttarakhand.
5. **The Kaziranga National Park** : This park is located in Assam on the southern bank of Brahmaputra River. It is one of the most picturesque wildlife parks in India. It is the natural habitat of the rare *Rhino,* elephants, gaur, wild bear, swamp deer and hog deer in large numbers. This park also has tigers, leopards, besides a rich bird life including ducks, storks etc. This has also been declared as a world heritage site.
6. **Kanha National Park** : This park is located in Madhya Pradesh, and offers a wonderful opportunity to observe wild tigers from elephant back. It is the only protected area, in which a subspecies of *barasingha* is found.
7. **Kevladeo Bird Sanctuary** : This is also known as Ghana bird sanctuary and is located in Bharatpur, Rajasthan. This is most popular water-bird sanctuaries in the world, thousands of ducks, geese, herons and other water birds can be seen here. This sanctuary is very famous due to large number of migratory very rare *Siberian Cranes*, and birds from China in winter season. This sanctuary has, therefore, been declared as a *Natural World Heritage Site.*
8. **Sariska National Park and Tiger Reserve** : This is located in Alwar district of Rajasthan. It stretches over an area of 866 sq km comprising of forests, dry deciduous forest, grasslands and rock hills. The area was declared a wildlife reserve in 1955, and given tiger reserve status of tiger reserve in 1978. This wildlife sanctuary was declared a national park in 1990.

   It is the first reserve in the world to have successfully relocated tigers. In 2003, there were 16 tigers living in the reserve, but in January 2005, it was reported to have no tiger left in Sariska. Poaching was blamed for the disappearance of tigers.

   In order to repopulate Sariska with tigers, a total of 5 tigers were relocated (these were shifted from Ranthambore National Park in Sawai Madhopur, Rajasthan) to this reserve. By August 2014 there were 13 tigers in total. As of October 2018 there are 18 tigers.
9. **Gir Sanctuary (Sasan Gir National Park)** : This is located in Gujarat and was established in 1965 with a total area of 1412 sq km. Because of hunting, at the end of the 19th century only about a dozen Asiatic lions were left in India, all of them in the Gir Forest. Today, it is the only area in Asia where Asiatic lions are found.

   Now, the number of Asiatic lions is increasing. It was 359 is 2005, 411 in 2010 and 523 in 2015 census.

The lions of Gir rarely attack people. They live in close proximity to people which makes their behaviour different almost as if they have a acquaintanceship with the neighbouring people.

10. **The Pariyar National Park and Tiger Reserve** : This is situated within the confines of the Western Ghats. This is developed around a large artificial lake, and is famous for its large population of elephants, which can be seen by boat from the lake. Other wildlife found in the area are *gaurs*, Indian wild dogs, nilgiri *langoors*, large tortoise.
11. **The Desert National Park** : This is located in Thar desert for the protection of the wildlife desert, like *black-buck, nilgai, chinkara* and the Great Indian Bustards.
12. **Kudramukh National Park** : This is a wildlife protected area belonging to a tropical wet evergreen type of forest in the Western Ghats. The area receives an average rainfall of about 7000 mm because of the evergreen vegetation. The wet climate and the tremendous water retention capacity of the grasslands and the forests have led to the formation of 3 major rivers of the region, namely Tunga, Bhadra and Nethravath, forming an important life line for the people of Karnataka and Tamil Nadu.

Its ecology has a diverse assemblage of endangered large mammals found in the area, namely tiger, leopard and wild dog. The important tiger prey base found within the park is *gaur,* sambar, wild pig, bonnet macaque, *langur* and the lion tailed macaque.

Kudramukh Iron Ore Co. Ltd (KIOCL) a government run company is mining iron ore over the years. Opposition to its activities by environmentalists and wildlife conservationists concerns about the threat to the regions, flora and fauna and farmers who were affected by the pollution of the streams due to mining. The company was then banned from operating in this beautiful natural reserve by the Hon'ble Supreme Court from open cast mining of low grade iron ore and other land surface disturbances caused by mining operations resulting in very high sediment discharge.

The company then changed its process of mining, and now with the high quality process adopted to mine, the flora and fauna remained intact, causing no adverse effects on the nature. The company is sending iron ore through pipes and converted to pallets at their plant. These pallets are then exported to countries like China, Iran, Japan etc. by ships.

4. Costal Protected Areas

It aims to protect and conserve the natural marine ecosystems in their original condition. These protect the vital life support processes of the sea and also ensures sustainable productivity and fish production. The Marine Protected Areas (MPAs) in India comprise of a 33 national parks and wildlife sanctuaries designated under the Wildlife (Protection) Act, 1972.

Two important sanctuaries are the *Chilka Lake* in Odisha, (already described in earlier paragraphs), and the most excellent coastal ecosystem provided by the *Sunderbans* (West Bengal), which is the delta region of the Great River Ganga. The Sunderbans protect the largest mangrove delta in India. The Sunderban have been declared as one of the World Heritage sites by the UNESCO. Besides, the Sunderbans is the home to the largest number of wild tigers in the world.

In addition, several protected areas have been created in *Andaman and Nicobar Groups of Island*. The diverse scenery includes *mangrove creeks, tropical rain forest, and coral reefs*.

5. Special Projects for Endangered Animal Species

In order to protect certain endangered animal species, the Government of India has started certain specific projects to protect, conserve, and enhance the number of animal species, like tigers, elephants, lions, snow leopards, crocodiles etc. These projects were respectively named as : Project Tiger, Project Elephants, Girdion Project, Snow Leopard Project, and Crocodile Breeding Project.

(A) **Project Tiger** : The project tiger was launched by the Government of India with the support of World Wildlife Fund (WWF) in 1973 with the aim to protect the endangered species of tigers and their habitats.

This project was initiated in 9 Tiger Reserves in different ecosystems of the country, covering an area of about 16,340 sq km. By 2001, the number of Tiger Reserves increased to 27, covering an area of about 37,760 sq km. The number of tigers increased from 268 in 1972 to around 1,500 in 1997 in 23 Tiger Reserves. From this project it was recognised that tigers cannot be protected in isolation, and that to protect the tiger, its habitat must be preserved.

(B) **Project Elephant** : Project elephant was launched in 1992 as centrally sponsored scheme to assist states to ensure long term survival of identified viable population of elephants in their natural habitats. The project is being implemented in 12 states.

The *objectives* of this project are :

1. To protect elephants, their habitats and corridors.
2. To address issues of man-animal conflict.
3. Welfare of domesticated elephants.
4. Ecologically restoring the natural habitats and migratory routes of the elephants.
5. Protecting the elephants from proachers and other unnatural causes of death and preventing illegal ivory trade.

5.12.2 Ex-situ Conservation

Ex-situ conservation involves maintenance and breeding of endangered plant and animal species under partially or wholly controlled condition *e.g.*, zoos, botanical gardens, seed banks, genetic resource centres, tissue culture etc. Human care eliminates the stresses which living organisms experience in natural habitats and provide conditions for a secure life and breeding. When species become critically endangered and close to extinction, then it may become necessary to conserve, breed and nurture these species outside their natural habitats, in artificially developed habitats under controlled conditions. This type of conservation is also important for conservation of crop varieties for further crop improvement or afforestation programmes.

These artificial habitats for plant life are known as *botanical gardens*; while the artificial habitats made for animal life are called as the *zoological parks* (or *zoos*). Modern zoological parks, like *crocodile parks*, *snake parks* etc. also promote breeding of specific critically endangered species to increase their numbers. These zoological parks provide enclosures to different animal species that simulate their wild habitats.

In India, the following genetic research centres cum gene/seed banks have been set up:

1. **National Bureau of Plant Genetic Resources (NBPGR):** Located at New Delhi is used to preserve seeds of various varieties of agricultural and horticultural crops and their wild varieties. This centre also provides the facility of conservation of varieties of crop plants and trees by using the techniques of *tissue culture*.
2. **National Bureau of Animal Genetic Resources (NBAGR):** Located at Karnal (Haryana) preserves the semens of domesticated bovine animals for future development of species of choosen genes, so as to produce more useful species.

5.12.3 Birds' Migration

Migration refers to the regular, recurrent and cyclical seasonal movements of birds from one place to other. The distance of migration may range from short distance to thousands of kilometers. But at the end of period, the birds will eventually return to the original place.

The main *reasons for migration* are :

1. To avoid adverse factors *i.e.* extreme climatic condition.
2. To manage food shortage.
3. To manage water shortage.
4. To have better breeding conditions.
5. Less competition for safe nesting places.

Examples of winter migratory birds of India are : Siberian Cranes, Eurasian Pigeon, Greater Flamingo, Common Teal, Yellow Wagtail, White Wagtail. Examples of summer migratory birds are : Asian Koel, Black Crowned Night Heron, Comb Duck, Blue Cheeked Bee Eater.

5.12.4 World Heritage Sites

World Heritage sites means "Sites any of various areas or objects inscribed on the United Nations Educational Scientific, and Cultural Organisation (UNESCO) World Heritage list."

The sites are designated as having outstanding universal value under the convention concerning the protection of the World Cultural and Natural Heritage. This Convention, which was adopted by the UNESCO in 1972, and enforced in 1975, provides a framework for international cooperation in preserving and protecting cultural treasures and natural areas throughout the world.

The value of biologically-rich natural areas, is now being increasing recognised throughout the world as the biowealth is of unimaginably high value. World Heritage Convention have executed in an attempt to protect and support biologically rich areas. India is a signatory to this convention, and several of its protected areas stand included as **"World Heritage Sites"**. Out of the 26 such Indian sites in this list, 5 are under the natural heritage category. These include;

(i) *Manas Wildlife Sanctuary* on the border between Bhutan and India.

(ii) *Kaziranga National Park* in Assam.

(iii) *Keoladeo National Park* (*Ghana* bird sanctuary) in Bharatpur, Rajasthan.

(iv) *Nanda Devi National Park* in Uttarakhand.

(v) *Sunderbans* in the Ganges Delta in West Bengal.

5.12.5 Export-Prohibited Goods

India has signed **"the Convention on International Trade in Endangered Species (CITES)"** which is intended to reduce the utilization of endangered plants and animals by controlling trade in their products and also the trade in pets.

The prohibited items are not permitted to be exported. The following common items prohibited in the flora and fauna categories :

- All wild animals, animal articles including their products and derivatives excluding those for education, scientific research and management under Wildlife (Protection) Act 1972.
- Beef of cows, oxes and calf.
- Meet of buffalo (both male and female).
- Peacock Tail Feathers and its handicrafts and articles.
- Savings of Chital and Sambhar and their arcles.
- Sea Shells.
- Wood and wood products including wood pulp.
- Fuel wood and wood charcoal.
- Sandalwood in any form (excluding finished handicraft products).

5.12.6 Laws to Protect Biodiversity

India has enacted the following legislations to exercise control on killing and exporting of wild animals and plants, and to promote production of biological resources:

1. The Wild Life (Protection) Act, 1972;
2. The Forest Conservation Act, 1980; and
3. The Biological Diversity Act, 2002.

QUESTIONS

1. Define the term biodiversity, and explain its levels.
2. Describe the importance and values of biodiversity.
3. Discuss the various biogeographical regions in India, indicating the types of species commonly found in these regions, with a special mention of their endemic species.
4. Write an essay on "India as a Mega-diversity nation.
5. Write a brief note on 'Global biodiversity', indicating the Indian status in this regard.
6. Write a detailed note on 'Indian biodiversity'.
7. Discuss characteristics of the biogeographical zones in India.

8. Discuss about the biodiversity hotspots, giving examples of world hotspots, and Indian hotspots.
9. Define and explain the threat perception and extinction of biological species.
10. Discuss the impacts of loss of biological diversity on the environment, in general and on humans in particular.
11. Discuss the different threats to biodiversity.
12. Write short notes on:
    (a) Habitat loss.
    (b) Poaching of wildlife.
    (c) Man-animal conflict.
13. What do you understand by conservation of biodiversity? Discuss the various methods of conservation of biodiversity in brief.
14. Distinguish between endangered and endemic species.
15. Differentiate between a national park, a wildlife sanctuary and a biosphere reserve.
16. Discuss the protected area concept for the conservation of biodiversity.
17. Differentiate between in-situ and ex-situ methods of conserving biological life. Explain both of them in detail.
18. (a) Enumerate the legal laws, enacted in India for conserving and protecting its flora and fauna.
    (b) Write a brief note on India's Wildlife (Protection) Act, 1972.
19. Write short notes on:
    (a) Red Data Book
    (b) Biosphere Reserves
    (c) Sanctuaries
    (d) National Parks
20. Explain the following in brief:
    (a) How to measure bio-diversity
    (b) Cold spots of bio-diversity
21. Write a small note each on 2 important National Parks and 2 sanctuaries.
22. Write short notes on:
    (a) Kevladev Bird Sanctuary
    (b) Gir Sanctuary
    (c) Sariska Tiger Reserve
    (d) Bird Migration
23. Explain the following in brief:
    (a) World Heritage Sites
    (b) Project Tiger
    (c) Project Elephant

PART III : NATURAL RESOURCES

6
Chapter

NATURAL RESOURCES

6.1 INTRODUCTION

Natural resources are those which belong to natural environment and include both, (*i*) abiotic (such as land, air, water, soil, minerals etc.), and (*ii*) biotic (such as plants and animals) resources. Natural resources can also be defined as wealth supplied by natural — mineral deposits, soils, forest, fuel, wind, saline water, hydropower, fish, wildlife etc.

The needs of people vary depending on the stage of cultural and technological development of the people. Everything that comes from nature has some utility for man but its utilisation is possible on the availability of appropriate technology. Although natural products existed during pre-historic time, but man had neither the tools nor the technology to use them. With time he could learn to cultivate the land, run the wind and water mills. Hence, these natural materials turned into resources only, when they could be used. Thus, the furniture, clothes, mechines and tools are more valuable than wood, cotton and metal respectively. Any material which can be transformed in a way so that it becomes more valuable and useful, can be termed as *resource.* From this, we can say that it is possible to obtain valuable items from any resource. Therefore, water, minerals, land, forests, wildlife and also human beings are resources.

Any material may be termed as a resource provided that an appropriate technology is available to transfer that into more valuable goods. Gifts of nature acquire value as a result of technological development. Rocks, minerals, soils, rivers, lakes, oceans, plants and animals are the gifts of nature or natural endowments. They become resources only, when man locates them or find a use for them or proposes to use them. For example, a river is a natural endowment, but it becomes a resource when man utilises the river for irrigation, power development, navigation etc.

6.2 CLASSIFICATION OF RESOURCES

Resources can be classified in following ways:

1. Biotic (Organic) and Abiotic (Inorganic)

These are classified on the basis of their origin. Biotic resources are also called as ecological resources. Plant or Floral resources include forests and their products, agricultural crops, horticultural and flowery plants etc. Animal and faunal resources comprise wild and domestic animals, birds and fishes and animal products. *Abiotic resources* include non-living things of the biosphere ecosystem, such as land, water, soils, rocks and minerals, sunlight, air, which are renewable and exhaustible abiotic resources while most of the mineral resources are exhaustible and non-renewable, such as coal, mineral oil and natural gas.

2. Renewable and Non-renewable

Renewable resources are those resources which have the inherent capacity to reappear or replenish themselves by quick recycling, reproduction and replacement within also known as *non-exhaustible resources.* Air, water, soil (land), animals and plants (including forests and all types of biodiversity) are the examples of renewable resources. *Cyclic resources,* like water, which can be used again and again also fall in this category.

Non-renewable resources are those resources which do not have the ability for recycling and replacement within a reasonable period of time. These are also known as oil and natural gas and minerals are the examples of non-take millions of years in their reformation. Thus, all non-living things taken out from the interior of the earth to satisfy human wants are non-renewable resources.

There are certain non-renewable resources that are perpetually available to us, naturally. Although these cannot be recycled, but they are perpetually available to us for times to come and their availability seems infinite. These resources include run, mind and ocean tides. Since these resources are perpetually available, these are usually considered as a part of renewable resources.

Here, it is pertinent to point out that even the renewable resources may become non-renewable, when over-exploited *i.e.* if used too rapidly and in excess quantities by improper management. Therefore, it is important to use our natural resources to just fulfill our minimum needs and preserve their stock to make them available for generations to come.

Thus, broadly speaking, renewable resources are those which are reproducible *i.e.* they are obtained from the biomass of living organisms. Non-renewable resources, on the other hand, ore of finite stock are not reproducible.

3. Agricultural and Pastoral Resources

Agricultural resources include both, agricultural lands and agricultural poducts and are also called as basic resources which meet the basic needs of human beings (food, clothing and shelter) and also provide fodder and solid food (grains) to the domestic animals. Agricultural resources also provide raw materials to agro-based indusries. *Pastoral resources* include food items (*e.g.*, meat, milk, butter, cheese, food processed items and by-products *e.g.*, hairs, wools etc.).

4. Minerals and Industrial Resources

Mineral resources may be metallic mineral ores, non-metallic minerals and fossil fuel minerals.

Industrial resources may be mineral resources, agricultural resources, pastoral resources and industrial by-products.

5. Classification Based on Utility

Based on utility, resources are mainly of following types:

(i) **Human Resources:** Human resource is the most important resource of a nation. For the development of a nation not only general education, but also imparting of skills in different folds is essential. All the developed nations have developed their natural resources only after the development of human skills and technology.

(ii) **Land Resources:** Since humans live on land, it is an important resource. He obtains most of their needs from the land. Land is put to different uses, such as construction of buildings (residential and non-residential), roads, railways, cultivation, grazing, mining, industries, forests etc. In recent times forests have been cleared for cultivation, urbanisation, mining and other uses resulting in a reduction in forest area. Proper planning of land-use with reference to the nature of land and the needs of community is essential for providing maximum returns. Improper use of land, such as clearance of forest for cultivation may result in soil erosion. Reclamation of lowlying land or marshes for construction of building is treated as a good proposal. Since total land is affixed asset, we must adopt integrated land-use planning.

(iii) **Soil Resources:** Soils is the most important natural resources, as it is indisposable for the growth of all plants on which human and other animals depend for their food. The soil consists of mineral matter, such as sand and clay as well as organic matter, such as decayed leaves, flowers, dead tissues of organisms etc. And contains varying amount of moisture and air between soil particles. Soil formation is a very low process and it may take thousands of years to form a thick soil layer. Soils are usually formed by weathering of rocks on the surface of the earth, soils may also be formed by deposition of material by agents of gradation.

Soil erosion is a natural process, occurring due to rain, wind and deforestation. When the top soil depletion due to erosion is more than the formation, the land loses its fertility. When farmers lose top fertile soil, they may affect the farming land productivity the use of chemical fertilizer or through irrigation. As soil is carried by farming from run off, it may end up in local streams, rivers, canals and hydro-electric reservoirs. This results in reduction of irrigation capacity, electricity generation and the navigability of waterways.

(iv) **Water Resources:** Water is essential for the plants, animals and humans. Water is needed for domestic needs of man, and for all economic activities, like cultivation, industries, power generation etc. Humans are affected not only by scarcity of water but also by its occasional abundance leading to floods. Floods may cause large scale destruction of human lives, property and loss of crops. Multipurpose prospects help in preventing soil erosion and floods, besides regular supply of water for irrigation, industries, navigation, power generation and other needs.

Apart from surface water from rivers, lakes and tanks, man has been tapping underground water through wells, construction of percolation ponds may increase seepage of water into water table.

(v) **Agricultural Resources:** Land, soil and water are the basic resources required for agriculture. Agriculture has been practised by man for over four thousands years. River valleys and coastal plains have relatively level land, fertile alluvial soils and abundant water supply which favour cultivation.

Availability of suitable land for cultivation is limited and due to increase in population, new areas are being cultivated by clearing forest land or grass land, terracing of hill slopes, reactors and other machineries are also being increasingly used to cultivate large areas with minimum human labour. In recent decades, productivity in agriculture has been increased by using fertilizers, pesticides, high yielding varieties of seeds.

Agriculture started with cultivation of foodgrains, like wheat and rice, needed by the people. Agriculture comprises the cultivation of variety of crops besides foodgrains. Other forms of crops are: fibre crops, like cotton, jute, commercial crops, like tobacco, sugarcane and variety of spices, plantation agriculture yielding tea, coffee and rubber. A variety of fruit trees are also grown. There are certain crops which provide raw materials for a variety of industries which encourage the growth of agro-based industries.

(vi) **Animal Resources:** Animals provide meat and milk products which supplement food obtained from agriculture. Animals are also a source of wool, hides and skins needed by man. In certain countries, animals supplement, human labour for work on the farms, especially in ploughing, drawing water from wells and pulling carts.

(vii) **Forest Resources:** Forests now cover about 30% of the land area of the world. Forests play an important role in the environment. Trees absorb carbon dioxide from the atmosphere and release oxygen, and add water vapour, prevent soil erosion, provide a safe habitat for wild animals, birds and other organisms. Forests also provide raw materials for the manufacture of wood, pulp, paper, synthetic fibres, rubber and timber.

Because of the increasing pressure of human population, forests are being cleared for cultivation, mining and other uses, like timber, firewood and other products. Destruction of forest cover in recent decades has resulted in environmental degradation causing soil erosion, floods, more siltation in the reservoirs, water table is going down and habitat of animals and birds getting destroyed.

(viii) **Wildlife Resources:** Plants, animals, birds and other organisms which live in natural habitats are known as *wildlife.* As we have already discussed that all organisms in the biosphere are interdependent, as they are part of a food web. There is a balance in the environment of the relative number of species is not disturbed. Therefore, conservation of forests and wildlife go side by side. Establishment of national parks and wildlife sanctuaries helps conservation of wildlife. India's criminate hunting of animals and birds for food, feathers or skins has already resulted in extinction of some species.

(ix) **Fisheries:** Fishing has been practised since ancient times. These may be fresh water fish and marine or sea-water fish. Fresh water fish may be obtained from rivers, lakes, tanks and other inland water. Marine fishes are available in abundant quantity in the coastal zone of shallow waters than in the open ocean. Fishes provide animal protein in the diet of people, oil to be used as tonic, and their residuals as fertilizers.

If properly managed and controls are exercised in fishing, marine fisheries provide renewable resources of food. There is significant decline in catches of marine fish in some of the fishing grounds because of the pollution of coastal water, oil spills and oil drilling in continental shelves and over fishing.

(x) **Mineral Resources:** Mineral is a substance that is found in the earth's crust and which generally has a definite chemical composition. Minerals include all materials derived from the earth by mining. Such materials may be of (*a*) organic origin, like coal, petroleum or natural gas, or (*b*) inorganic in origin which include both, metallic and non-metallic materials.

Coal and lignite occur as seams or layers between sedimentary rocks. Petroleum and natural gas metallic minerals, like iron, copper, zinc, lead, silver etc. occur as ores. Mineral ores contain a number of mineral occurring in the area varies from place to place. These have to be processed to remove impurities and obtain pure mineral.

6.3 CONSERVATION OF RESOURCES

Conservation is defined as "the planning and management of natural and man-made resources so as to secure their judicious use and continuity of supply while maintaining their quality, value and diversity." The expanding human population and increasing per capita consumption due to scientific progress and technological development has resulted in great demand for the limited resources available on the earth. This has created a situation when the non-renewable resources may come to an end after sometime.

Conservation has also been defined as, "the management for the benefit of all life including human kind of the biosphere so that it may yield sustainable benefit to the present generation while maintaining its potential to meet the needs and aspirations of the future generations."

Land has become a scarce resource in densely populated countries. Valuable cultivated land is being used for the construction of houses or industries. Forest land is being used for cultivation and other uses. Land use should, therefore, be optimum. An important part of soil conservation is prevention of soil erosion. The main agents of soil erosion are water and wind. Soil erosion can be protected by adopting following guidelines:

(i) protection of soil from impact of rain drops,

(ii) to slow down the water from concentrating and moving down the slope in narrow path,

(iii) to encourage more water to enter the soil,

(iv) reduction in the mind velocity near the ground by growing vegetation cover, redging the land, and

(v) to grow the vegetation cover to catch and hold the moving particles of soil.

Water resources need to be conserved by tapping the available surface water. Lowering of water-table due to excessive use of groundwater may be checked by improving surface storage so that water-table gets recharged by seepage. Modern methods of sprinkle irrigation or drip irrigation will consume less water. Water can also be conserved by creating dams and reservoirs.

Biotic resources (forests) can be conserved by protecting them from encroachment, fire and cutting. Afforestation is also very helpful. National parks and sanctuaries will preserve wildlife (*i.e.* wild plants animals and birds) in their natural habitat.

Minerals are non-renewable resources and the metallic minerals like iron, copper, aluminium etc. and non-metallic minerals, like coal, petroleum, natural gas etc. are in great demand. Metallic minerals can be conserved by recycling the scrap. While non-metallic minerals should be judiciously used and their alternative resources should be found out. Coal, petroleum and natural gas are used in generation of energy. Efforts must be to use non-conventional renewable sources of energy *i.e.* Hydropower, solar, wind etc. and by reducing transmission and distribution losses, and improving the efficiency of generation, automobiles, electric motors etc.

6.4 ENVIRONMENTAL DEGRADATION

There is severe pressure on natural resources and environment because of simultaneous increase in population and economic growth, sharp rise in agricultural production, rapid urbanisation and industrialisation are increasingly affecting the natural environment. Though environment is keeping man alive but he is steadily destroying it. Man chops down millions of trees everyday; for every 10 litres of petrol consumed, he releases 20 kgs of carbon dioxide. He dumps tonnes of chemicals and nuclear waste into the sea; the food eaten by man is spread with lethal pesticides, which on long run can have serious side-effects, like paralysis, cancer, kidney problem etc.

Man, being the most intelligent animal of all animals and technologically most advanced, tries to make use of the environmental resources and to make the environmental influences favourable for him. In doing so, man affects and modifies the environment and thus environmental conditions are changed by human by man exceeds critical limit environmental degradation starts, but when the environmental residence is such that it can withstand the environmental changes, environmental stability is maintained and the natural environment is protected from being degraded.

Over-exploitation of exhaustible resources, use of those resources whose generation is a very slow process (like soil), continuous exploitation of those natural resources whose reserves have reached a critical limit and are at the average of exhaustion etc. are a few examples of rapacious use of natural resources by greedy man. Such human actions make the environment sick and degrade the environmental quality to such an extent that it endangers the very existence of human beings. Clean air and water, fertile land, rich biotic community (plants, animals, wilderness and recreational areas), sufficient mineral resources, ideal weather etc. are the elements of the environment which are needed for the maintenance and development of human society.

The environmental crisis being faced by the earth, can be attributed to the manner in which the industrial and technological development has taken place in the world.

Man's impact on the environment has resulted in pollution of environment. Pollution affects air, water, land and also the organisms in the biosphere. The natural ecosystem has the capacity to decompose dead organisms or excreta and recycle them. When harmful substances contaminate the environment in large quantities, the ecosystem is unable to absorb them and they accumulate in the system, resulting in degradation of environment.

Reasons for Environmental Degradation

The main reasons for environmental degradation and resultant environmental crisis are:

- Population growth
- Accelerated rate of scientific and technological developments
- Accelerated growth of production potential

• *Population Growth*

The growth of world population at exponential rate is the main reason of environmental degradation while other reasons are offshoots of the factor of over-population. Increased population growth is responsible for increased air and water pollution, increased demand for food and natural resources, such as minerals, forests and water, increased urbanization and insufficient and inadequate housing, schools and health services as well as loss of agricultural land, increase in means of transport and communication etc. People wants to generate better facilities for their existence and better life. This has necessitated the development of scientific techniques and advanced technologies in order to save the humanity from hunger.

The problems for environment are further enhanced due to the desire of man for a greater share of resources and considerably higher standard of living as compared to that enjoyed by his forebearers. Because of this reason, the rate of exploitation of natural resources increases more rapidly than the rate of population growth. This has caused and is causing exhaustion of certain valuable non-renewable.

• *Science and Technological Development*

Industrialisation due to advancement in science and technology in the beginning blindly followed the race of industrialisation and did not care for its adverse impacts on the natural environment. No doubt rapid rate of industrial development has given economic prosperity to human society and has provided material comfort to the people but it has also created many fold environmental problems.

Rapid rate of industrialisation has resulted into (a) rapid rate of exploitation of natural resources, and (b) increased industrial output. This has created environmental degradation and ecological in balance at global, regional and local levels in a variety of ways.

Exploitation of natural resources to meet the industrial demand of raw materials has resulted into:

(i) reduction of forest cover due to reckless felling of trees,

(ii) excavation of land for mining purposes,

(iii) reduction in cultivable land due to industrial expansion,

(iv) lowering of groundwater table due to excess withdrawal of groundwater,

(v) collapsing of ground surface due to withdrawal of minerals, oil and groundwater etc.

(vi) soil pollution due to excessive use of chemical fertilizers, pesticides and insecticides for over-utilisation of soils for increasing production of sugarcanes (for sugar industries), cotton (for textile industries) etc.

There are several undesired outputs from the factories, such as industrial wastes, polluted water, toxic gases, chemical precipitates, as gas and smokes etc. which pollute air, water, land, soils etc. and thus, degrade the environment.

• *Accelerated Growth of Production Potential*

Earlier technologies were less destructive because the main emphasis was to provide minimum requirements for the sustenance of life, whereas the modern technologies have become more destructive because these are meant for accelerated rate of exploitation of natural resources and their production into various forms to raise the material standards of human beings.

Some examples of modern technologies used for accelerated growth and creating environmental crisis are:

(i) Modifying weather by *cloud seeding* to stimulate clouds and induce precipitation through crystallisation of supercooled drops through the application of solid carbon dioxide and certain compounds of iodine. In such cases enormous heat energy employed to induce precipitation, to prevent hail storms, and to disperse and clear clouds and fogs is released and it causes additional heating of the lower layer of the atmosphere and thus, changes the natural meteorological processes.

(ii) For increasing the agricultural production, modern techniques demand increased application of chemical fertilizers (to increase the productivity of the soils) and pesticides, insecticide and herbicides (to remove plant diseases and destroy unwanted weeds). These chemicals pollute the soils, surface water of ponds, lakes, rivers, and ground water because of gradual accumulation of toxic chemicals.

(iii) The manufacture and use of several luxury items, like, refrigerators, airconditioners, hair driers and the operation of fire extinguishers etc. release substantial quantity of chlorofluorocarbons and halons into atmosphere which depletes the life saving ozone layer of the stratosphere.

(iv) Burning of fossil fuels in the thermal power stations and in the factories release huge quantity of carbon dioxide in the atmosphere. This intensifies greenhouse effects and thus, raises earth's temperature.

(v) Leakage of toxic gases from chemical plants not only pollute the air but also causes deaths of human beings, plants and animals (*e.g.*, Bhopal gas tragedy).

Interaction between the environment and society depend largely on the social and political systems within which arise the environmental and ecological problems. The capitalistic and socialistic systems perception and reactions to the environment are quite different. Inherent ideology of capitalism is to increase the production with minimum production cost for maximum profit leading to accumulation and hoarding of wealth and to shift the responsibility of environmental degradation on the shoulders of developing and underdeveloped countries on the other hand. Whereas the socialistic system of government lays more emphasis on the social importance of natural resources and environmental and ecological problems.

6.5 EQUITABLE USE OF RESOURCES FOR SUSTAINABLE LIFESTYLES

The concept of **sustainable development** is to meet the needs of the present without compromising the ability of future generations to meet their needs. Sustainability also means, conservation of natural resources and management of biosphere for human needs and use for an indefinite period of time.

Sustainability is also for, wise use of nature and its resources, to provide the greatest sustainable benefit to present generation while maintaining the potential to meet the needs and aspirations of future generations.

Following are some of the *interpretations of sustainability:*

— Living in harmony with nature.

— Living off the income of nature, without eroding the capital.

— Must give back to the earth what we take from it.

— Realisation that the biosphere is both, for us and for our future generations.

— It is an intergenerational concept seeking equity over time, and minimisation of disparities among generations.

— Survival of human beings as well as of all other life forms.

— Maintenance of bio-physical productivity.

— Preservation of environmental quality and ecosystem.

— Inter and intra-generation equity.

— Human survival can only be ensured by preserving the basic resource base.

Considering all the above mentioned facts, we can define the *sustainable development* as that development, which is free from environmental degradation, poverty, and depletion of natural resource base. To use our resources efficiently and without depletion, we must first understand what our resources are and what role do they play in our ecosystem. This is essential to manage our resources. The word commonly used in the content of resources today is the *sustainable utilization,* which in simple terms means optimum utilisation.

Till about four decades earlier, it was believed that all the resources are abundant and we can exploit them as much as we can. Little thought was given to the quantum of minerals used or forests cleared or the capacity of our environment to absorb pollutants. Converting forests into cropped and urban land, over-fertilizing the land resource, over-harvesting forests or avoiding pollution controls are the harmful practices to meet the needs of the growing population. They are the only short term solutions and harm in long run.

Human environment is deteriorating day-by-day to the point of affecting not only the equality of life, but also even the very existence of life. Now, at this point, the man has realised the need to preserve and improve his environment. We have to build a sustainable world–a world which should last forever. Everyone in this world should get at least the basic requirements of everyday life *i.e.* food, clothing, drinking water, shelter etc. in such a way that there could be no damage to other life-forms and the environment. This would only be possible when man learns to sustain the natural resources.

6.6 ROLE OF INDIVIDUAL IN THE CONSERVATION OF NATURAL RESOURCES

During last some decades, humans are endlessly exploiting the earth's ecosystems and natural resources. However, in the last few decades it has been emphasized that the global ecosystem has the limited capacity to sustain. The overuse or misuse of resources cannot be replenished. The resources may be classified as 'renewable' or 'non-renewable'. The non-renewable resources will be rapidly exhausted if we continue to use them as intensively as at present. Even 'renewable' resources are being degraded by overuse and may be permanently destroyed.

As discussed in previous chapters, the most damaging factors are increasing consumerism for 'so called' improve lifestyle (by affluent section of society), and rapid population growth.

The solution to these is the conservation, reuse and recycling of resources, and practicing the principles of sustainable development. For the environmental conservation, it is important to be aware of the various problems that the earth is facing, the underlying causes, and the various means by which the problems can be minimized and mitigated 'World Population Day', 'Earth Day,' 'Water Day' etc. are all celebrated to create wide-spread awareness about the environment.

Various aspects about awareness, sustainable development, conservation of resources have been discussed at relevant places in the book.

Everyone of us should realise that we have to build a sustainable world—means a world which should last forever. Everyone in this world should get atleast the basic requirements of everyday life *i.e.* food, clothing, drinking water, shelter etc. in such a way that there is no damage to other life-forms and environment.

Everyone of us and society as a whole must focus attention on the following:

1. Maintain clean and hygienic living and working conditions.
2. Encourage afforestation according to type of soil and environmental conditions.
3. Improve awareness on environmental issues.
4. Encourage use of pollution-free non-conventional and renewable sources of energy.
5. Use organic fertilizers and other ecofriendly biotechniques.
6. Encourage use of pollution free and environment friendly products.
7. Encourage conservation of natural resources.

8. Encourage reuse and recycling of products.
9. Maximum attention be given to waste management.

QUESTIONS

1. What are the natural resources?
2. How will you classify the natural resources?
3. Distinguish between biotic and abiotic resources.
4. Distinguish between renewable and non-renewable resources.
5. What do you mean by human resources?
6. Explain the following in brief:
   (i) Land resources
   (ii) Soil resources
   (iii) Water resources
   (iv) Food resources
   (v) Agriculture resources
   (vi) Animal resources
   (vii) Forest resources
   (viii) Wildlife resources
   (ix) Fisheries resources
   (x) Mineral resources
7. Discuss about the conservation of resources.
8. Explain the meaning of environmental degradation.
9. Explain the reasons for environmental degradation in brief.
10. Discuss the equitable use of resources for sustainable lifestyle.
11. Discuss the role of individual in the conservation of natural resources.
12. Discuss the need for conservation of resources.

7

Chapter

FOREST RESOURCES

7.1 INTRODUCTION

Forest is any land managed for the diverse purpose of forestry. The Indian word *'jungle'* describe a collection of trees, shrubs etc. that are not grown in a regular manner, as contrasted with 'forest' which is any vegetation under a systematic management.

Forest can be defined as an area set aside for the production of timber and other forest produce, or maintained under woody vegetation for certain indirect benefits which it provides, agroclimatic or protective.

Forest can also be defined as a plant community predominantly of trees and other woody vegetation, usually with a closed canopy.

Forestry is the theory and practice of all that constitutes the creation, conservation and scientific management of forests and the utilization of their resources to provide for the continuous production of the required goods and services.

A forest is a natural ecosystem having diversified species of plants, shrubs, small and large trees in different stages of life-span. Forest vary greatly in composition, density and pastures. Their scenic effect changes with the seasons. Certain forests are evergreen like *Deodar* forests, while others are deciduous, like *Teakoroak* forests, becoming leafless either before the forest of intense dry summer or before the advent of winter when vegetative activity almost ceases.

The forests of a country are a natural asset of immense value. The forests, if of adequate extent, ideally dispersed, scientifically managed and judiciously utilized can be kept perpetually productive and useful. Forests are, therefore, treated as a renewable natural resource.

We have already discussed in earlier chapters that forests are essential for ecological balance of an area, and they are an important components of our environment and economy. We have also discussed earlier about the forest types and biogeological zones of the country. Forests contribute to the economic development of our country by providing goods and services to the people and industry. They also play an important role in enhancing the quality of environment by influencing the life supporting system.

Forests cover about one-third of the earth's land surface, of which about 50% is occupied by tropical forest. Of the total geographical area of India, only about 22% forest is available as against a minimum of 33% forest cover prescribed under National Forest Policy. Due to large scale deforestation carried out by humans, there is a continuous decrease in the forest cover all over the world.

Every forest has various layers of plants. The five basic forest strata (layers) from highest to lowest level are:

(i) The canopy (also known as upper canopy).
(ii) The under-story (also known as lower canopy).
(iii) The shrub layer.
(iv) The herb layer.
(v) The forest floor.

Forest Cover in India

Class	Area in sq km	Percentage of Geographical Area
Very dense forest	99,278	3.02
Moderately dense forest	3,08,472	9.38
Open forest	3,04,499	9.26
Total Forest Cover*	**7,12,249**	**21.67**
Tree Cover	95,027	2.89
Scrub	46,297	1.41
Non-forest area#	25,28,923	76.92
Total Area	**32,87,469**	**100.00**

* Includes 4,975 sq. km under mangrove cover

\# Non-forest included tree cover (percentage round off)

Constraints and Threats to Indian Forests

1. **Population Explosion** : At the time of independence India had a population of about 30 crores, which is now over 130 crores, resulting into large scale land use changes.
2. A large number of major projects have affected forests and wildlife.
3. Rural development imbalances.
4. Market pressures led to collection and removal of both timber and non-timber forest produce.
5. Lack of effective implementation of legislation and policy and lack of political will.
6. A large area of forest land is under-encroachment of forest land.
7. Fragmentation of wildlife habitats owing to loss of forest connectivity is a serious limiting factor for wildlife. The main reason for this is diversification of forest areas for the development activities and loss of forest cover due to illicit felling, excessive fires and grazing.

8. Poaching and illegal trade of wild flora and fauna.
9. Damage from mining.
10. Degradation and diversion of water bodies and wetlands.
11. Natural and man-caused fires cause damage to nests of ground-laying birds, death to reptiles, soil compaction, damage to plant regeneration.
12. Adverse impact of tourism.

7.2 USES OF FOREST

Forests are invaluable property of a nation because they provide raw materials to modern industries, timber for building purposes, habitats for numerous types of animals and micro-organisms, nutrient-rich soils having high content of organic matter, offer protection to soils by binding the soils through the network of their roots and allow maximum recharge of groundwater.

The forests not only produce innumerable material goods, but also provide several environmental services, which are essential for life. The various **economic and environmental services and products that are provided by forests are** discussed hereunder.

Commercial and Economical Services

Forests provide us a large number of goods that have a huge commercial value:

(i) Forests are the source of timber, firewood, nuts, flowers, fruits and seeds, medicinal plants, bamboos, canes.

(ii) They also provide facility for the grazing of their live-stock and yield a variety of industrial products, such as charcoal, raw materials for making paper, newsprint, rayon, plywood, *bidi* leaves, gums, resin silk, lacquers, railway sleeper etc.

(iii) They also provide spices (used in cooking of food), essential oils, matches, sport goods, matches and packing articles.

Ecological (or Environmental) Services

Main ecological services provided by the forests are explained hereunder:

(i) Forests play an important role in the environment, as trees absorb carbon dioxide from the atmosphere and release oxygen to the atmosphere. Thus, the forests help in renewal of the atmosphere, and therefore, rightly called as the "lungs of the Earth".

(ii) It maintains the natural ecological balance by providing habitat for various species of plants and animals.

(iii) Forests prevent soil erosion and desertification as the roots other trees bind the soil particles. Forests also protect the soils from direct impact of falling raindrops. They encourage and increase infiltration of rainwater and thus, allow maximum recharge of groundwater resources, minimize surface sunoff and hence reduce the frequency, intensity and dimension of floods.

(iv) Healthy forests play a key role in regulating water quality and quantity, and can reduce the risks of landslides, floods and droughts.

(v) Forests add water vapour to the atmosphere by the process of transpiration. Thus, the humidity and temperature of the air and soil is maintained favourably by the forests.

Forests serve as wind breakers, which reduces soil erosion and check the spreading of deserts.

(vi) Forests absorb several toxic air pollutants, and help in cleaning the air.

(vii) Forests also absorb noise, and help in reducing noise pollution.

(viii) Fodder from the forests form an important source for cattle and other grazing animals.

(ix) Forests improve soil environment by thorough its composition, structure, chemical properties, water content etc. Forests also play an important role in biogeochemical cycles of water, carbon, nitrogen, phosphorous, sulphur etc. Some species of trees have the ability to return nitrogen to the soil through root decomposition or fallen leaves, such plants help in improvement in soil characteristics.

(x) Forest biomass satisfies the basic requirements of the tribal people in its immediate vicinity.

(xi) Forests also provide natural, beautiful, peaceful and congenial environment to people. Many people visit these forests for camping, hiking, to enjoy scenery and to relax. Thus, the forests help in promoting recreation and tourism.

7.3 OVER-EXPLOITATION OF FORESTS

It is a matter of serious concern that the present money-minded man has forgotten the ecological and environmental significance of natural vegetations, mainly forests and grasslands, and has destroyed the forests so rapidly and alarmingly that the forest areas at global and regional and local levels have so markedly decreased that several serious environmental problems have cropped up.

Excess cutting of trees for commercial and industrial purposes is referred to as *'over-exploitation'* with the growing population, demand for forest produce has increased resulting in large scale logging (cutting of trees for specific use of logs of wood as firewood), mining, road building and clearing of forests. This results in removal of nutrients from forests in the form of wood harvested, or soil erosion or rainwater leaching.

Inspite of so many life supporting functions performed by the forests, man due to his needs and greeds is over-exploiting and is wiping out the forests for expanding his agricultural, industrial and residential areas.

7.4 DEFORESTATION AND AFFORESTATION

Each day as more and more trees are full for timber, forests cleared for agriculture and settlements, fuelwood extracted for the rural hearths, fodder removed for cattle, the condition of forests continue to decline.

Deforestation

Deforestation is defined as the destruction of forests by cutting of trees. The problems caused by deforestation have already been discussed in earlier paragraphs. Deforestation results in decreased soil fertility soil erosion, land slides in hilly slopes, increased silt build up behind dams effecting their capacity. In drier areas, deforestation contributes to the formation of deserts.

Deforestation contributes to the extinction of many species. Migratory species, including birds and butterflies, also suffer from deforestation. Deforestation poses great dangers, not only to the local climates, but also to the regional climates.

The loss of forest in our country, like in other parts of the world, is directly associated with the growth of population and need of timber for housing, furniture, manufacturing of pulp and paper, burning of fuelwood. Expansion of agricultural and residential areas have also led to deforestation. Large scale unauthorised felling of trees have also resulted in loss of forest area.

Our present forest cover is about 20% as against the desired 33% under the National Forest Policy. On the contrary, the area under forestland had been reducing due to industries residential settlements, rehabilitation of landless people construction of roads and river valley projects, mining and such other activities.

The deforestation, overgrazing etc. being about changes in rainfall, temperature, wind velocity etc. and also lead to soil erosion. Such changes then lead to *desertification* (process leading to desert formation) of the area.

Afforestation

As we have discussed till now, the forests re-store ecological balance of all ecosystems (including desert), maintain biological diversity, act as catchment for soil and water conservation and prevent floods. In order to meet such needs, we, therefore, need to develop massive *afforestation* programmes of indigenous and exotic fast growing species for production and protection forestry.

The major goals for forestry should be (*i*) supply of goods and services to people and industry by a well thought out plan of production, and (*ii*) long term ecological security through conservation of forests cover and its restoration. A massive social forestry programme is needed to meet demands of local people for fuel, fodder, timber etc.

Afforestation means planting and covering the soil surface by vegetative cover. These vegetative covers try to reduce the securing effects in the following ways:

(i) It provides a cushion to falling rain-drops and thus, reduces their impact and consequent erosion.

(ii) The roots of the plants go deep into the soil and bind the soil for stabilising it against water and wind erosion.

(iii) Plants offer obstacle to the flowing rain, for reducing its flow velocity.

Ravines of rivers, like Yamuna and the Chambal need to be re-claimed to utilise the land for growing wood by afforestation and fodder, and where possible fruit and food crops.

Impact of Deforestation

As already discussed, the major impact of deforestations are:

(i) Soil degradation and erosion.

(ii) Changes in climate conditions.

(iii) Destruction of natural habitats.

(iv) Destruction of valuable sink for removing the environmental pollutants.

Causes of Deforestation

Deforestation is a consequence of over-exploitation of our natural ecosystems for space, energy and materials. The major seasons for deforestation are:

(i) *Conversion of forest land into agricultural land* for expanding the agriculture.

(ii) *Shifting or Jhuming Cultivation.* This is a cyclic cultivation being practiced chiefly in certain tribal areas where the soil is not fertile. To keep the soil fertile, the cultivators cut the trees and grass, burn it, and raise field crops for one or more gears and then move onto another site for repeating the process.

(iii) *The transformation of forests into pastures.* This is responsible for the soil degenerates within a short span of time due to over-grazing (also called as *cattle ranching*). This also results into soil erosion.

(iv) *Firewood collection.* Contributes much to the depletion of tree cover.

(v) *Timber harvesting.* Timber resource is an important asset for a country's prosperity. As a consequence to this, natural forests are being mercilessly explored.

(vi) *Forest fires,* whether natural man-made, destroys forest covers, atmospheric lighting is the major source of natural forest fires. Besides, man causes forest fires through his intentional or unintentional actions. Throwing of *'bidies'* especially in summer causes forest fires. This fire hardens the ground surface which decreases the porosity of the soil and results in surface runoff water. Besides, forest fires kill all the micro-organisms living as decomposers, and in the soils and plant roots. Thus, forest fires not only destroy natural vegetation and retard the regeneration of trees but also cause tremendous damange to the biological communities.

(vii) *Multipurpose river projects* require large areas to be submerged for the storage of huge volume of water in the reservoirs. The complete submergence of forested riverine areas completely destroys the natural forests.

(viii) *Construction of hill roads* most of which are in fragile belt of Himalayas. Road construction affected the stability of hill slopes, damaged the protective vegetation cover both, above and below the roads, resulted in debris covering forests vegetation, orchards and agriculture fields, blocked natural damage and damage to property.

Thus, we have seen that principal causes of deforestation and denudation have been the population explosion in man and livestock leading to enhanced requirement of timber and fuelwood, and grazing respectively. Another major reason for deforestation are agriculturalisation, urbanisation and industrialisation. Preservation of forests has been given a very low priority because they have long been looked upon as revenue generation sector.

7.5 NATIONAL FOREST POLICY, 1988

Basic Objectives of the Forest Policy are:

(i) Maintenance of environmental stability through preservation and, where necessary, restoration of ecological balance that has been adversely disturbed by serious depletion of the forests.

(ii) Opposition to reckless extension of agricultural land at the cost of forestland.

(iii) Checking the soil erosion and denudation.

(iv) Checking the extension of sand-dunes in the desert areas.

(v) Increasing substantially the forest/tree cover through massive afforestation and social forestry programmes, so as to being 33% of the country's geographical area under forest.

(vi) Encourage efficient utilization of forest produce and maximise substitution of wood.

(vii) Creating a massive people's movement with the involvement of woman, for achieving these objectives and minimise pressure on existing forests.

(viii) Arrange the requirements of fuel wood, fodder, minor forest produce and small timber of the rural and tribal populations.

(ix) Ban on the cutting of trees without permission of the government.

(x) The forest should not be diverted to non-forestry used. If it is very essential and unavoidable, it should be done with the approval of the central government, and the loss of forests due to diversion should be suitably compensated by planting trees in other proportionate areas.

7.6 EFFECTS OF MINING ON FORESTS

Mining is the extraction of minerals and ores from the earth's crust. It provides us coal, ores, clays, stones, precious stones etc. To meet the increased demand due to industrialisation, urbanisation and modernisation, the exploitation of mineral resources has increased manifolds, leading to large scale mining activities. Mining, whether it is at the surface or underground, alters the whole landscape and the ecosystem existing at the place of mining. It also upsets the existing land conditions.

Since mining activities are essential for the development and cannot be prevented, it is necessary to carry out mining activities in a planned way.

Mining has a serious impact on forest areas. A large areas have been clear-felled and laid barren as a result of open cast mining. Impact of mining include loss of production (forests, agriculture, pastures), loss of top soil, surface water pollution, lowering of ground water table, ore transport hazards (which include damage to vegetation), fire hazards and air pollution.

Major environmental impacts of mining can be classified as under:

1. **Air Pollution:** Air pollution is caused by (*i*) emition of *dust particulates* to the atmosphere during ore/coal handling, blasting, and transportation; and (*ii*) addition of

harmful *gaseous pollutants* which are originated in mining activities. Mine fires also pollute the atmosphere.

2. **Water Pollution:** Mining causes water pollution in two ways: (*i*) It discharges acid-mine-waters into surface water bodies, like ponds, rivers etc. (*ii*) It also changes the hydrology of the area, like change in river course and flow discharge. It may also lower the water-table of the area.
3. **Degradation of Land:** The land degradation effect of mining can be categorised as under:
   (a) **Land Subsidence:** Due to underground mining, the land surface subside, which causes demage to agricultural land, forests, buildings and surface drains, sometimes they also damage highways, bridges, water and sewage lines etc. Land subsidence has also taken a tell of several lives in many disasters. Land subsidence incidence have also been reported due to excessive mining of oil and natural gas. Land subsidence problems have mostly been arisen due to unscientific practices followed in underground mining.

   (b) **Land Degradation:** Open cast mining results into removal of top soil and causes deep and large excavations, pits and cuts. Thus, large areas become unusable and denuded of vegetation, causing continuous soil erosion, and silting of streams and water bodies. The degraded land also lead to landslides which are sometimes disastrous.

   (c) **Deforestation:** Large scale felling of trees from forest areas, may sometimes, be required for extraction of mineral deposits. The tree covers are also destroyed for making in-roads into the running area. Besides, thirst, dust and overburden, during mining, spreads on the nearby areas causing large scale damage to the vegetations in that area.
4. **Noise and Ground Vibrations:** Mining operations create large scale noise and vibrations which interferes with the hearing and speech communications, causes annoyance, headache, high blood pressure, and ultimately hearing loss. Heavy ground vibrations caused by heavy slashing may also damage the nearby structures.
5. **Health Hazards:** In all underground mining, rock falls, roof-falls, flooding and inadequate ventilation are the major hazards. Use of explosives had also been disastrous in mines. Mining also poses several long term health hazards.

7.7 EFFECTS OF DAMS ON FORESTS

For harnessing the precious water resources, dams are constructed across the rivers. Dams are constructed where heavy rainfalls occur only during a few rainy months of an year. If no dam is constructed, then very heavy flood flows will pass through a given river during such heavy rainfall spells, causing floods and the precious water will flow as waste to join the sea. After the expiry of rainy season, the river will become dry, and there will be no water to fulfill various demands of water.

Since a dam accumulates water on its upstream side, it submerges lowlying land (generally forest areas) and villages, displacing people. A valuable forest resource is lost permanently alongwith a number of animal species. The downstream portion, on the other hand, suffers due to lack of sufficient water, sediments and nutrients. Due to such reasons, the forests in downstream flood plain and adjoining area will be replaced in due course of time by new species adapted to the modified environment.

Due to construction of dam and change in water use pattern and seepages from reservoir water-logging and salinity problems are cropped up. This causes the land to be infertile, which adversely affects the nearby forests and agricultural lands.

The implementation of project i.e. construction of dam and filling of reservoir, results in flourishing growth of trees and crops in the command area. Such production may far exceed the loss of wood and crops caused by the submergence of forests and agricultural land.

Construction of dams need large quantity of quarry material and roads etc. for transportation. These activities adversely affect the forests.

One of the most profound effect of dams is displacement of people from submergence area and area covered by the project. Due to submergence, the residents are forced to vacate their ancestral homes, agricultural fields and other properties. These people are re-settled in other areas, where they come across numerous problems and have to re-adjust to a totally new type of living. These people loose their society community, religious sites and old environment to which they are accustomed to life and economy of the displaced people, generally tribal people, are absolutely dependent on forests. The rehabilitation of tribal people to a different habitat makes them lose their identity and culture.

The reservoirs created due to construction of dam causes breeding of mosquitoes with possibility of onset of vector borne diseases, like malaria and dengue.

The dams have several other negative impacts, besides several positive effects. These are discussed in the section "water resources".

Sardar Sarovar Project

Sardar Sarovar Project is a multipurpose project and built on Narmada River in Gujarat.

Its foundation stone was laid on 5th April 1961. Initially the project was facing financial problems. It picked up work in 1979 when the World Bank loan of US$200 million was sanctioned for the project to increase irrigation and produce electricity. The dam was inaugurated on 17th Sept. 2017.

This dam is the largest of the 30 dams planned on river Narmada. The project is planned to irrigate 18000 sq km area, most of it in drought prone areas of Kuchch Saurashtra. Total installed capacity of the power facility is 1450 MW (including 250 MW from Pump Storage Plant).

Delay in completion caused lot of sufferings to the people and submergence and resettlement caused adverse impact on forest flora and fauna. Against submergence of 4523 ha of forest in Gujarat, compensatory afforestation was done in 4600 ha in non-forest areas of Kuchch. In addition, compensatory afforestation in 93 ha (double the area of submergence) has been done in degraded forest land in nearby districts.

In order to mitigate the rate of soil erosion in the catchment and also to check the silt inflow in the reservoir Catchment Areas Treatment measures have been carried out in 27,204 ha. In addition, treatment in 1953 ha of non-forest in the catchment has been carried out. Besides 5732 ha of canal sides and borrow area plantation have been done.

For the conservation of wildlife, bear sanctuary in the catchment area is extended to cover an area of 607 sq km for prime habitat.

There had been a strong movement against the construction of the project resulting its delay for several years. The questions raised by the movement were related to social economic equity as well as the enormous environmental losses, including loss of the biological diversity of the inundated forests in the Narmada Valley.

Reasons for long agitation were : displacement of project affected people, dispute on compensation package, effects of submergence on biodiversity. All these factors were mainly due to not carrying out proper Environment Impact Assessment (EIA) and Environment Management Plan in time (in the initial stages and lack of knowledge of project developers). Delay in execution caused long sufferings, and increase in family size resulted in lack of R&R package. Further, the people residing in upstream area did not want sufferings for the benefit of downstream people.

Tehri Dam Project

Tehri dam project built on Bhagirathi river in Uttarakhand is a multi-purpose project. Detail project report was prepared in 1969 which was technically cleared in June 1972 and its administrative approval was accorded in July 1976. Stage-1 (1000 MW) of the project become fullly operational in March 2007.

The Tehri Hydro Power Complex (2400 MW) has 3 components, namely Tehri dam and hydropower plant (1000 MW), Koteshwar hydro-electric project (400 MW) and Tehri pumped storage plant (1000 MW). Tehri complex has two lakes formed by 260.5 meter high Tehri dam and 103 meter high Koteshwar dam and have a spread of 42 sq km and 2.65 sq km respectively at full reservoir levels.

Due to submergence caused by these dams and project colony, Tehri town and 125 villages were affected. A total of about 1800 (5300 urban and 5500 rural families) were displaced. In addition about 4100 rural families were partially affected, who were not shifted but compensation were made.

In addition to power generation, the project is providing additional irrigation facility in 2.70 lac ha, and 300 cu secs of drinking water for Delhi, which is able to meet the requirement of about 40 lac people.

Against 4193.813 ha forest land diverted for the project including land for submergence land used in construction of project and project colonies, project has completed the **compensatory afforestation** in an area of 4586.07 ha in Lalitpur and Jhansi districts (in a different ecosystem, which is not ideal). In addition 1358.20 ha forest land diverted in second phase for resettlement, the compensatory afforestation has been carried out in 2716.40 of degraded forest land in Haridwar district.

In order to reduce soil erosion the project has carried out **Catchment Area Treatment** (CAT) in 52,204 ha of entire degraded catchment. In CAT works, the species as recommended by BSI (Botanical Survey of India) based on flora study of the area, have been planted. A Botanical garden in an area of 14.28 ha has also been established and plantation of special species coming under submergence was carried out, so as to preserve important flora of region.

Faunal studies were got conducted through ZSI (Zoological Survey of India). As per studies no adverse impact was envisaged on mammals, birds, reptiles, (Snakes and Lizards), frogs and toads, and fresh water fisheries. As suggested by ZSI, action plan for possible mitigation of *Mahseer fish* was prepared. The *Mahseer Fish* Hatchery and Fish Farm have been developed with the help of National Research Centre on Cold Water Fisheries (NRCCWF, Bhimtal).

Besides above, a green belt have been developed along the rim of reservoir, so as to check soil erosion and resultant siltation of the reservoir, to protect and regenerate the vegetation in the rim area and to increase the natural beauty of the landscape.

The project took about 38 years for completion. During this period there had been lot of agitations. The main reason for all these problem was lack of proper planning, insufficient ESA and Environmental Management Plan (EMP) and probably due to lack of knowledge of project developers. In fact, the knowledge and interest in various environment aspects is gaining in last about 2 to 3 decades. Now the project has adopted sufficient mitigation measures.

7.8 CONSERVATION OF FORESTS

Conservation of forest resources are necessary for economic development and maintenance of environmental and ecological balance from local through regional to global levels. Forest conservation measures include:

1. Maintenance of *ecological balance*
   (a) Protection of remaining forest covers; and
   (b) Increase in the forest cover.
2. Supply of goods and services to meet the needs of people and industry by a well thought out plan of *production/commercial forestry.*

The second measure is necessary to meet the need of judicious relationship between the environment and the development.

Protection of Forest Cover

The first and the foremost task to conserve the forests is to protect the existing forests from merciless, reckless cutting of trees by greedy economic man. This task may be achieved through government legislation and by arousing public interest in the importance of the forest resources.

It is recognised that commercial exploitation can be allowed in the reserve forest areas, and these need protection from fuel-starved villagers and fodder-starved cattle. The reserve forest include the areas of Himalayas, Western and Eastern Ghats, National Parks, Sanctuaries, Biosphere reserves. For this, public support must be generated to achieve the real goal of eco-development.

During 'first Earth-Summit' (also known as *'Rio-Summit'*) held in 1992, a deep concern was expressed at the reckless cutting of forests. There was emphasis in the summit for imposing restrictions on cutting of tropical rainforests in order to use them for medical purposes and as natural carbon drinks for the absorption of emitted carbon dioxide.

Chipko Movement

Originally, about 300 years ago, a ruler in Rajasthan decided to fell the Kejri trees to create lime. This decision of the ruler was challenged by the local women led by Amrita Devi, a Bishnoi woman and these women clung to the trees to prevent the felling of trees, as these are the scarce resources on which they were dependent. The women agitators were ruthlessly massacred.

The story was revived in 1970s when tree felling for timber in Uttarakhand prompted local women to prevent deforestation by timber contractors. The movement of these local women was supported by Chandi Prasad Bhatt and Sunder Lal Bahuguna to make it a peoples' movement. This movement was named as "Chipko Movement" or "Chipko Andolan".

Chipko movement primarily begun and supported by the local women in the hilly areas of Garhwal region in Uttarakhand, where the women who were traditionally collecting the fuelwood realised that their fuelwood and *fodder* resources have receded due to deforestation. The chipko activists made long *padyatras*, for creating awareness, across the Himalayas, protesting against deforestation.

The movement was successful and made to realise that the forests of the hills are the life-support system of local communities and it was realised that there is need for empowering local womens' groups, who are most seriously affected.

Increasing the Forest Cover

There is a need, as per forest policy, to increase the forest cover so as to cover 33 percent at the total geographical areas of the country by forests through (a) *afforestation* in the open wasteland, and (b) *reforestation* of already deforested areas. This need to be done particularly in those deforested areas which are not suitable for cultivation, such as mountainous and hilly areas.

Afforestation: As already discussed earlier, afforestation is necessary for exchanging the area under forest. In afforestation, efforts should aim to generate forests and not plantation. A forest as a whole with its three storeys (tall trees, smaller trees or shrubs, and the ground cover of small shrubs and/or herbs) together with soil and microflora.

Reforestation: Reforestation can reduce and check accelerated rate of soil erosion mainly over the hills which have been deprived of their dense cover of original forests through extensive deforestation.

Forest Conservation Act, 1980

The Act was enacted with a view to check, indiscriminate dereservation and diversion of forestland to non-forest purposes. As per the provision of this Act, prior approval of Central Government is required before any reserved forest is declared as dereserved, or forest land is diverted to non-forest purposes. If diversion is permitted, compensatory afforestation is insisted upon and other suitable conditions imposed. Where non-forest lands are available, compensatory afforestation be raised over equivalent area of non-forest lands. Where non-forest lands are not available, compensatory plantations be raised over degraded forests twice in extent to the area being diverted.

7.9 PRODUCTION/COMMERCIAL FORESTRY

The forests maintained to obtain from them production of economic importance to earn revenue by meeting the commercial and industrial requirements for the manufacture of products of commercial value. Production forests must be managed on strictly scientific lines to drive the maximum value from them on a sustained basis.

Local supply forests should be managed to meet the domestic and agricultural needs of the local people to the maximum extent without impairing future productivity of the forests.

The basic aim of production/commercial forestry is to supply goods and services to meet the needs of people for firewood, fodder, fibre, timber, medicine etc. or for industrial purposes as timber of all types, plywood, matchwood, fibre board, paper and pulp, rayon, silvichemicals etc. This can be done by: (*i*) intensive plantations, and (*ii*) production/captive plantation.

1. **Intensive Plantation:** This is done by planting on all the available land from villagers' fields, to community land, to road/rail sides every available space. Indigenous and/or exotic species can be used for plantations that removes pressure on natural forests. Social/participative/agro-forestry programmes are included in this category. This would meet all needs of fuelwood, fodder, fertiliser, fibre, medicine and timber and thus, prevents denudation of natural forests.
2. **Production/Captive Plantations:** This is commercial forestry developed to meet the need of forest based industry that requires large quantity of raw materials. These plantations are done to follow (barren) land not being used for agriculture, mostly on free grazing lands. A part of such plantations may be used to generate fodder for cattle. Moreover, short rotations of indigenous or exotic species are to be preferred. The major strategy should be to increase productivity of wood per unit area per unit time.

7.10 FORESTRY AND SILVICULTURE

Forestry is the scientific management of forests for continuous production of goods and services. It deals with the theory and practice of all that constitutes the creation, conservation and scientific management of forests and the rational utilisation of their resources. The primary objective of maintaining forests in the ideal condition is the ultimate satisfaction of the desires of the beneficiaries—the present as well as the future generations. Forestry is of 3 types, namely: (*i*) protection forestry (to protect land against wind and water erosion, hazards of floods etc); (*ii*) commercial forestry to produce timber and (*iii*) other forest produce for business; and social forestry (this is the forestry other than conventional forest areas).

Silviculture is the art and science of cultivating forest crops with particular reference to environmental factors. Cultivation of forest crops consists of *regeneration,* namely, renewal of forest crop after the standing, old or decrepit crop is removed; and *tending*, that is carrying on operations from time to time for the benefit of growing the crop, which include wedding cleaning, pruning, climber cutting and girdling or unwanted growth.

Silviculture, in short, can also be defined as the science of managing forests on sustainable basis, *silviculture* can also be defined as that branch of forestry which deals with the establishment, development, care and reproduction of stands of timber.

1. Social Forestry

Social forestry can be described as forestry of the people, by the people and for the people. Social forestry means the management and protection of forests and afforestation of barren lands with a purpose of helping in the environmental, social and rural development as against the traditional objective of securing revenue. Its principal aim i[illegible]vert the pressure away from traditional forest areas by raising plantation of fuelwood, fodder and grasses.

There are two main *objectives of social forestry*:

(i) use of public and common land to produce in a decentralized was firewood, fodder and small timber for the local poor men and also to manage soil and water conservation, and

(ii) to relieve pressure on conservation of forests.

Economic Benefits from Social Forestry

Social forestry has many economic benefits, some of which are:

(i) It generates direct and indirect employment potential.

(ii) It helps to create an economic capital of increasing value.

(iii) Increased production of food and fodder.

(iv) It opens up a vast store of raw materials that may be used for a number of cottage industries, such as furniture baskets, wood carvings, sports goods, match-sticks, handmade paper, pencil and walking sticks etc.

(v) It provides raw materials for important forest based industries, *very*, paper and pulp; match-box and match-sticks; pencils; packing boxes; gums, honey, lac, essential oils, kattha etc.

(vi) It plays an important role in regulating water cycle.

(vii) It helps in flood control.

(viii) It provides fuelwood, thereby reduces the pressure on natural forests.

Participatory Management Programme (Joint Forest Management)

Participatory forest management has been considered since long by many as the only long term solution to the problems of deforestation and environmental degradation in the country. Since National Forest Policy of 1988 lays down the need to involve local forest dependent villagers particularly tribals and women, in forest management, Joint Forest Management (JFM) programmes are being increasingly adopted by several State Forest Departments. The JFM approach involves negotiating partnerships between local institutions of forest users and State Forest Departments on the basis of sharing forest protection responsibilities in exchange for a share of the income from forest products.

By adopting Community Forest Management, it is possible to protect these lands through mutual co-operation to increase their productivity and also put an end to their gradual decline into barren wastelands.

Social forestry programmes have mainly three components:

(i) *Farm Forestry*—encouraging farmers to plant trees on their own farms;

(ii) *Wood lots*—planted by the forest departments for the needs of the community, especially along road side, canal banks and other such public lands; and

(iii) *Community wood lots*—planted by the community themselves on community lands, to be shared equally by them.

Farming and forestry had for the past many centuries been developed as isolated enterprises. It was only recently complementary role of forestry realised and applied practice. According to the principle of *Agro-forestry*—raising trees and crops simultaneously, involving the farmer.

2. Agro-Forestry

It is the raising of trees and agricultural crops either on the same land or in close association in a way that all the land inclusive of the waste patches is put to good use. In this, crops are grown in between the rows of trees, or alternatively, trees are raised along the bunds of fields. In this way food, fodder, fuel, timber and fruit can be raised on the farm itself. This involves the maximum production from a particular patch of land, and provides a boost to the rural economy. In addition, agro-forestry has the ability to provide employment all through the year.

3. Community Forestry

Community forestry involves the raising of trees on public or community lands as compared to form forestry, in which trees are raised on privately owned lands. The main feature of this programme is to provide benefit to the community as a whole. In this, tree planting is done, on community lands or state owned land meant for the local community, with the help of seedlings and fencing provided by the forest department. However, the community is responsible for protecting the plantation.

4. Urban Forestry

Urban forestry is the raising and management of trees on public and privately owned lands in and around urban centres. Urban forests and trees have more aesthetic value as compared to that of forests and trees in rural areas. The urban forests include green belts roadside avenues, parks, reserved lands, industrial and commercial green belts etc.

7.11 FOREST MANAGEMENT

Forest management is of significant value for the mankind, as if used on sustainable basis, forests can continue to provide forest goods and services indefinitely. *Forest Management* deals with the organization of a forest property for its proper maintenance and utilization.

For successfully managing, the forester must have adequate knowledge of all the branches of forestry and allied sciences. The forest management is, in fact applied forestry, and has its object of *economic* (maximise the yield of desired forest products); *protective* (preserve the physical features of the soil and regulation of flow of matter); and *bio-aesthetic* (enhancing the scenic beauty and preserving the fauna and flora).

Essentials of Forest Management

1. Protect the existing forests and forestlands.
2. Increase the forest and vegetable cover rapidly with special considerations on hill slopes; in catchment areas of rivers, lakes and reservoirs, ocean shores, and on semi-arid and desert tracts.
3. Improve the productivity of forests and forestlands.
4. Diversion of good and productive agricultural lands to forestry should be discouraged in view of the need for increased food production.
5. The network of national parks, sanctuaries, biosphere reserves and other protected areas should be strengthened and extended adequately. This is necessary for the conservation of total biological diversity.
6. Provision of sufficient fodder, fuelwood, and pasture, especially in areas adjoining forest is necessary in order to prevent depletion of forests.
7. Minor forest produce provides sustenance to the tribal population and to other communities residing in and around the forests. Such produce should be protected, improved and enhanced with a view to the generation of employment and income.

Strategies for the Forest Management

Following strategies may be employed to obtain continuous forest yields, on sustainable basis, from year after year:

1. Forest management requires to harvest only that much quantity of timber and other products annually that can grow during that period. Over-harvesting of forest products, especially timber, will destroy and vanish the forests completely. Hence, *restrained felling* and *selective cutting* of trees is required to be adopted.
2. *Block cutting methods of trees* should be adopted, in which trees are harvested in blocks only. This method is very useful where forest is having almost even-aged trees of a fewer number of species.
3. Reforestation of the denuded area is essential to ensure sustained yield.
4. *Alternative sources of fuel wood* must be made available to the people cheaply, so as to discourage them from using wood. *Planting of fuel wood providing trees* on waste land can also help in this regard.
5. Provisions of strict punishment be made for *unauthorised felling of trees* on poaching of animals.
6. *Re-cycling* in the use of forest products like paper etc. can help in reducing the demand on forests.

7. Declaring forest areas as national parks, sanctuaries and other protected areas.
8. No developmental project be allowed to be executed without thorough considerations about forest and ecological losses.

QUESTIONS

1. Describe the importance of forest resources.
2. What is the importance of forests in maintaining ecological balance of nature?
3. Explain the importance of forest in providing economical and commercial services and goods.
4. Differentiate between forests and open forests.
5. Describe the structure of a forest, indicating its various strata.
6. Define forest. Explain the different types of forests.
7. Define and explain the term deforestation and afforestation.
8. Discuss various impacts of deforestation.
9. Differentiate between deforestation and afforestation.
10. Write a detailed note on conservation of forests.
11. Discuss sustainable management of forests.
12. Explain the different uses of forests.
13. What do you understand by "over-exploitation of forests"?
14. Write short notes on:
    (a) Impacts of deforestation
    (b) Causes of deforestation
    (c) National Forest Policy
15. Discuss the various effects of mining on forests.
16. Describe the different impacts of dams on forests.
17. Write a detailed note on "protection of forest cover."
18. Explain in brief:
    (a) Forest Conservation Act
    (b) Commercial Forestry
    (c) Forestry and Silviculture
19. Explain different types of forestry.
20. Write a detailed note on 'Joint Forest Management'.

8

Chapter

WATER RESOURCES

8.1 INTRODUCTION

Water is the most important resource of the society as a whole, since no life is possible without water. Water is essential for existence and survival of plants, animals and humans. Water is one of the five basic elements constituting the life. It is the major component (70–90%) of organism's body.

About 97% of the total available water on the earth is contained in oceans, and is saline in nature. Out of the balance 3%, which is available as fresh water, about 2% is contained as ice on poles, and 0.75% as groundwater. Out of the remaining about 0.25%, only 0.75% as groundwater. Out of the remaining about 0.25%, only about 0.01% is found available in lakes and rivers at any given time, as the rest occurs as glaciers and snow. Total water contained in the atmosphere is still less, and is of the order of 0.001% of the total available water.

It can, therefore, be easily understood that the surface water, which can really be utilised by the society, is generally very-very small, being quantity contained in lakes, or the one-flowing in rivers, caused by rain storms and melting of snow. It is also estimated that about 96% of the total annual surface run-off, flowing in the rivers, goes and joins the oceans, and is, thus, not put to any worthwhile use.

We have already seen a brief account of the circulation of water from the oceans to the atmosphere and then to the lithosphere and back to the oceans. The total quantity of water available in the hydrosphere, atmosphere and lithosphere remains constant, though the quantity available in each one of these realms may vary from place to place and according to the seasons. We know that distribution of fresh water resources from precipitation, rivers and underground water is highly uneven. While in the rainy season, water may run-off rapidly leading to floods, surface water may dry up in the dry season. Such seasonal contrasts in volume of water in rivers may be regulated by construction of dams to store the water on reservoirs and utilising in dry season.

As we have seen that global supply of water is fixed and finite, the average annual rainfall and run-off are also fixed, while the population is increasing day-by-day, and at the same time demand is also increasing due to modern lifestyle. Similarly, the irrigation as well as industrial demand is also increasing.

Global Estimates of Water Quantities

Segment	% of total water	% of fresh water
1. Oceans	97.3 (1,34,000 × 10^4 B.m^3)	—
2. Fresh water	2.7 (3750 × 10^4 B.m^3)	100.0
Total water	**100.0 (1,38,550 × 10^4 B.m^3)**	**100.0**
Fresh water available in various sources		
(a) Polar ice and Glaciers	2.03	75.2
(b) Ground water upto 800 m depth	0.27	10.0
(c) Ground water 800 m to 4000 m depth	0.34	12.6
(d) Lakes and Rivers	0.01	0.3
(e) Other sources—soil moisture, biogeological water and atmospheric vapour	0.05	1.9
Total fresh water	**2.7**	**100.0**

The only possible way to increase our available water supplies seems to be the storage of excess flow during rainy season by building dams; or by groundwater recharge by rain-water harvesting. Besides, we should curtail our demands by conserving and judicious use of water. Water resource projects are planned, designed, executed, operated and managed with specific objectives of providing irrigation, domestic and industrial water supply, power generation, flood control etc. for helping the society to develop and improve its socio-economic status.

8.2 WATER RESOURCES

Precipitation is the primary source of water in the form of rainfall in hot regions and as rainfall/snow in cold regions. These give rise to secondary sources, which are broadly classified as:

1. Surface water
2. Ground water

Microsources of water are fog, moist and dew.

8.2.1 Surface Water

Surface water is available from rivers, (either direct run-off from rains or snow melt). Streams, tanks, springs, lakes or reservoirs. Precipitation of water from the atmosphere results in water falling on the land surface as well as in oceans. This water runs down the slope of the land in the form of streams and rivers. Water may also get stored on the land in the form of ice-sheets. While extensive and thick ice-sheets occur in Antarctica, Greenland and other regions around poles, less extensive ice-sheets occur in the high altitudes of mountain regions. Of the total water available on the surface of the land, about two-thirds is in the form of ice-sheets. The quantity of surface water in any region depends on the total precipitation and its seasonal distribution and also on the nature of rocks and soils on the surface of the land.

Much of the water falling during early phase of monsoon is accumulated in the surface depressions, such as lakes, pools, marshy places etc., and the remaining part flows downslope. Run-off is the concern treated part of the precipitation that flows from higher to lower elevations on or below the ground surface and subsequently appears at the outlet of a basin.

The total **surface run-off has following components:**

(a) *Surface run-off:* It is that part of the run-off which travels over the ground surface and through the streams to reach the outlet of a basin.

(b) *Subsurface run-off:* It is that part of the run-off which infiltrates into the soil zone and then moves laterally downslope *i.e.*, towards the streams, mainly in the form of seepage.

(c) *Groundwater run-off:* It is that part of the run-off which infiltrates into the zone of saturation by percolation and moves downslope laterally and then emerges out in the form of spring.

8.2.2 Underground Water

A part of the rainfall, which falls down on the land surface, seeps below the surface through the soil layer, pore spaces, joints and fissures in the rocks. The quantity of seepage depends on the nature of rocks and soils. The seepage of water through soils and rocks produces a zone of saturation below the surface. The upper limit of this zone of saturation is called the *water-table*. The depth of water-table varies from place to place and also with season. Water stored below the surface of the land is result of accumulation over a long period. In some places, under-ground water may not be potable on account of dissolved salts.

Underground water is being used not only for domestic needs but also for irrigation and industries. Underground water is drawn from open wells or bore wells generally in localities where surface water is not freely available.

Thus, the underground water is that water which occurs in the saturated zone of variable thickness and depth, below the earth's surface. Cracks and pores in the existing rocks and unconsolidated crystal layers, make up a large underground reservoir, where part of precipitation is stored.

In India, groundwater has played the pivotal role in fulfilling the demands of domestic, industrial and agriculture sectors. At present groundwater in India contributes more than 58% drinking water, 52% for agriculture production, and 50% for urban and industrial sectors. Indiscriminate and unscientific exploitation of this resource has led to multiple problems of decline in groundwater level, sea-water ingress, inland salinity, groundwater pollution, land subsidence etc.

The measures that need to be adopted in the country to meet the increased water demand would include exploration of deeper aquifers, groundwater recharge, development of aquifers in flood plains, conjunctive use of surface and groundwater in canal command area and regulation of groundwater development.

8.3 STATUS OF INDIA'S WATER RESOURCES

As we have already discussed, distribution of water available through precipitation is not uniform with respect to space as well as with respect to time. The annual precipitation including snowfall, which is the main source of water in the country, is estimated to be of the order of 4000 billion cubic meters (BCM) (1 BCM = 1 km^3 = 10^9 m^3 = 0.10 million ha m). The estimated precipitation during the monsoon season (June to September) is of the order of 3000 BCM. Out of the total precipitation of 4000 BCM, we get an average annual run-off of about 1869 BCM, and the rest about 2131 BCM goes back to the atmosphere through evaporation. The 1869 BCM of average annual flow for a population of 1.1 billion gives an annual per capital total water availability = $\frac{1869 \text{ BCM}}{1.1 \text{ B}}$ = 1700 m^3. The scientists stated that a country whose renewable fresh water availability on annual per capita basis is about 1700 m^3, would suffer from occasional local water problems. Below this, the country begins to experience a period of frequent water stress. Since our replenishable water quantity is fixed, the per capita availability will continue to fall below the present limiting value of 17000 cum/capita/year.

Thus, India is not a rich country in the water resources. Further, India has only 4% of world's average annual supplies of 47000 BCM, as against the 16% population of the world of 6.5 billion. Moreover, about 85% of this fresh water supply of 1869 BCM, is produced only during the 3-4 monsoon months, and that too in 4 to 5 flood spells.

As against average annual utilisable flow of 1122 BCM, our actual utilisation, at present, is only about 640 BCM. Even this could be achieved by constructing large number of dams. The entire average annual flow in the country flows through its 14 major river basins. The availability of water is river. *Brahmaputra* and *Barak* basin is excessively high with average annual per capita availability of about 19650 m^3, whereas it is as low as 380 m^3 in some of the East flowing rivers of Tamil Nadu. These values are further decreasing with increasing population.

The above facts emphasize that if we failed to take the following immediate and effective measures: (*i*) Exploitation if the large amount of run-off presently going waste to the oceans, (*ii*) We will soon be facing a severe crisis curtailment of the wastage of water.

8.4 STATUS OF WORLD'S WATER RESOURCES

On an average, 1,09,000 BCM of precipitation falls annually on the total land area of about 1,45,000 billion square meters. Out of the total precipitation volume of 1,09,000 BCM, about 62,000 BCM goes back to the atmosphere through evaporation, leaving 47,000 BCM of average annual run-off volume. This available water is very unequally distributed over various countries and continents, and majority of this flow comes within 3 to 4 months of the year.

Thus, the world's per capita average annual availability is about $\frac{47{,}000\,\text{BCM}}{6.5\text{ B}} = 7200\text{ m}^3$, against India's average annual per capita availability of 1700 m^3. The precipitation is very uneven in various areas, which can be seen by the example that South America with relatively uninhabited *Amazon* river basin gets 12,200 BCM (about 26% of the world's total flow). Similarly, the per capita average annual flow is also very uneven, for example, in Canada, it is 100,000 m^3, while it is just 1000 m^3 in Egypt.

Due to the existence of excess available flows at various places, a large quantity of available run-off goes waste to the oceans. The total utilisable water in the world has been estimated to be only 9000 BCM, against the total fresh water supply of 47,000 BCM. Out of this 9,000 BCM, the present exploitation and utilisation is hardly 4,500 BCM. *i.e.*, about 50% of the utilisable quantity. This is because several countries have yet not exploited their entire utilisable waters due to non-creation of sufficient storages.

With the present utilisation of about 4,500 BCM by a total world's population of about 6.5 billion, the per capita annual consumption is only $\frac{4{,}500\,\text{BCM}}{6.5\text{ B}} = 700\text{ m}^3$.

Since, even the average figure of 700 m^3 per capita annual is on much lower side than the minimum suggested value of 1000 m^3, this shows the extraordinary poor condition with regard to availability of water in various countries.

Below 1000 m^3 water availability is considered to be water stress condition.

8.5 USES OF WATER RESOURCES

Water is important in almost every sphere of human activity including drinking, irrigation, thermal power generation, industries, livestock management, hydroelectric generation, fisheries, navigation, recreation, etc.

Water is required by the community for fulfilling its several needs, and for surviving not only human life but also for animal and plant life. Water should not only be in sufficient quantity, but should also be of good quality, so as not to cause any harmful effects on such lives. Water is generally required for the following important uses:

1. Domestic Use

This includes the water required for drinking, cooking, bathing, washing, lawn sprinkling, gardening, sanitary toiletary uses etc. The amount of domestic water consumption per person vary according to the living conditions (lifestyle) of the consumers. On an average, the domestic consumption under normal conditions in an Indian city is expected to be around 135 litres per day per person against those of the developed countries, like U.S.A. to be around 340 litres per day per person.

2. Industrial and Commercial Use

This includes the water used in factories, hospitals, hostels, hotels, restaurants, offices, schools etc. In thermal power plants, water is required for boiler make-up water, for cooling in cooling towers etc.

3. Public Uses

This includes the water required for public utility purposes, such as watering of public parks, gardening, washing and sprinkling on roads, use in public fountains, fire fighting purposes.

4. Irrigation Needs

Irrigation is generally defined as an application of water to soil for the purpose of supplying moisture essential for plant growth. The main purpose of irrigation are: (a) supply essential moisture for plant growth, (b) save crops against short duration droughts, and (c) wash or dilute salts in the soil.

Every crop requires a certain quantity of water after a certain fixed interval, throughout the period of growth. If the natural rain is sufficient and timely so as to satisfy both these requirements, no irrigation water is required for raising that crop.

5. For Hydropower Generation

Hydroelectric power is generated by running the turbine blades by the energy of the flowing water, and thus, producing electricity from the generators coupled to the turbines. In order to ensure the continuous production of hydropower, it is essential to store water to a certain height and then let a discharge flow through the turbines. Production of hydropower is an excellent use of water, as no water is consumed in it.

6. For Navigation and Recreation

Navigation is another important use of water, which provides us a cheap means of transport without any consumption of water. Although no water is consumed in this use of water, however, sufficient quantity of water must remain available throughout the year, so as to enable the ships, motorboats, steamers etc. to safely float over it with minimum drag.

8.6 OVER-UTILIZATION OF WATER

Water as a resource is under relent pressure due to population growth, rapid urbanisation, large scale industrialisation and agriculture. Increasing water pollution resulting in water quality deterioration, due to contamination of surface and groundwater. Today, in almost all spheres of human activity, more water is drawn than what is actually needed. Due to carelessness, much of the water is wasted and this is resulting in water scarcity at many places, where enough clean water is no longer available. Everywhere we tend to use more water than is actually necessary, often because it is available in plenty or because we can afford it. Such an attitude results into over-consumption and wastage. A little care can result in substantial savings and longer life for resource base.

Pressure of demand on underground water resources has gone up considerably. Every year more and more water is drawn from sub-surface layers, whereas recharging of underground water has slowed down. In a number of localities where underground water is regularly drawn, the water-table is receding deeper and deeper because output exceeds the input.

Water level declines as a result of injudicious exploitation of groundwater resource may lead to several problems. Some of these problems are; reduced well yields, land subsidence, intrusion of salty water, especially in coastal areas, leakage of highly mineralized water into the aquifer. In order to overcome these serious implications, the recharge potential of groundwater resource is most essential.

The water crisis could precipitate faster than expected, unless integrated land water use strategies were implemented, and suitable water storage and conservation practices were adopted. Failure to evolve appropriate strategies and implement suitable measures to keep pace with the actual demand of water might throw the whole process of water resource development out of gear and precipitate serious crisis. Since, the water resources projects usually have a fairly long gestation period, it is important to underline the need for having a long-term planning for water resources development and management.

Groundwater is a significant source of supply for nearly 50 percent of India's irrigated area. Groundwater irrigation began to expand rapidly with the spread of rural electrification. Presently, about 40% of agricultural output comes from areas irrigated with groundwater.

Rapid pace of groundwater as well as surface water development has resulted in a number of problems. In many arid and hard rock areas, overdraft and associated water quality problems are increasing. The overdraft has resulted in failure of wells, shortage of water supplies, and salinity ingress in coastal areas.

Careful crop planning is required where water is scare. Drought resistant, deep-rooted, less water consuming and short duration crops are the needs of these areas. Traditional surface irrigation generally achieves only about 40–50% water use efficiency. Modern irrigation techniques, such as drip and sprinkler are very efficient and useful in water scarce areas.

Water sources that are close to urban sources are now either exhausted or will get exhausted soon due to over-utilisation and urbanisation. Efforts are, therefore, needed to conserve water by: (a) leak detection and preventive maintenance, (b) demand management, and (c) wastewater management:

Large areas, particularly in the command areas of major and medium irrigation projects, suffer from water-logging and soil salinity or alkalinity. High intensity of irrigation without adequate drainage has resulted in rapid upward movement of water-table and increased chances of development of water-logging conditions.

An agricultural land is said to be water-logged, when due to high water-table, the root zone of the plants gets flooded water and thus, become ill-aerated. The supply of oxygen gets cut off, when the land becomes ill-aerated resulting in reduction in the plant growth, which reduces the crop yield. Other **problems created by water-logging are:**

(i) The normal cultivation operations cannot be easily carried out in wet soils. In extreme cases, the free water may rise above the surface of land (known as swampy land), making the cultivation operations impossible.

(ii) Certain water loving plants, like grasses, weeds etc. grow profusely in water-logged lands. Thus, affecting and interfering with the growth of the crops.

(iii) Water-logging also leads to salinity in the root zone of the crops which checks the plant growth, and the plant ultimately fades away.

The **water-logging** may occur due to the following *reasons*:

1. Over and intensive irrigation.
2. Seepage of water from the adjoining high lands.
3. Seepage of water through canals.
4. Impervious obstruction in the water seeping below the soil, which may not allow water to go deep into the ground.
5. Inadequate drainage.
6. Submergence due to floods.

The water-logging can be controlled by adopting following measures:

(i) Living of canals and water courses.

(ii) Reducing the intensity of irrigation.

(iii) Introducing crop-rotation.

(iv) Providing an efficient drainage system.

(v) Introducing lift irrigation.

8.7 CONSERVATION OF WATER

Conservation of water is mainly in the context of freshwater. As already discussed earlier, with the increasing industrialisation and urbanisation and consequent growth of human population, water is being used in enormous quantities. Agriculture also consumes a lot of water.

To conserve water, we must reduce domestic as well as industrial utilisation. At domestic level one can think of better toilet designs, prevention of wastage by checking and repairing pipe and tapes. Re-cycling of water is another way of conserving water. Sewage effluents can be used to grow plants and vegetables. Industries can also partially purify used water and re-cycle it. Better effluent treatment technologies will not only conserve water but also prevent water pollution caused by these effluents.

Agriculture accounts for more than 75% of water consumption. It is, therefore, imperative that new and more economic ways of irrigation, like drip and sprinkler system, are developed and used.

If dams and canals are kept free of weeds, it will further reduce water loss due to evapo-transpiration. Storing rainwater in especially created lakes is one good way of preventing useful water from draining away. Such storage also helps to recharge the groundwater.

If we want to meet our future requirements of food security, water supply, energy, flood modernisation, regulation of river flows, dilution of pollution in the rivers, it would be necessary to construct storage reservoirs to conserve the flood flows in our major rivers. The water crisis could precipitate faster than expected, unless integrated land-water use strategies were implemented, and suitable water storage and conservation practices were adopted. In addition, promoting innovative water management techniques, including rainwater harvesting, watershed management, revival of traditional wisdom of water storage and conservation practices is important.

At present, huge quantity of water is wasted in irrigation sector, for conservation of water, we have to eliminate this wastage. The wastage in irrigation sector is caused mainly by two factors; *i.e.*

(i) Due to huge *seepage losses* in canals, which can be reduced by living,

(ii) Due to *evaporation losses* that takes place when irrigation supplies flow from a long network of open earthen canals, which can be reduced by providing conduit canals. Since this is costly option, it is not common.

(iii) Due to use of excess quantity of water by the cultivators, as they are not, normally, paying revenue through volumetric measurements of consumed water. Since the farmers has to pay a fixed sum of money, based on the area of crop grown, and the rate of water supply for irrigation is heavily under priced. It is, therefore, desired that cultivators must switch to the modern techniques, and adopt sprinklers and drip irrigation methods.

Promoting *rainwater harvesting*, both for recharging the groundwater as well as to collect the rainwater in underground tanks is very useful to increase the availability of water.

Recycling of water is another way of conserving treated municipal water. The treated sewage effluents can be used to water the lawns and plants in houses. This will also help in controlling pollution of our rivers.

Industries can also purify their effluents and re-cycle it, thereby reducing their requirement of fresh water and reduce pollution load on rivers.

8.8 FLOODS

A flood may be defined as an overflow coming from some river or from some other water body. A river may get flooded due to excessive rainfall or excessive melting of snow or due to some other form of obstruction. Whenever the water overflows the banks of the river, the river is said to be flooded. On important rivers, the *warning* and *danger levels* are generally assigned at important locations. Apart from the overflow of rivers, the floods may be caused by the failure of some dam or with a sudden release of huge amounts of water, causing considerable damage to life and property, and even carry off the top-soil, leaving the land barren.

Flood occurs when soil and vegetation cannot absorb all the water, the water then runs off the land in quantities that cannot be carried in river channels or retained in natural ponds or dams.

Flood in simple terms means "*inundation of extensive land area with water for several days continuously*. Rapid run-off due to flood causes soil erosion and sediment deposition problems downstream, the lowlying area adjoining the deep river channel, within which a river water can spread, is called the *flood plain*. A river and its flood plain forms a single ecological unit. The man, however, by his greed to occupy more area for his living, commercial and other activities, has been encroaching upon the flood plain, thereby reducing the width of the flood plain, and making the inhabitated area more vulnerable to floods.

When river overtops, it causes serious damage not only to public but also to its own ecosystem. Plant and animal life of the riverine environment will be lost, and the river ecosystem will be considerably changed.

Floods have been regular features of India, especially in *"Brahmaputra River Region"* and in *"Ganga River Region"*. Assam, Bihar, Odisha, Uttar Pradesh and West Bengal are the worst affected states.

Causes of Floods

Rapid increase in population and consequent increase in all round human activities, has led to occupation of the *'flood-plains'*, resulting in increased flood damage. Actually the man has gone to the flood and not the other way. Deforestation, reclaiming areas for occupation and shifting cultivation, all have added to the fury of floods.

Other major causes of floods are:

1. Flooding by the inadequate capacity within the banks of the rivers to contain the high flows brought down from the upper catchment due to heavy rainfall.
2. Areas having poor drainage characteristics get flooded by accumulation of water from heavy rainfall.
3. Flooding is caused by erosion and silting of the riverbeds resulting in reduction of carrying capacity of river channels.
4. Earthquakes and landslides leading to changes in river courses.
5. Obstructions to flow of water in river channels.
6. Synchronisation of floods in the main and tributary rivers.
7. Retardation due to tidal effects.
8. Cyclones in the coastal areas, which often are accompanied by heavy rainfall leading to flooding.
9. Prolonged rainfall on the catchment area of the river.
10. Dam failure.

Impact of Floods

Apart from loss of life (human and cattle) and damage to property, the devastating effects of floods, sense of insecurity and fear in the minds of people living in the flood plains is enormous. The after-effects of floods, like the agony of survivers, spread of epidemics, non-availability of essential commodities, and medicines, loss of their dwellings make floods most feared natural disaster being faced by human kind. These also damage crops, wells of the water, public utilities etc.

8.8.1 Flood Control Measures

As we have already discussed, flood in a river is caused by excessive precipitation in the catchment area of that river. Since the excessive precipitation cannot be controlled by man, the only alternative left for him is to somehow manage the excessive waterflow of the river. The methods adopted to manage such heavier river discharges are known as "flood control measures." Inspite of making his best efforts, man has not been able to fully eliminate floods, and has been only to control them or to keep them away from intensive habitations, or to check and reduce their devastating effects by pre-warning the area residents.

Various measures for flood management can be classified as under:

(A) Structural Measures

These measures include:

(i) *Construction of embankment, flood walls and sea walls:* These are planned to restrict the river in its existing course and are designed to avoid over-flowing of banks by increasing the channel capacity to pass the probable floods.

(ii) *Dams and reservoirs:* These store excess water during heavy rains at appropriate places in the river reach, so as to reduce the river flow downstream.

(iii) *Natural detention measures:* The most effective measure of flood control lies in reducing the run-off and allowing the water to percolate into groundwater through increasing the permeability of soil. The simplest, cheapest, most effective and highly beneficial method of reducing water flow is to artificially induce infiltration of water and recharging the groundwater so as to allow this water to infiltrate. The dug wells are sunk along the beds of ephemeral streams.

Other measures are, adopting soil and water conservation technique to help increase the vegetative cover in the catchment area of the river, and constructing terraces to help increase ground infiltration, thereby reducing river run-off. Water and soil conservation methods are based on the slogan, *"stop the little rain drops were they fall"*. Such conservation measures are planned to evolve an integrated programme of watershed development by construction of check dams, retarding basins, terrace cultivation, plantation of trees, providing vegetable cover to barren hilly lands, encouragement to pasture land, better forest management, discouraging shifting cultivation etc. These measures, apart from preserving land fertility, also improve infiltration capacity of soil, reduce run-off and retard the peak flow.

(iv) *Drainage Improvement:* Maximum damage due to floods, especially to crops, is on account of drainage congestion. The quicker the drainage, the lesser will be the losses. Highest priority in flood prone areas should be given to increase the capacities of road and rail bridges, culverts, cross-drainage works, drainage channels etc.

(v) *Diversion of flood waters.*

(B) Non-structural Measures

Non-structural measures contrary to the structural measures for aiming to mitigate the flood damage by trying to keep the flood water away from the people, strive to keep the people away from the flood water, bearing in mind the fact that the flood plains belong to the river. It contemplates use of flood plains judiciously, simultaneously permitting vacating of the same for use of the river whenever the situation calls for. This technique allows the use of flood plains reducing the hazards, while retaining its beneficial effects.

The non-structural measures can be broadly grouped under two categories, *i.e.*:

(i) Flood forecasting; flood proofing including disaster preparedness; and flood plain zoning.

(ii) Disaster relief; flood fighting; and flood insurance.

Flood Forecasting: It enables forewarning as to when the river is going to use its flood plain, to what extent and for how long. With reliable advance information/warning about impeding floods, loss of human lives and moveable properties can be reduced to a considerable extent. People and cattle etc. can be shifted to safer places.

Flood Proofing: These measures help greatly in the mitigation of distress and provide relief to the population in flood prone areas. These techniques include providing raised platforms for flood shelter and raising the public utility installations. These include: quick drainage potable drinking water and sanitary arrangements, human dwellings and animal shelters, storage facilities for food, fodder and other essential commodities, communication links—telephone/wireless/rail/road/boat.

Flood Plain Zoning: The basic concept of flood plain zoning is to regulate the land use in the flood plains in order to restrict the damage by floods which are bound to occur the damage by floods which are bound to occur from time to time. Flood plain zoning, therefore, aims at determining the locations and the extent of areas likely to be affected by floods of different magnitudes/frequencies and to develop those areas in such a fashion that the resulting damage is reduced to minimum. For the purpose of flood plain zoning, we can broadly classify the areas into three zones *i.e.*:

(a) *Prohibitive zones:* In which no building activity to be permitted, as this area would be highly susceptible to flooding.

(b) *Restrictive zones:* Which are moderately flood prone area, and the flood waters would enter only once in 5 or 10 years.

(c) *Warning zones:* Which would be safe to floods upto the design flood. These areas are very less liable to flooding.

Flood Preparedness and Relief: Flood disaster relief is rendered as soon as flood damage occurs. The flood preparedness and relief has, therefore, to be planned in advance to meet any disaster. Such preparedness generally include measures like: (*i*) flood forecasting, (*ii*) flood warning, (*iii*) flood fighting, (*iv*) evacuation, (*v*) community shelter for people, and (*vi*) relief camps etc.

Flood Fighting: Flood fighting covers construction of temporary dykes along the river, closing breaches immediately. Stagnant water becomes the breeding ground for mosquitoes, affecting public health. Public health operations should ensure availability of supplies and equipment, co-operation with other organisations engaged in disaster relief, for immediate mobilisation of personnel to eliminate health hazards.

Flood fighting measures normally involve:

- Strengthening of Central, State and District flood control rooms.
- Evacuation of flood victims.
- Air dropping of food pockets.
- Supply food and other rations including kerosene etc.
- Supply of fodder.
- First-aid and health operations.

- Restoration of road/rail links.
- Restoration of tubewells etc.
- Pumping out water from ponds and low-lying areas in cities/villages.
- Restoration of public assets and structures, power installation, sewerage and water supply schemes, roads and drains etc.

Most of the above activities are of immediate nature. Participation by voluntary organisation and local people should be encouraged for flood fighting.

Flood Insurance: Flood insurance has several advantages as means of modifying loss burden. The insurance does not reduce the flood loss potential directly, but it provides a mechanism for spreading the loss over large number of individuals.

8.9 DROUGHTS

A drought is said to be occurring at a place, when the place does not get as much water as we expect or need, over a significant period of time. As a drought occurs through a slow process, we only see its effects. A drought originates from a deficiency of precipitation (rainfall or snowfall) over an extended period of time, resulting in water shortage at same area. Availability of water in the right quantity and at the right time is very essential for plant growth and yield. If water is not available especially during the critical crop growth stages, yield is seriously affected. Hence, due to scanty or erratic rainfall, this essential requirement of plant growth is seriously disturbed thereby creating drought situation. During drought, ground water depletes for quite a long period of time.

Losses due to Droughts

As droughts do not cause too many fatalities (deaths) as the people migrate to other places, the people do not realise that a drought is very serious and costly natural hazards. As and when a drought occurs, it can have far-reaching impacts. Not only biotic, even the abiotic world, like land and soil, need water to stay intact and uneroded by weather.

The losses caused due to droughts may be of following types:

(a) Economic Losses

(i) Loss of crops which means loss to farmers and increase in price.

(ii) Loss of timber causing loss to timber industry.

(iii) Loss to the business dealing in fisheries, farm equipment, fertilizer, food processing etc.

(iv) Loss incurred due to transportation of water.

(v) Reduced hydropower generation.

(b) Environmental Losses

(i) Loss of fish and wild life habitat.

(ii) Lack of food and drinking water for humans and wild animals.

(iii) Migration of human beings and wild animals.

(iv) Increased stress to endangered species, leading to their extinction and loss of biodiversity.
(v) Loss of trees and vegetation.
(vi) Loss of reservoirs, lakes, ponds and wetlands.
(vii) Increase ground water depletion and land subsidence.
(viii) Reduced water quality.
(ix) Increased soil erosion due to winds.
(x) Degraded landscape.
(xi) Increased number of fires and loss of forests and human habitats.

(c) Social Effects

(i) Large scale migration of people, resulting in loss of income, and metal and physical stress.
(ii) Health and nutritional problems due to loss of food, vegetation and water, and due to reduced quality of air and water.
(iii) Famine and hunger and thereby resulting in all round chaos.
(iv) Loss of human and animal life due to heat, hunger, depression etc.

Causes of Drought

The major causes of drought are poor rainfall, poor quality and excessive depth of groundwater limiting its use for irrigation and drinking; absence of perennial rivers, poor water holding capacity of soil.

Drought Control Measures

Drought proven areas can be made less vulnerable to drought associated problems through soil-moisture conservation measures, water harvesting practices, minimisation of evaporation losses, development of the groundwater potential and transfer of surface water from surplus areas, where feasible and appropriate pastures forestry. The relief works should also be undertaken for providing employment to drought sticken population preferable for drought proofing.

Important measures for controlling the drought are:

(i) Creation of surface storage.
(ii) Planning for less dependable yield.
(iii) Reduction in conveyance losses.
(iv) Maintenance of irrigation system.
(v) Better irrigation practices.
(vi) While designing the canal capacities, peak demand of water in critical periods by the high yielding varieties of crops should be kept in view.
(vii) Conjunctive use of surface and groundwater.
(viii) Watershed development which include construction of soil conservation works on crop lands; construction of structures, like check dams, Nalla bunding, contourbunds, Gully plugging, percolation tanks, development of rainwater harvesting etc.

(ix) Prevention of evaporation losses from reservoirs.

(x) Long term solution to drought problem may be found in the basic principles of transfer of water from surplus.

(xi) Help of 'long range forecasts' for rainfall be taken to assess the likelihood of occurrence of drought. This help in taking suitable measures well in advance to mitigate the consequences of drought.

Some of the Major Droughts in India

India has suffered from innumerable natural and man-made disasters. Droughts and famine are major natural disasters which occurred in 20th and 21st century. Some of the major droughts occurred in India are given below in chronological order :

1. Drought of 1900, killing 1.23 million people.
2. Drought of 1942, killing 1.5 million people.
3. Drought of 1943, killing 1.9 million people.
4. Drought of 1965, killing 1.5 million and affecting 100 million people.
5. Drought of 1972, affecting 200 million people.
6. Drought of 1982, affecting 100 million people.
7. Drought of 1987, affecting 100 million people.
8. Drought of 2000, affecting 50 million people.
9. Drought of 2007, affecting 310 million people.
10. Drought of 2009, affecting 400 million people.

8.10 CONFLICTS OVER WATER

As we have already discussed, with the development of civilization, our water demand is going on increasing, while the supply remains almost the same, therefore, it becomes evident that the available fresh water will become scarce in future, even if sufficient water is available today. By "sufficient water", we mean sufficient quantity of water to meet all our demands including domestic demand, irrigation demand, industrial demand etc. There are various places in the world, where water is in great scarcity even today. Even in our own country, shortage of fresh water is creating serious problems in major parts of Rajasthan, Gujarat and Tamil Nadu.

Shortage of water is one of the causes of the conflicts between two countries/states. Conflicts can arise from the use of common water resources or through polluting these common resources. Conflicts are increasing over rivers and river basins shared by many countries around the world. For example, tussle between India, Nepal and Bangladesh is simmering over the best uses of water. India and Nepal want to exploit the basin's huge hydroelectric power generating potential, whereas Bangladesh wants the water managed in such a way also minimise flooding during monsoon months and water shortages during dry months.

Similarly, water conflicts are between Karnataka and Tamil Nadu, which border the Cauveri river. There had been agitations over water.

Upstream versus downstream disputes are becoming more common. We always criticize those living upstream for all our water problems, little realizing that someone lives downstream of us also, criticizing us for their problems. People abstracting water from water courses and those polluting than ignore the basic tenet that others also live downstream. Upstream industrial dumping, diversion of water for irrigation, damming for hydropower and the increased consumption of growing human population mean that downstream water supplies are becoming more polluted and less available.

In India, several rivers are under dispute between states with regards to sharing of water.

Causes of Conflicts

Water conflicts may arise due to following two reasons:

1. *Conflict on use of water:* Activities, such as construction of a dam or the channeling of the river flow leads to conflict. Most running water conflicts are similar, where the upper-lying state controls the quality and quantity of the flow water through hydrological means.
2. *Conflict on pollution:* Besides production of electricity irrigation and shipping, rivers and lakes also serve an industrial purpose. Not only do they act as reservoirs for the supply of fresh water, but also as a means of disposing waste water and industrial rubbish. With the increasing decline in the quality of the water crossing borders, the problem of cleaning the water takes the shape of a conflict.

8.11 DAMS—BENEFITS AND PROBLEMS

Water resources development is essential in developing countries, where population is increasing at a fast rate and economy needs accelerated growth to provide basic needs as well as high standard of living to its people. Accordingly, construction of a number of large dams has been taken up the earth a view to exploit the water resources for multiple benefits. But some detrimental effects are also associated with them. We shall now discuss benefits and problems of these large dams.

A. Benefits (Positive Impacts) of Large Dams

- Irrigation.
- Hydropower generation—source of power.
- Domestic and industrial water supply.
- Flood control.
- Navigation.
- Help in afforestation.
- Fisheries development.
- Tourism and recreational facilities.
- Employment generation.
- Development of agro-based industries.

– Improving general socio-economic standard of people.
– Reduces the impact of drought.
– Provide habitat for animals and birds due to increased availability of water and more vegetation.

B. Problems (Negative Impacts)

– Submergence of land and property causing displacement of people.
– Spread of water related diseases, like malaria and other waterborne diseases.
– Loss of flora and fauna in submergence area.
– Deforestation.
– Submergence of historical monuments.
– Reservoir sedimentation.
– Water-logging and soil-salinity in command area.
– Land loss and change in landuse.
– Downstream impacts.
– Landslides.

It is desired to have detailed studies conducted to assess the extent of impact and analysing mitigating measures, with reference to above mentioned possible negative impacts. Remedial actions are then taken to mitigate these impacts.

8.12 RAIN WATER HARVESTING AND WATERSHED MANAGEMENT

Rainwater Harvesting is the collection and storage of rain water from roofs or a surface catchment for future productive use.

A **watershed** can be defined as an area from which run-off resulting from precipitation flows past a single point into a large stream, river, lake or pond.

Watershed management is the management of a single unit of land with its water drainage system.

'Rainwater harvesting' as well as 'watershed management' have been discussed in detail in the chapter on "Social Issues" in the book.

QUESTIONS

1. Explain the significance of water resources.
2. Discuss the problems involved in utilizing underground water.
3. Discuss about the inventory of the world's water resources, with specific status of availability of water in India.
4. Discuss the necessity of conserving water and its resources. Explain the methods of their conservation.
5. Explain the different types of water resources.

6. Write short notes on:
   (a) Surface water
   (b) Underground water
   (c) Uses of water resources.
7. Discuss impact of over-utilization of water resources.
8. What is water-logging? Mention various reasons of water-logging. How it can be controlled?
9. Write a detailed note on 'conservation of water'.
10. Define floods. What are the causes of floods?
11. Discuss the various impacts of floods.
12. Discuss the different flood control measures.
13. Explain in brief:
    (a) Structural measures for flood control.
    (b) Non-structural measures for flood control.
14. Write short notes on:
    (a) Flood proofing.
    (b) Flood plain zoning.
    (c) Flood preparedness.
    (d) Flood fighting measures.
15. Explain the term 'Drought'. Discuss the various losses due to droughts.
16. What are the major causes of drought? Discuss the various drought control measures.
17. Write a detailed note on 'Conflicts over Water'.
18. Explain benefits of large dams.
19. Discuss problems due to large dams.
20. Write short notes on:
    (a) Rain water harvesting.
    (b) Re-cycling of water.
    (c) Positive and negative impacts of large dams on environment.
21. Enumerate the various techniques that are used for rain-water harvesting.

9
Chapter

MINERAL RESOURCES

9.1 INTRODUCTION

The *mineral* is a naturally occurring homogeneous solid that is an inorganic substance, having an orderly atomic structure and definite chemical composition and characteristic physical properties. The earth contains several rocks, which are of great utility to humans in their day-to-day use. These rocks in turn, are made of inorganic substances, called minerals.

Mineral resources are broadly defined as elements, chemical compounds, minerals or rocks that are concentrated in a form that can be extracted to obtain a usuable commodity. The finite stock of minerals on the earth is non-renewable and the geographical distribution of minerals is unequal.

Some mineral deposits are formed by biological processes and many minerals are formed under conditions of biosphere that have been greatly altered by life. The minerals are used to make various things required for the use of human beings which essentially come from the earth's crust. The minerals are the most common solid materials found on the earth. The land as well as oceans on the earth, all rests on a layer of rock, made of minerals.

There are about 3000 minerals, but only about 100 of them are common, as most others are most difficult to find. Many people use the term mineral for any substance taken from the earth. Such substances include coal, petroleum, natural gas and sand. Certain substances in food and water, such as calcium, iron and phosphorous are also called minerals. But mineralogists (the scientists who study minerals) do not consider any of them as minerals.

As per mineralogists, mineral is a substance which has following features:

(i) A mineral is found to occur in nature.

(ii) A mineral is made up of substances that were never alive. Therefore, coal, petroleum and natural gas are not minerals, because they were formed from the remains of animals and plants.

(iii) A mineral has a definite chemical formula, irrespective of wherever it is found.

(iv) The atoms of a mineral are arranged in a regular pattern, and form solid units, called crystals.

As, composition and structure of a mineral is fixed, its physical properties are fixed. Minerals vary greatly in appearance and properties. Some minerals have glass-like surface that sparkle with colour; others look dull and feel greasy. The softest ones can be scratched by a finger nail (*e.g.*, talc); while the hardest ones can scratch and even cut glass (*e.g.*, diamond). The main characteristic of minerals are: external appearance and internal structure, lustre, cleavage, hardness, colour etc.

9.2 TYPES OF MINERALS

Generally, minerals are classified as:

1. Metallic Minerals, and
2. Non-metallic Minerals.

1. *Metallic Minerals.* These are those minerals from which we get metals, like iron, copper, aluminium, zinc, lead, manganese etc. These minerals are found in the earth's crust mixed with various impurities. Such materials so mined from the earth are known as *ore.* Minerals are extracted from the ores. Some of the important ores are: *Magnetite* (producing iron), *Bauxite* (producing aluminium), *Chromite* (producing chromium) etc.
2. *Non-metallic Minerals.* These are those minerals which do not produce metals. These include minerals, like rock salt, clay, talc, native sulphur, gypsum asbestos, limestone etc.

   Out of the above minerals, some minerals are known as *economic minerals,* which include all those minerals which are of significant economic value. These include all the important metal producing minerals and other minerals having large industrial value, such as *mica* (used as an insulator in electrical goods), *asbestos* (used as heat insulator), *corundum* (used as an abrasive), *chromite* (used as refractory material) etc. Rocks, which are made of minerals are also used as construction materials, and building stones, like granite, marbles etc. Economic materials also include minerals which are the sources of economic energy, like *uranium*, thorium etc. The *fossil fuels* like coal, petroleum and natural gas are also mined from the Earth's crust and are valuable resources. These fossil fuels, though strictly are not minerals, but for all practical purposes, are considered as economic minerals.

From the economic importance point of view, mineral resources can be divided into four main groups; namely: (*i*) *essential resources,* like soil and water; (*ii*) *energy resources,* like fossil fuels and nuclear fuels; (*iii*) *metal resources,* like iron, aluminium, ornamental metals, such as gold, silver etc.; and (*iv*) *industrial* minerals, such as salt, asbestos, sand etc.

Mineral deposits have two geological characteristics that make them a real challenge to modern civilisation. First, almost all the minerals are non-renewable resources. Second, minerals deposits have a place value.

9.3 AVAILABILITY OF MINERALS

1. In the World

As already mentioned, about 3000 minerals, but only about 100 of them are common, as others are difficult to find. About 118 elements are known to be existing in nature, but only eight of them constitute about 98.59% (including 46.60% oxygen) portion (by weight) of the Earth's crust, the remaining about 110 elements are found only in trace concentrations, altogether constituting only 1.41%, as shown in the table below. The table also shows the estimated quantities of available deposits of these minerals alongwith their average annual extraction (production) and also the life expectancy of the available stocks considering 2% annual growth.

S.No.	*Mineral / Metal*	*Reserves (estimated deposits)*	*Average annual extraction*	*Life expectancy in years*	*Remarks*
1.	Bauxite ore (for Aluminium)	25 Billion Tons (Bt)	0.0124 Bt.	81 years	Reserves measured in terms of Bauxite ore
2.	Copper	340 Bt	0.012 Bt	22 years	
3.	Iron	74 Bt	0.56 Bt	65 years	
4.	Lead	64 Mt	3.07 Mt	17 years	
5.	Nickel	46 Mt	0.016 Mt	15 years	
6.	Zinc	190 Mt	7.75 Mt	20 years	
	Fossil Fuels				
1.	Coal	987 Bt	4.56 Bt	84 years	
2.	Crude oil (Petroleum)	1145 Billion Barrels	23.7 B Barrels	31 years	
3.	Natural Gas	177 trillion cu. m.	2.28 trillion cu. m.	41 years	

The important producers of bauxite in the world are: Australia, Jamaica and Brazil. Copper is found mainly in Chile, USA, Russia, Canada, Zambia, Poland and Peru. Major producers of Iron ore are Russia, China, Australia, Brazil, USA and India.

2. In India

India has significant mineral deposits and produce large variety of minerals including fossil, metallic and non-metallic minerals. The metallic production is accounted for iron-ores, copper ore, chromite and/or zinc concentrates, gold, manganese ore, bauxite ore (producing aluminium), lead concentrates. Among the non-metallic minerals, more than 90% of the aggregate value is shared by limestone, magnetite, dolomite, gypsum, phosphorite, fluorite etc.

India is world's largest producer of mica. India ranks 3rd in production of coal, chromite and barytes; 4th in iron ore; 6th in bauxite (aluminium ore) and manganese ore; 10th in aluminium metal; and 11th in crude steel in the world. India, therefore, has vast mineral resources, like mica, chromium, iron, aluminium, manganese, titanium, coal, thorium, magnesium etc. But we have deficient in some minerals, like nickel cobalt, antimony, tin, lead, zinc etc. while some others are highly deficient, like petroleum. Lot of emphasis is being given on exploration of petroleum and natural gas, especially in our off-shore exploration programmes.

South Bihar and Odisha, situated on the north-eastern parts of peninsular India (south of Vindya mountains), have large concentrations of mineral ores, accounting for nearly three-fourths of the country's coal deposits as well as rich deposits of iron ore, mica, copper, aluminium and manganese, which are also found over the rest of the peninsular India and in parts of Assam and Rajasthan.

Important fuel minerals, metallic minerals, and non-metallic minerals of India and the states where they are mainly available, are shown in the table below:

S.No.	*Mineral*	*Indian States where Found*	*Remarks*
(A)	*Mineral Fuels*		
1.	Coal including Lignite	Bihar, West Bengal, Tamil Nadu, Gujarat, Odisha and Andhra Pradesh	
2.	Crude oil (Petroleum)	Assam, Gujarat, Rajasthan	Petroleum and Natural gas are available in large quantity from off-shore
3.	Natural Gas	Assam, Gujarat	Petroleum and Natural gas are available in large quantity from off-shore
(B)	*Metallic Minerals*		
1.	Bauxite	Bihar, Odisha, M.P., Maharashtra, Tamil Nadu, Karnataka, Gujarat	Bauxite is used for producing aluminium besides for manufacture of chemicals, and also, used as refractory and abrasive material.
2.	Chromite	Odisha, Karnataka, Bihar, Tamil Nadu, Maharashtra	Used for producing chromium. Also, used as refractory material.
3.	Copper	Bihar, Rajasthan	In Singhbhum and Khetri Copper belts.

(Contd...)

4.	Gold	Karnataka	Some Gold is also found in Andhra Pradesh and Sikkim
5.	Iron	Bihar, Odisha, MP, Karnataka, Gujarat and Goa.	Main ore is *Hematite.* Other ores are Magnetite, *Siderite* and *Limonite.*
6.	Lead and Zinc	Rajasthan	The Lead and Zinc ores occur in nature in close association with each other. Small deposits are also available in T.N., AP, HP and J & K.
7.	Maganese	MP, Maharashtra, Kerala, Bihar, Odisha and Karnataka.	This is used for making alloys.
8.	Nickel and Cobalt	Rajasthan, Karnataka, J & K and Odisha.	
9.	Silver	Karnataka, A.P., and Rajasthan	
10.	Titanium	Kerala, Maharashtra, A.P., Odisha.	India is the second and largest producer.
11.	Tungston	Rajasthan, M.P., W.B., Tamil Nadu and Maharashtra	
12.	Thorium	Kerala	
13.	Uranium	Rajasthan, A.P., and Bihar	Uranium is required for producing atomic energy.
(C)	*Non-metallic Minerals*		
1.	Asbestos	Rajasthan, Bihar, Andhra Pradesh and Karnataka.	Used for heat insulation
2.	Arsenic	West Bengal, Uttaranchal, J & K.	
3.	China Clay	Rajasthan and W.B.	
4.	Diamond	MP	This is the hardest known minerals.
5.	Graphite	Odisha, Rajasthan, M.P., Kerala and A.P.	It is used as lubricant, in the pencils, crucibles and electro-typing works.
6.	Gypsum	Rajasthan, TN, U.P. and Maharashtra	Used as fertiliser, for making plaster of paris, for fire-proofing works and as a retardant in cement.

(Contd...)

7.	Limestone	M.P. and T.N.	
8.	Mica	Bihar, Rajasthan, A.P. and Kerala	
9.	Magnesium	T.N., Karataka	Used for making light-weight alloys.
10.	Sulphur	T.N.	
11.	Quartz	A.P. and Karnataka	
12.	Silica sand	U.P. and Gujarat	
13.	Talc	Rajasthan, MP., Bihar, AN, Karnataka and Odisha	Used in manufacture of talcum powders, ceramic material and paints.

9.4 USES AND EXPLOITATION OF MINERAL RESOURCES

Extraction, processing and disposal of minerals affects the environment adversely. Mining destroys the large stretches of land, as most mineral deposits are in the form of ore and are taken out by mining. Extraction of one element usually wastes a number of other elements. Worldwide smelting of minerals for extraction of metals introduces an enormous quantity of sulphur, nickel, mercury etc. into the environment.

The mining and extraction activities produce dust and many toxic gases which are inhaled to cause a number of lungs and other diseases. These activities also create pits and sears on the earth surface leading to destruction of biotic community. Besides, formation of large cavities beneath the earth's surface, which may cause surface collapse and subsidence. Disposal of certain waste materials into water and soil pollutes them which may be hazardous to organisms.

There is a mad race going on amongst various countries to extract and sell as much mineral resources as one can, without considering the interests of future generation, although it is the well known fact that materials are finite in quantity and are non-renewable sources, since their formation has taken millions of years. Over the past century, our consumption of various metals have tremendously gone high. The developed rich countries are the biggest consumers of minerals. The details given above indicate that just after next 25 to 50 years, we may be left with no worthwhile stocks of any important mineral or metal to meet our future requirements.

Here, it is also pertinent to point out that, mining has created some of the largest environmental disaster zones in the world. The mining and processing activities result into large quantities of waste materials. The mining activities consumes large amount of energy in handling overburden and ore which require the consumption of huge amount of energy, produced generally by thermal power houses, resulting into air pollution.

9.5 ENVIRONMENTAL EFFECTS OF EXTRACTING AND USING MINERAL RESOURCES

Mineral resources involve mining, processing and use activities. The impact of all these activities on environment are discussed in this article. Mining helps to extract costly and useful resources from the earth. Mining provides as fuels, metals, chemicals fertilisers, clays, rock and stones.

1. *Environmental damages caused during mining include:* Water pollution, air pollution, land subsidence, and land degradation including deforestation, ground vibrations and noise pollution, mining accidents, health hazards etc.
2. Most of the ores and minerals are required to be processed before they become usable. This processing involves transportation, purification, and manufacturing of usable metals or materials. All these *processes cause several adverse environmental impacts,* including air pollution, water pollution, soil pollution, solid waste generation, noise pollution, heat generation, and several health hazards.
3. Even the use of materials, involve their transportation to individual users, eventual use and ultimate discard and solid waste generation. This *process of use also causes environmental damages* which include air pollution, noise pollution, generation of solid wastes etc.

9.6 MAJOR ENVIRONMENTAL IMPACTS OF MINING

Mining whether (*i*) surface or open cast mines, or (*ii*) underground mines, involve following major types of environmental impacts:

1. *Air pollution:* Mining adds *gaseous pollutants* to the air. These pollutants, like sulphur dioxide, oxides of nitrogen, carbon monoxide etc. are originated in the mining activities. The *dust particulates* are produced during ore/coal handling, blasting, and transportation, *mine fires,* which are generally associated with the underground mining and also in solid waste disposals, also pollute the atmosphere.
2. *Water Pollution:* Mining causes discharge of acid-mine water into surface water bodies, like ponds, rivers etc. Thus, it contaminates surface and groundwater, leading to loss of aquatic life. Mining also adds toxic radioactive substances, like selenium, nickel, fluorine etc. to the water bodies. Sometimes, mining changes the hydrology of the area by changing the river course and flow discharge, which affects the agriculture and flora and fauna of the area. Mining activities, sometimes, lower the water-table of the area.
3. *Land Subsidence:* Land surface subsides due to the underground mining, which may cause damages to the buildings, surface drains, highways, water and sewage lines etc. existing on the ground.
4. *Land Degradation:* Open cast (surface) mining removes top soils and results into deep and large excavations and pits. This results into the unusable and denudation of vegetation of large areas and causes continuous soil erosion and its ill-effects like silting and degradation of streams and water bodies. Further, the waste dumps,

created by the surface as well as underground mining could lead to erosion and slips (landslides), if not attended and treated timely. Such landslides and flowing debris may sometimes cause serious damage to life and property and may also result in disasters.

5. *Deforestation:* For extraction of valuable mineral deposits, large scale felling of trees from forest areas have to be adopted. During mining operations, large amount of dirt, dust and overburden spreads on the nearby areas, thereby resulting into large scale damage to the nearby vegetations. Large scale deforestation, besides causing oxygen deficiency, drastically changes the climatic conditions, like rainfall, temperature, humidity etc. of the area which may result in large scale deaths of wild animals and birds living in the forest, causing imbalance to the natural ecology of the area.
6. *Noise and Ground Vibrations:* Mining operations cause large scale noise and vibrations induced into the ground, which interferes with the hearing and speech communication and results into distraction, annoyance and ultimately hearing loss, headache and high blood pressure.

   Heavy ground vibrations introduced by heavy blasting of areas may damage the nearby structures, buildings etc. besides causing irritation to the residents. Noise, vibration and dust pollution affects women the most during pregnancy.
7. *Mining Accidents and Health Hazards:* As mining is an hazardous occupation, the safety of the workers working in mines is an important environmental consideration. In all underground minings, rockfalls, roof falls, flooding, inadequate ventilation and fires are the greatest hazards. Therefore, underground mining is more hazardous than surface mining. Communities living around the mine sites are constantly vulnerable to mining accidents, as mining companies use explosives for blasting, which results in houses cracking and collapsing.

   Mining also poses several health hazards to the miners as well as to the persons residing in the nearby areas. The dust produced during mining operations is highly injurious to health and causes diseases related to lungs not only to the miners but also to the nearby residents. Some other common diseases found in the area caused due to unchecked pollution and high levels of toxicity are tuberculosis, cough and cold, malaria, diarrhoea, skin diseases, staining of teeth, joints pain etc. The fumes generated by incomplete dynamite explosives and methane gas emanating from coal strata are also very harmful. In mining of radioactive metals, like uranium, radiations are always life threatening.

9.6.1 Sand Mining

Sand is an important mineral for our society. It helps in protecting the environment as it acts as buffer against tidal waves and storm, habitat for crustacean species and marine organisms. Sand is used for making concrete for making roads and buildings etc.

Sand mining is the process of removal of sand and gravel. When the demand for sand increases, the sand mining becomes an environmental issue.

Hon'ble Supreme Court has prohibited sand mining without the requisite clearance from the concerned authorities, who places the limits on the quantities that can be mined at a particular site. But the sand is being illegally mined to meet the increasing demand of construction industry.

The ban on sand mining was imposed, because sand is useful from numerous enviromental reasons, such as:

(1) Sand acts as an aquifer.
(2) It acts as natural carpet on the bottom of the river, as stripping of this layer may cause downstream erosion and may result in change in channel bed and habitat type.
(3) It forces the river to change its course.
(4) Revenue loss to the exchequer.
(5) It acts as a buffer to the river bed.
(6) Depletion of groundwater tables.
(7) Sand mining has adverse impact on the habitats of micro-organisms.
(8) Increased river erosion.
(9) Might damage to roads and bridges. With the removal of sand, boulders may come down with the river water and strike the bridge pillars and may damage the pillars or side of the dam.
(10) Change of river course is a threat to agriculture.
(11) Damage to coastal ecosystem. Illegal mining in beaches may lead to erosion along the shoreline.
(12) Lesser availability of water for agricultural, industrial and drinking purposes.

Sustainable Mining

To decide, where to mine and where to prohibit mining, district survey report for each district is prepared, taking the river in that district as one ecological system. For this purpose, remote sensing data of ISRO is taken into consideration. For sustainable mining, it is allowed to mining out material only that much which is deposited annually. In this matter the District Collector is to be assisted by the district level expert appraisal committee. This committee is authorised to grant environment clearance upto 5 hectare of mine lease area.

9.7 CONSERVATION OF MINERAL RESOURCES

Material resources as exhaustible and are non-renewable. Therefore, it is very essential to conserve these resources. Once exhausted, the minerals cannot be replaced over long geological period which may be in millions of years. Our consumption of various metals have tremendously gone high due to fast urbanisation and modern lifestyles.

There is a mad race going on amongst various countries to extract without looking into the future, in spite of the well known fact that the minerals are finite in quantity. In this modern era of development, minerals are being over-extracted from inside the body of the earth, through mining activities, which are extremely harmful to human health as well as the environment in general. Advancement in mining technology is also responsible for over-exploitation of mineral resources at an accelerate rate.

The developed rich countries are found to be the biggest consumers of minerals. Such a consumption pattern is on excessively high side, and if things move like this in other developing countries also, we will be consuming the entire stocks of minerals of the world in just 25 to 50 years, and we may not be left with any worthwhile stocks of any important mineral or metal to fulfill needs of our future generations. The present rate of extraction of minerals is absolutely non-sustainable.

Mineral resources can be conserved by adopting following measures:

1. Economy in use of mineral resources.
2. Making finished products of more durable and long lasting quality.
3. Reuse and re-cycling of metals as this helps in reducing the waste.
4. Scarce minerals can be saved by substituting them with minerals which are cheaper and are found in abundance.
5. Search for new deposits.
6. Protection and economy of existing mineral deposits.

QUESTIONS

1. What is a mineral? Define the terms: mineral, elements and compounds.
2. What are the common elements of the Earth's crust?
3. How are minerals classified for a common man?
4. Write a detailed note on 'use and exploitation of mineral resources'.
5. Write down a brief note on the availability of important mineral in the world as well as in India.
6. Describe in detail the necessity of conserving mineral resources.
7. How does the mining activities effect our environment?
8. Discuss the environmental effects of extracting and using the mineral resources.
9. Explain the various environmental impacts of mining.

10

Chapter

Land (Soil) Resources

10.1 INTRODUCTION

Land is an important resource as humans live on it and obtain most of their needs from the land. Land is put to different uses, such as cultivation, grazing of animals, industries, buildings, roads and railways, mining, forest etc. The proportion of land put to different uses varies from one region to another and also from time to time in the same region. For example in recent few decades, forests have been cleared for cultivation, mining etc. resulting in reduced forest area.

Use of the land depends upon its physical features. The slope of the land, soil cover, availability of surface or underground water, the nature of rocks and minerals present are some of factors which determines the type of land use in an area. Relatively level land with adequate water facilities favours cultivation of crops, while occurrence of valuable minerals encourage their mining.

Proper planning of land-use with reference to the nature of land and the needs of community would provide maximum returns. Improper use of land, such as clearance of forest for cultivation may result in soil erosion. Reclamation of low lying land or marshes has resulted in making such land available for construction of buildings or other uses.

The land should be used in a prudent and scientific manner to obtain optimum benefits. The damage of the land usually results in irreversible condition. Therefore, this needs for an appropriate landuse planning based on suitability and capability adjudged through determination of carrying capacity. As the land suitable for practising agriculture or for growing trees or for maintaining as pastures and grasslands is limited, therefore, the management strategy must involve the prevention of (*i*) encroachment on forestland and (*ii*) use of fertile agricultural lands for non-agricultural purposes.

Careless use of land, damages soil that results into reduction in quality and quantity of woodland, grassland, cropland, and soil erosion and degradation of watersheds and catchments, deforestation and desertification.

Due to pressure of population growth, industrialisation and urbanisation, the land is under stress and good agricultural land is going to industrial estates and for urban development. Thus, valuable cropland is being lost for ever.

Land Use Pattern in India

Land Use pattern	*Area in Mha*	*Percentge*
1. Agriculturable land	140.23	41.88
2. Forest land	74.84	22.35
3. Permanent pastures and grazing land	12.47	3.73
4. Miscellaneous tree crops and grooves	3.96	1.18
5. Non-agricultural uses (settlements, industries etc.)	17.53	5.24
6. Barren and Unculturable land	21.94	6.55
7. Culturable wasteland	17.15	5.12
8. Others	46.69	13.95

10.2 SOIL RESOURCES

Soil is a layer of unconsolidated material at the surface of the earth which has been derived from rocks and organic matter through agencies of decay and disintegration. Nature takes thousands of years to build up soils and various factors of environment are responsible for their formation. A deep soil cover with a high degree of fertility may result into rich and prosperous agricultural economy, which would ultimately sustain a large human population.

The soil consists of mineral matter, such as sand and clay as well as organic matter, such as decayed leaves, flowers, dead tissues of organisms, minute bacteria and earthworms. It also contains varying amount of moisture and air between soil particles.

Soils are formed by physical, chemical and organic changes which take place continuously in the soil layer. Soil formation is a slow and long term process of complex interactions between climate, topographic and biological factors. Bare rock surfaces lead to physical and chemical disruption, called *weathering.* Physical processes of weathering include action of water, temperature and processes as wetting, drying, heating, cooling, freezing, solution and sand blast etc. The chemical process of weathering include hydration, hydrolysis, oxidation—reduction, carbonation etc. about 100 to 200 years may be needed for the formation of about 2 to 2.5 cm soil, depending upon the climate and soil type.

Parent rocks play a definite role in the case of derived soils. For example, soils derived from lava rocks are of black colour. Similarly, sandy soils are derived from sandstone, and alluvial from silty limestone. The renewal of soil is a very slow process, the soil becomes a non-renewable resources, although, usually soil is considered to be a renewable resource.

As a matter of fact, alongwith sunlight, air and water, a soil supports all living beings on the earth. Plants and vegetations have their roots in the soil, from which they absorb life-sustaining water and nutrients.

Similarly, all animals including humans, in turn, depend on these plants for food either directly or indirectly. Foods, vegetable oils and fats, fodder for livestock, wood, fibre and various other indispensable agricultural products come from soils.

Properties of Soils and Fertility

The **productivity** of the soil depends upon a number of physical properties, namely the size, shape and arrangement of its particles, the volume and form of its pores, and the effective depth of the soil. The flow and storage of water, the movement of air and the ability of the soil to supply nutrients to plants depend upon its physical properties.

Permeability of the soil depends upon its structure. Sandy soils have rapid permeability while clay and silt have slow permeability. Coarse soils have lower water holding capacity while black soils have higher capacity. A soil with good drainage is more productive than a soil with poor drainage. Steep slopes are more vulnerable to erosion than gentle slopes.

The **fertility** of soil depends upon the presence of nutrients. A large number of elements, namely carbon, hydrogen, oxygen, nitrogen, phosphorous, potassium, sulphur, magnesium, calcium, iron, manganese, zinc, copper, and molybdenum are required by a plant. The first three are supplied to the plant from air and water, and the others needed in small quantities are usually present in the soil. The proportion of nitrogen, phosphorous and soil. These elements, therefore, are added to the soil in required quantities for higher production. Besides, a large number of soil organisms also play an effective role in the fertility of the soil. These include bacteria, fungi, earthworms, ants and other insects and animals. They multiply very quickly under favourable conditions and add nutrients to the soil.

10.3 SOIL CLASSIFICATION

Brief descriptions of the major soil groups are given hereunder:

1. *Alluvial Soils:* This is the largest and most important soil group which contributes the largest share to its agricultural wealth. They are depositional soils transported by the streams and winds. They are largely sandy loam in texture and in certain areas, they are mixed with silt and clay. These are conductive for cultivation of variety of crops especially cereals, pulses, cotton, sugarcane and jute.
2. *Black Soils:* These are popularly known as 'black cotton soils,' and are the typical soils developed on basaltic rocks of the Deccan plateau. These cover about 30% of the total area of the country. The colour of the soil ranges from deep black to grey, and contains clay minerals. These are very fine grained and contain a high proportion of calcium and magnesium carbonates. These soils are sticky and swell when wet, and strink when dried large and deep cracks are formed on drying.

   These soils are rich in iron, lime, magnesia and alumina, and poor in phosphorus, nitrogen and organic matter. They have low permeability, high water-holding capacity, and are quite fertile and supportive to crop growth. In these soils, cotton, cultivation is widespread and hence, the name black cotton soil.
3. *Red Soils:* These soils are sandy and reddish yellow in colour. Red colour is largely because of the presence of iron oxides, but it looks yellow when it occurs in hydrated form. Very often the surface layer is red and the lower horizon yellow. The texture

ranges from sandy to clayey. The fine grained red and yellow soils are normally fertile, whereas coarse grained found in dry upland areas are poor in fertility. These soils are more wide-spread in peninsular India. Wheat, millet, cotton, potato and many other coarse grains are widely grown in these soils. Besides these, chillies and groundnuts are also grown in these soils.

4. *Laterite Soils:* These are typical soils of the tropical regions with heavy seasonal rainfall, which promotes leaching of the soil. With rain, lime and silica are leache away and soils rich in iron oxide and aluminium compound are left behind. The organic matter, nitrogen, phosphate and calcium are low in these soils, while iron oxide are enough for crop growth. Application of large quantity of manures and fertilizers are required for making the soils useful for crops.

   Laterite soils are widely cut as bricks for use in house construction. Cotton, rice, wheat, pulses, tea, rubber and coffee are grown in these soils.

5. *Desert Soils:* These soils are blown sand with arid climate. These areas receive little rain. A large part of the arid region, belonging to western Rajasthan, Haryana, Punjab and lying between the Indus river and the Aravali range, constitutes desert soils. The fertility of these soils improve from north-west to north-east. These soils range from red to brown in colour and are generally sandy and alkaline. In some areas, salt content is so high that common salt is obtained by evaporating the saline water. Due to dry climate, high temperature and accelerated evaporation, they lack in moisture and humus content. They also lack in organic matter, and are generally used for the cultivation of coarse grains, like jowar, bajra, ragi etc. and oilseeds.

6. *Saline Soils:* These contain a large proportion of sodium, potassium and magnesium, and thus, they are poor and infertile and do not support any vegetative growth. They acquire more salts largely because of dry climate and poor drainage. They occur in arid and semi-arid regions, and water-logged and swampy areas. Their structure ranges from sandy to loamy. Saline soils are more widespread in western Gujarat, deltas of the eastern coast and Sunderban areas of West Bengal. In the Rann of Kutchh, the southwest monsoon brings salt particles and deposits there as a crust. Seawater intrusions in the deltas promote the occurrence of saline soils.

10.4 LAND DEGRADATION

Land degradation is the change in the quality and fertility of the soil due to natural and anthropogenic (man-made) activities. One important constituent of a soil is *humus*. Decaying remains of plants and other organic matter form the humus. It helps in retention of moisture and provides space for oxygen to circulate through the soil. Soil and its humus takes thousands of years to develop. Anything that damages the humus, or any action that loosens the soil, eventually leads to degradation of soil.

Due to increasing population, more and more land is required for agriculture and forestry, but good land is shrinking, both in quantity and quality. Increasing misuse of land resources through shortsighted development policies has resulted into wasteland.

The **major factors** responsible for soil (land) degradation are:

– Landslides
– Soil erosion
– Desertification
– Soil pollution

10.5 LANDSLIDES

A *landslide* is the rapid sliding of large masses of bed rocks or regoliths. Landslide is a natural hazard and is defined as "any downward movement of mass of bedrock or regolith under the influence of gravity." It is a rapid sliding of large rock masses beginning their descent as unit of blocks, without internal flowage. The landslides often occur in the hilly areas.

Whenever mountain slopes are steep, there is possibility of large disastrous landslides. Landslides are triggered by sudden rock failures or earthquakes. They can also result when the base of a slope is over-steepened by excavation or river erosion. The landslide becomes dangerous especially when there is heavy rainfall or snowfall on the steep hill sides which makes the rocks to slip and break. It is the rocks that break and carry with it the soil and debris. A major cause which tiggers off the landslide is the weight of the overlying material and the presence of a lubricating material, like water. Freezing and thawing of the rocks on mountain slopes cause them to break and roll down the slopes. The overbearing weight of snow or water which has seeped into the soft permeable rocks also lead to slipping and breakage of hill slopes. The rivers and roads often get blocked by the debris brought by landslides.

The extent of landslides depend on the steepness of the slope, the bedding plane of rock, the amount of vegetation cover and the extent of folding and faulting of the rocks.

Therefore, the main factors responsible for landslides are:

(a) *Natural Factors.* As discussed above these include heavy and prolonged rainfall, wind, temperature, steep slopes, soil structure and texture, earthquakes, volcanic eruptions etc.

(b) *Anthropogenic (Man-induced) Factors.* These include:
- Clearing of vegetation on slopes
- Excavation of slopes
- Overloading of slopes by wastes and construction materials
- Pumping of ground water
- Mining activities
- Blasting for construction and mining activities
- Change in land-use

10.6 SOIL EROSION

Soil erosion may be defined as the detachment and removal by transportation of the top soil layers from a soil mass. The transportation of soil takes place due to the action of wind, water in motion or by the beating action of the raindrops.

Soil erosion refers to a series of processes leading to soil depletion *in-situ* and the export of rediment towards downstream areas. Soil erosion reduces the fertility of the left out ground soil, because it is the top soil layer, containing humans, which is most fertile. When the top soil depletion due to erosion is more than the formation, the land loses its fertility.

Types of Soil Erosion

Soil erosion may be broadly classified into following two groups:

1. *Geological (Natural) Erosion:* It takes place under the cover of vegetation through the action of wind, water, gravity etc. It is a natural (or normal) and slow process and the loss of soil is compensated by the formation of new soil under the natural weathering process. It is the natural physical phenomena which have helped in shaping the present form of the earth's surface. It is a continuous, slow, but constructive process, which has resulted in building up of flood plains and coastal plains, and formation of some of the most fertile valleys of the world, such as Indo-Gangetic valley. This erosion does not disturb the equilibrium between the soil loss and the soil formation.
2. *Accelerated Soil Erosion:* Human activities, like deforestation, brick making overgrazing, shifting cultivation etc. usually cause deterioration of the top surface of the land, due to which the soil erosion caused by the natural erosive factors, like water and wind, gets accelerated. Thus, the natural balance existing between the soil formation and soil erosion is disturbed. The removal of soil (erosion) takes place at a much faster rate than it can be built-up by the soil forming process. This is known as *accelerated erosion* or abnormal erosion. Due to removal of top fertile soil, the left out ground soil may be severally degraded, making it an infertile wasteland.

Classification Based on Agency

The soil erosion, whether normal or accelerated requires: (a) detatchment of soil particles, (b) transportation of these detached particles, and (c) deposition of these particles of some other place. In order to carry out all these things, certain medium or agency is required. These agencies are water, wind and gravity.

(a) Water Erosion

The water erosion process starts as soon as the rain starts. This erosion action of the water has been found to be most destructive at the places where the natural vegetative cover have been removed and the bare land surfaces are exposed to the direct action of the storms. The removal of soil particles depend on the magnitude, intensity and duration of rainfall. It is also affected by:

(i) the amount and velocity of surface flow,

(ii) the nature of soil,

(iii) the nature and extent of vegetative covers,

(iv) slope of the land surface, etc.

The two **principle erosive agents** that become active during rainstorm are:

(a) falling raindrops, and

(b) flowing run-off. Raindrops falling on the land surface cause detachment of soil particles and are subsequently transported with the flowing water.

The energy of the falling water is applied from above and is utilised in detaching the soil particles. When a rain-drop strikes the surface, the soil particles are dislodged, and these dislodged soil particles are scattered in all the directions. The amount of material which is transported from a hilly catchment is always more than that from a flat catchment.

The energy of the surface runoff is applied parallel to the surface and is utilised in transporting the dislodged soil particles.

(b) Wind Erosion

Wind erosion is the detachment and transport of soil particles by the forces of wind. (Wind is the movement of atmospheric air in a direction parallel to the earth's surface.) In arid regions of low rainfall and high day temperatures, the wind velocity is invariably very high, which causes enormous soil erosion. Like water erosion, wind erosion also undergoes three stages, *viz:* (*i*) detachment and removal of soil particles from the weakened or denuded ground soil surface, (*ii*) transportation of eroded soil particles, and (*iii*) deposition of sediment particles when the wind velocity subsides.

Such a removal of loose sand and dust-sized particles, may at certain places leave a denuded surface. The blowing winds when pick up sand particles and move with them as sand laiden winds causes choosen by grinding and rubbing action of area within 30–60 cms in its way.

The larger and heavier particles, such as sands and gravels, which are moved by the winds, but not lifted a more than 30 to 60 cms of the earth's surface, constitute the *bed load* and produces *sandstorms.* Whereas, the finer clay or dust particles which are lifted by the moving winds by a distance of hundreds of meters above the earth's surface, constitute the *suspended load* and produces *dust storms.*

As and when the velocity of the wind is reduced, or the wind is obstructed by some object, it causes the wind to drop the sediment, and form the (*i*) *sand dunes* by the deposition of bed load from the sandstorms, and (*ii*) *loess* by the deposition of suspended load from the dust storms. *Loses* are the blanket deposits of silt and clay formed by the deposition of such material from the suspended load of winds, under the action of falling rains. Initially, these sheet like deposits are porous and friable, but with the passage of time, they become stronger due to the protective grip of the grasses that grow over them.

Wind erosion is primarily caused by the human mismanagement of land resources, such as intensive farming over-grazing by increasing cattle population, deforestation, etc. Such human practices destroy the cohesive properties of the particles of the ground soil, making them susceptible to wind erosion.

Effects of Soil Erosion

Some of the major harmful effects of soil erosion are:

1. Loss of productive soil which affects crop production.
2. Silting of reservoirs and lakes, thus, reducing their capacity and useful life.
3. Damage to agricultural lands, roads and railways due to soil deposition.
4. Loss of nutrients available in topsoil which is removed by erosion.

10.6.1 Measures to Control Soil Erosion

1. Protect the soil surface with a vegetative cover, as plant roots bind the topsoil.
2. Roughen the land surface to reduce wind velocity and trap the drifting soil.
3. Establish barriers of trap strips at an interval to reduce wind velocity and soil drifting. Strip cropping is done perpendicular to the direction of the prevailing erosive winds.
4. Wind breaks and shelter belts should be provided. A *wind break* is a type of barrier for protection from winds. These may be mechanical or vegetative barriers. The barriers provide maximum reduction in wind velocity.

   A *shelter belt* is a longer barrier than a wind break and consists of a combination of shrubs and trees intended for the conservation of soil and moisture and for the protection of field crops.
5. *Biological Measures.* These methods used for controlling water erosion consist of: (*i*) contour cultivation, (*ii*) strip cropping, to be carried out in combination with other farming practices, like crop rotation, (*iii*) maintenance of soil fertility.
6. Sanddune stabilization.
7. *Engineering Measures.* These measures are adopted where the landslopes are more than 2%. These measures are:
   (a) *Bunding* (constructing small embankments or bunds across the slope of the land).
   (b) *Bench Terracing.* This is the most popular soil conservation measure for cultivation on sloping and undulating lands.
   (c) *Checkdams.* These are the structures that are constructed to control the velocity of running water along the gullies. These reduce the steep uniform gradient of the gully by constructing a series of checks. They break the velocity of running water and make it deposit the silt carried by it. They also increase the time of concentration which increases the scope of percolation of water into the soil.

Restoration of the Fertility of Soil

Following are the important means of re-storing the fertility of soil:

1. Animal manuring.
2. Green manuring. Green plant material is called the green manure.
3. Application of inorganic fertilizers. Nitrogen, phosphorus and potassium are largely added to the agricultural fields in the form of inorganic fertilisers.

10.7 DESERTIFICATION

Desertification is a process in which soil loses its productivity as a consequence of degradation of land. The process involves loss of regenerative capacity of soil due to extreme depletion or absence of plant supporting factors. Thus, desertification can be defined as a process through which fertile land is converted into infertile land. In simple words, we can say that desertification is a process which leads to desert formation.

Desertification may result either due to a natural phenomenon linked to climatic change or due to abusive landuse. Removal of vegetative cover brings about marked changes in the local climate of the area. Thus, deforestation, overgrazing etc. bring about changes in rainfall, temperature, wind velocity etc. and also lead to soil erosion. Such changes then lead to desertification of the area. Desertification often starts as patchy destruction of productive land, and increased dust particles in atmosphere lead to desertification.

As a gross mismanagement of natural resources including land, certain irreversible changes have triggered the breakdown of nutrient cycles and microclimatic equilibrium in the soil indicating the onset of desertic conditions.

Controlling Desertification

Following measures should be taken to control desertification:

(i) Ban on cutting of desert vegetation.

(ii) Plantation of ecologically suitable plants in affected areas.

(iii) Development of pasture lands.

(iv) Control of over-grazing.

(v) Stabilization of sanddunes.

(vi) Proper use of available ground water resources.

(vii) Management of land use and farming practices.

(viii) Creating awareness.

(ix) Planting of shrubs in areas where some water is available for irrigation. This decreases the wind velocity near the base of the dune and prevents much of the sand from moving.

(x) Planting of tree fences/grass belts oases and farmlands in windy regions to protect them. Small plots of trees may also be scattered inside oases to stabilise the area.

(xi) Developing windbreaks made from trees and bushes.

(xii) Use of 'solar ovens' in order to relieve the pressure for cutting trees for firewood.

10.8 LAND POLLUTION

Soil is also getting degraded due to pollution or contamination, caused by the discharge of industrial effluents, municipal and domestic sewage-sludge used as a manure, and by chemicals used in agriculture, like pesticides and fertilizers. The disposal of industrial wastes, like flyash or heavy metals are also creating significant problems of land pollution.

Readers may also refer "soil pollution" for details under chapter on "Environmental Pollution".

10.9 OTHER SOIL DEGRADATION FACTORS

Other common factors for soil degradation are:

(i) Salination.

(ii) Refuse for domestic, municipality and industries.

(iii) Water-logging.

(iv) Shifting cultivation.

Salination means soluble soil contamination of the soil. As land use intensifies and water becomes more limited, salivation increases devastatingly. Mismanagement of agriculture induces secondary salination of the soil. In arid and semi-arid regions which practices irrigation, accumulation of salts and alkalinity reduces the productivity of soils.

The refuse from domestic and municipal sources and industries becomes a serious factor of soil degradation. Sewage effluent is also becoming a soil degrading source.

Desertification is the destruction of the biological potential of the soil, which can ultimately lead to desert-like condition. About 3.2 lakh km of land of the four states of Rajasthan, Punjab, Haryana and Gujarat falls in desert environment, largely man-made. It is spreading as a result of the exploitation of limited natural resources by over-grazing pastures, cutting of forests, and poor cultivation—all contributing to wind erosion.

Irrigation from the major dams generally leads to the rise in the water table, culminating eventually in *water-logging* and associated problems. The main causes of water-logging are: deep percolation, run-off and seepage losses from canals.

Shifting cultivation is an age old practice and is still very widespread in the tropical regions. In this, forest is cleared and the debris burnt off and a subsistence crop grown for about two years with the minimum effort by using hand tools.

10.10 SOIL CONSERVATION

Soil conservation is a methodology to maintain soil fertility, prevent soil erosion and exhaustion and improve the degraded condition of the soil. Soil degradation can be checked to a great extent by controlling the factors responsible for it. Thus, the term 'soil conservation' mean 'conserving various properties of the soil' and, thus, protecting the soil in its best possible form.

The first step in any rational solution to soil erosion is to check open cultivable lands on slopes from farming. Lands with a slope of more than 15 per cent should not be used for cultivation. If at all, the land is to be used for agriculture, terraces should be carefully made.

Over-grazing and shifting cultivation have seriously affected the natural cover of land and given rise to extensive erosion. They should be regulated and controlled by educating villagers about the consequences.

Contour bunding, contour terracing, regulated forestry, selective weeding, cover cropping, mixed farming and corporation are some of the remedial measures which should be adopted to reduce soil erosion.

Efforts should be made to prevent gully erosion and control their formation by terracing. In larger gullies, the erosive velocity of water may be reduced by constructing a series of check dams.

In arid and semi-arid areas, efforts should be made to protect cultivable lands from encroachment by sanddunes through developing shelter belts of forests. Efforts should be made to stabilise sanddunes. Thick hedges can be planted on the edges of cropped land.

Lands not suitable for cultivation should be converted into pastures for grazing.

Integrated land use planning is the best technique for proper soil conservation. The land which is deep, level, fertile, well-drained and having good texture is highly suitable for intense cropping. The land which is steep, shallow and stony may be suitable for only for wildlife and other uses that cause no erosion. Landuse can be broadly classified into cropland, pasture land, woodland, wildlife, and recreational land and miscellaneous use. Management of landuse can alter the effect of single type of landuse on a wide range of erosive tendencies. For example, a pasture may have grasses and legumes that were especially selected to provide ground cover and forage (fodder crops as pasture). If the livestock number per hectare is not controlled, over-grazing may kill the vegetation and expose the soil for erosion.

Crop rotation should be adopted, which means to use a series of different crops with some providing more income and some giving more soil protection. Choices of landuse crop rotation and cover crops are all vegetative means of erosion control. Extensive mono-culture *i.e.* growing of one type of crop in thousands of hectare of land at a time year after year is harmful to the soil, especially with crops, like maize, since maize fields loose soil very fast.

Tree crops grown for many purposes ranging from wood and paper products and ornamental uses to fruit products to help restore the fertility of tropical soil. The raindrops are repeatedly intercepted by the leaves and branches at different levels of the tree before falling on the ground, and finally absorbed by the litter layer without allowing the raindrop to strike the soil surface. Thus, the volume and the speed of run-off water are arrested considerably, thereby preventing soil erosion considerably.

Afforestation should be adopted especially in the upper reaches of river basins.

Use of organic manures should be encouraged, while use of chemical fertilizers should be discouraged.

Flood irrigation should be replaced by sprinkler and drip irrigation.

Salts and alkalinity can be reduced in soils by providing: (i) adequate drainage, and (ii) sufficient quantity of good-quality irrigation water to meet crop use and also to leach the salt below the root-zone in the soil, are made available. Besides, addition of specific chemicals like sulphur and gypsum and raising green manure crops can improve saline and alkaline soils.

The areas affected by *water-logging* conditions can be reclaimed by surface drains. In canal areas, water-logging can be prevented by proper lining of the canals and by constructing strong embankments.

QUESTIONS

1. Why the soil is considered as a natural resource?
2. How does the soil degradation affect the country's economy?
3. What are the causes of soil degradation? Explain.
4. What is meant by soil erosion? Explain the factors which cause the soil erosion.
5. Enumerate the different types of soil erosion caused by water.
6. Discuss the various problems created by water erosion.
7. Differentiate between 'water erosion' and 'wind erosion'.
8. Describe the different types of wind erosion.

9. Write a note on soil resources of India and on the extent of soil degradation existing in the country.
10. Discuss the methods adopted for conserving the soils to protect them from water erosion, wind erosion and salinity.
11. Describe the measures that you would suggest for preventing and controlling desertification of land.
12. Write short notes on:
    (a) Land as a resource
    (b) Landslides
    (c) Properties of soil and fertility
    (d) Soil erosion
    (e) Land degradation
    (f) Salination
13. Explain the different types of soils.
14. Discuss the measures to control soil erosion.
15. Write a detailed note on 'Land Pollution'.
16. Explain the different methods of soil conservation.
17. Explain different factors of soil degradation.

11

Chapter

ENERGY RESOURCES

11.1 INTRODUCTION

Energy is required for doing any type of work. Energy can, therefore, be defined as the ability to do work. The unit of energy is Newton-meter (Nm) known as Joule (J). Its higher units are KJ (Kilo joule), MJ (Mega joule), GJ (Giga joule) etc. When energy is converted into electricity, it is called power. The unit of measurement of power is J/sec, known as w (Watt). 1000 Watt is one kilowatt (KW), one megawatt (MW) is equal to 10^6 watt. 10^9 watt or 1000 mw is 1 Gigawatt (GW).

Energy is required for cooking, heating, cooling, lighting etc. in our homes. It is also required to run our machines and other mechanical equipment in industries. It is used in transport to run locomotives, ships, aeroplane, automotive vehicles. The amount of consumption of energy by a nation is usually considered as an index of its development. This is because almost all our developmental activities are directly or indirectly dependent upon the energy consumption.

The energy is the primary and most universal measure of all kinds of work by human beings and nature. Everything what happens in the world is the expression of flow of energy in one of its forms.

An energy source may be defined as the one that can provide us with adequate amount of energy in a usable form over a long period of time.

11.2 CLASSIFICATION OF SOURCES OF ENERGY

11.2.1 Primary and Secondary Sources

(1) *Primary energy sources* are those which provide a net supply of energy. Coal, oil, uranium etc., are examples of this type. The energy required to obtain these fuels is much less than what they can produce by combustion or nuclear reaction. Their energy yield ratio is very high. The energy yield ratio is the ratio of energy fed back by the material to the

energy received from the environment. Though these (primary fuels) contribute considerably to the energy supply, their supply is limited and hence it becomes very essential to use them sparingly.

Secondary energy sources do not produce net energy, though, these are necessary for the economy. Intensive agriculture is an example wherein terms of energy, the yield is less than the input. These sources require higher investment in terms of energy.

Coal, natural gas, oil and nuclear energy using breeder reactor are net energy yields and are primary sources of energy. Examples of secondary sources are: solar energy, wind, water energy, etc. It may be necessary in future to develop the secondary sources: like solar, wind, tide wave, geothermal and ocean thermal sources.

Examples of primary and secondary fuels are given in the following table.

Table 11.1: Primary and Secondary Fuels.

Type of fuel	*Primary fuel*	*Secondary fuels*	
		Manufactured	*By-product*
Solid	Wood coal	Coke Charcoal Briquettes Pulverised coal	Wood refuse Biogas Coke breeze,
Liquid	Petroleum	Diesel Petrol Kerosene Alcohol Fuel oil Naphtha	Tar Pitch Benzol Paper pulp mill waste
Gaseous	Natural gas	Producer gas Water gas Coal gas Oil gas Gobar gas Butane Propane Acetylene Hydrogen LPG	Blast furnace gas Coke oven gas Oil refinery gas Sewage gas.

11.2.2 Commercial and Non-commercial Sources

The **commercial sources** include fossil fuel (coal, oil and natural gas), hydro-electric power and nuclear power, while the *non-commercial sources* include wood, animal waste and agricultural wastes. In developed countries, most of the energy requirements are met from commercial sources, while in developing countries, the use of commercial and non-commercial sources is almost equal. World's energy supply comes mainly from fossil fuels. The dependence of developing countries on non-commercial sources is likely to continue till it is replaced by other alternative sources of energy.

11.2.3 Major and Minor Sources of Energy

Major sources of energy are:

(i) Fossil fuel *i.e.*, solid fuels (mainly coal including anthracite, bituminous, brown coals, lignite and peats), liquid and gaseous fuels including petroleum and its derivatives, and natural gas.

(ii) Water power *i.e.*, energy stored in water

(iii) Energy of nuclear fission.

Minor sources of energy include: Sun, wind, tides in the sea, geothermal, ocean thermal electric conversion, MHD, fuel cells, thermo-ironic converters, thermo-electric generators etc.

11.2.4 Renewable and Non-renewable Sources of Energy

The renewable energy sources are those which are perpetually available in nature (such as heat of the sun, power of the winds, power of tidal sea waves, thermal energy of the oceans, heat present in the earth's crust etc.); or are generated continuously in nature (such as the potential energy of the stored water; and energy of the biomass, such as wood, animal dung, agricultural waste, kitchen waste and leafly materials, and organic municipal solid wastes).

The *non-renewable energy sources* are those energy sources which have accumulated in nature over a long span of time (millions of years) and cannot be replenished even in hundreds of years. Therefore, these are the sources which are finite in quantity and cannot be reproduced. They are sure to be exhausted after some years. These energy sources include: fossil fuels, such as coal, petroleum and natural gas and nuclear fuels like Uranium-235.

In the age of development, mainly through industrialisation, and due to urbanisation these sources of energy are mainly used in the form of electrical energy. The electricity is the only form of energy which is easy to produce, easy to transport, easy to use and easy to control. Energy in the form of electricity is, therefore, used for transmission and distribution. As electricity is the main form of energy being used, its consumption per capita is considered as the index of living standard of people of a place or country. Electricity in bulk quantities is produced in power plants by using various sources of energy.

Thermal power plants generate about 80% of the total electricity produced in the world. Fossil fuels, *viz.* coal, fuel oil and natural gas are the energy source, and steam is the working fluid. Hydraulic power plants, besides generating power, also cater for irrigation, flood control, fisheries, afforestation, navigation, etc.

11.2.5 Classification According to use of Fuel

Depending upon the usage, fuels are classified as domestic fuel, illuminating fuel, industrial fuel, rocket fuel, etc.

Domestic fuels supply heat for cooking and space heating. Illuminating fuels, *e.g.*, Kerosene provide light in darkness. Industrial fuels supply heat for process heating, steam and electricity generation, etc. Rocket fuels *e.g.*, hydrazine are used for producing enormous thrust for the propulsion of rocket.

11.2.6 Chemical Fuels and Nuclear Fuels

As per this classification, the fuels which generate heat by combustion are called 'chemical fuels' and the fuels which generate heat by nuclear fission or fusion are known as 'nuclear fuels'.

Nuclear fuels are of two types: (*i*) fossil fuel, which release heat by its fission by neutron bombardment *e.g.*, uranium-233, uranium-235, plutonium-239. (*ii*) There are some nuclear fuels *e.g.*, thorium-232 and uranium-238 which cannot be fissioned directly, but they can be converted into fissile material *e.g.*, uranium-233'and plutonium-239 respectively. Such nuclear fuels are called fertile fuels.

11.3 FOSSIL FUELS

Fossil fuels (coal, oil and natural gas) are the remains of organisms (flora and fauna) that lived 200-500 million years ago. During that stage of the earth's evolution large quantity of dead organic matter had collected and buried under layers of sediment and converted by heat pressure into coal, oil and natural gas.

After discovering them, we started consuming them faster and faster. Today, our consumption rate is extremely high. Though, fossil fuels continue to be formed, but at an extremely slow pace, that is why fossil fuels are considered as non-renewable.

Several researches have suggested that an oil shortage is imminent. Very soon, the oil from wells will be insufficient to meet the demean and there will be massive disruption in economy. Further, we do not have sufficient time to switch over from oil to other available sources.

At current rates of use, the world's coal reserves will likely to last not very long. Burning of fossil fuel are the prinicipal sources of air pollution and associated effects.

Production and Consumption of Fossil Fuel in India

Fossil Fuel	*Production*	*Consumption*
Crude oil	700,000 barrels per day	1 million barrels per day
Natural gas	33,000 m cum	54,000 m cu m
Coal	610 million tons	680 million tons

The fuels are broadly classified as:

(a) Solid fuels, like wood and coal

(b) Liquid fuels, like petroleum products

(c) Gaseous fuels, like natural gas.

Fuel is defined as any material that combines chemically with oxygen and liberates heat. The primary fuels which are burned to release heat and generate steam in boilers are fossil fuels in the form of coal, fuel oil and natural gas, which represent the remains of plant and animal life that are preserved in sedimentary rocks. Besides these, industrial wastes like blast furnace gas, coke oven gas, refinery gas, sugar factory refuse, saw mill wood dust, rice husk, etc., are also used.

The most important fuels which are generally used are coal, petroleum oil and natural gas.

11.4 COAL

Coal is a very important fuel used in homes, as well as furnaces in industries, power plants and locomotives. It forms the backbone of all schemes aimed at the industrial progress of the country and hence, it is of utmost economic value. Coal is formed by the transformation of buried vegetation matter into the carbonaceous product, over a long periods of time. In composition, therefore, it consists of carbon with moisture, volatile matter and some mineral residuals.

The relative proportion of fixed (or pure) carbon, in relation to volatile matter and moisture, decides the quality of coal. The greater the percentage of fixed carbon present in a coal sample, the better will be its quality.

The initial decomposition of the buried accumulated vegetation matter into carbonaceous product, takes place through several biochemical processes. With this initial decomposition, starts the process of transformation into coal. The effect of heat and pressure due to further consolidation results in the expulsion of moisture and volatile constituents, giving rise to coal of higher grades. As this process of dynamo-chemical transformation continues, improved varieties of coal get produced.

Soil pressure, heat and movement of the earth's crust distilled off some of the bog's gaseous matter to form brown coal, or lignite. Continuing subterranean activity reduces the coal's gaseous content progressively to form different ranks. These are:

(*i*) Peat (*ii*) Lignite (*iii*) Bituminous and (*iv*) Anthracite.

The larger deposits of coal in the country are in Bengal, Bihar, Jharkhand, Madhya Pradesh, Chhattisgarh, Odisha and Tamil Nadu. The average ash content in Indian coal is as high as 20%. The high ash contents of coal reduces thermal efficiency, as loss of heat through unburnt carbon, excessive clinker formation and heat in ashes is considerably high. The difficulties arising due to high ash contents are overcome by using such coals in pulverized form.

11.4.1 Classification of Coal

Following are the main types of coal:

1. **Peat.** This is the first stage in the transformation of vegetation matter into the coal. It has only partly carbonised vegetation matter. Chemically, it contains a large percentage of moisture, and other ingradients including carbon, sulphur, nitrogen and hydrogen. Peat is generally used as a fertiliser, or as a fuel in dried forms. The high moisture content greatly reduces its calorific value, to about 730 kcal/kg as against

3000 to 5600 kcal/kg when it is dried, and therefore, is not suitable for power plants and mostly used for domestic and other purposes. It is sundried to remove the greater part of moisture and converted into briquettes before being transported.

2. **Lignite.** Lignite is the next stage in the development of coal. It also contains high percentage of moisture (30 to 45%) but can be dried by exposing to air, and can bring down the moisture upto 6%. It is brown in colour and exhibits a woody structure. It can be used as fuel in pulverised form. Its heat value range from 7000 to 11500 kJ/kg.

3. **Bituminous Coal.** This is the real coal formed from lignite, variety and is most popular form of coal used for all purposes. It has low moisture content and has high caking power (caking is the process in which the volatile matter is driven off from the coal when heated, leaving behind particularly pure carbon and formed product is known as coke). Bituminous coal is a black coloured, brittle substance, in which plant residuals are not visible. In chemical composition, bituminous coals show considerable variation, with carbon from 40 to 90%, moisture from 5 to 20%, ash from 6 to 12%, and volatile constituents from 10 to 45%. The bituminous coal shows a typically banded appearance.

   Low volatile matter and high caking coals are prepared for metallurgical industries, while high volatile matter and low caking bituminous coals have great utilities for gas making purposes. Its heat value varies from 9,500 to 16,000 kJ/kg.

   Inferior quality of bituminous coal, also known as sub-bituminous coals are similar to lignite but have lost the woody structure and contains 50% or less moisture. This type of coal is also known as black lignite and is also used in the form of briquettes or pulverised state.

   Higher grade of bituminous coals are also known as *Semi Bituminous Coal*. These when burnt generate little smoke and are of low volatile variety, which is intermediate between bituminous and anthracite coal in properties. It is most widely used for steam raising in thermal power plants. It has low percentage of moisture, ash and sulphur, and has high available hydrogen. It has carbon content ranging between 90–93%.

4. **Anthracite Coal:** Anthracite is the last stage in the process of transformation of buried vegetation into coal and it contains higher percentage of carbon 92–98% and the percentage of volatile matter is below 8%. It burns at high temperature with little or no flame, and pulverisation of anthracite is very difficult. It is clean, dense, hard, and non-caking.

11.5 LIQUID FUELS

Liquid fuels include petroleum and its derivatives. Petroleum is generally believed to have been formed during past geological ages from decayed marine life, both vegetable and animal. Dead marine animals and vegetable matter accumulated for millions of years ultimately got transformed into oil, mainly in sedimentary rocks, by pressure and heat. Crude petroleum oil deposits accumulate in voids in the rocks and sands below the earth's surface. Fairly dense earth strata cover most deposits. Oil has an underlying body of water and pockets of pressurized natural gas above. Oil-well drilled through the layer earth strata penetrate the deposits. The pressure forces the gases and oil to the surface. After the pressure has diminished, the oil should be pumped.

There is also possibility that some oils were formed through the decomposition of vegetable matter mainly, particularly where coal has been formed in adjacent territory. Liquid fuels are an excellent energy source, as they are clean, easy to handle, easy to store and easy to burn. These are primarily a mixture of hydrocarbon compounds, which may also contain nitrogen, oxygen and sulphur. The most of the hydrocarbons belong to the paraffin series, like methane (CH_4), ethane (C_2H_6), propane (C_3H_8) and butane (C_4H_{10}) which are gaseous, and pentane (C_5H_{12}), hexane (C_6H_{14}) and octane (C_8H_{18}) which are liquid at STP.

The crude petroleum, after it is extracted from the oil fields is taken to refinery where it is heated, vapourised, and cooled at different heights. Each separated oil is further refined into various products, such as bitumen, diesel, kerosene, petrol, etc. The wastes of the crude oil is also taken to other parts of a refinery, where it is used to manufacture other materials like plastics, fertilizers, paints, rubbers, wax, etc.

India is quite deficient in production of oil and natural gas, and a large quantum of crude is being imported to fulfill our domestic requirements.

Classification of Oil

Liquid fuels are hydrocarbons and are classified as light oils and heavy oils. Some commonly used liquid fuels for power engineering are:

(i) Gasoline or petrol
(ii) Diesel oil
(iii) Paraffins
(iv) Heavy fuel oil

Liquid fuels are commonly employed in I.C. engines, and for steam generation. The fuel oil used for industrial or domestic purposes are obtained by refining the crude oil. A refining process produces gasoline, kerosene, lubricating distillate and a residual oil, tar or petroleum coke.

Main **advantages of liquid fuels** over coal are:

(i) Higher calorific value
(ii) Economy in space
(iii) Cleanliness
(iv) Easy control of consumption
(v) No loss of heat during storage
(vi) Less operational labour requirement
(vii) The storage and handling in easy and economical
(viii) Elimination of wear and tear of great bars
(ix) Combustion losses are less
(x) Liquid fuels leave very little ash after burning, hence, it saves from ash handling problems.

11.6 GASEOUS FUELS

The gaseous fuels are of two types, namely, natural gas and manufactured gas. Since manufactured gas is costly, only natural gas is used for steam generation. **Manufactured gases** are: blast furnace gas, coal gas, producer gas, water gas etc. As discussed earlier, **natural gas** may be considered as an associate product of petroleum which exists naturally under the surface of the earth. Natural gas comes out of the gas wells and petroleum wells. About one-third of the natural gas produced is found with petroleum.

Natural gas contains 60 to 95% of methane (CH_4), with small amounts of other hydrocarbons, such as ethane (C_2H_6) upto 16% napthene, CO_2, and N_2.

The natural gas is carried through pipes to distances which are hundreds of kilometers from the source. The natural gas is a colourless, odourless and non-poisonous gas. These gases have calorific value between 8000 to 12000 kcal/m^3.

Commonly used manufactured gases are:

Blast furnace gas. This gas is obtained as a by-product from blast furnace used for producing pig iron. Its approximate composition is CO = 30%, N_2 = 52%, H_2 = 3%, CO_2 = 2%, CH_4 = 3%.

The calorific value of this gas is 970 kcal/cubic metre.

Coal gas. Coal gas is a by-product obtained during the destructive distillation of coal. Its calorific value is 7600 kcal/cu metre. Its approximate composition is CH_4 = 35%, H = 45%, CO = 8%, N = 6%, CO_2 = 2% and other hydrocarbons = 4%. This gas is also known as **town gas** or **city gas.**

Producer gas. Producer gas is produced during incomplete combustion of coke in current of air. Its main constituents are nitrogen and carbon monoxide. Its approximate composition is N = 62%, CO = 23%, H_2 = 6%, CO_2 = 5%, CH_4 = 4%. Its calorific value is 1200 kcal/cu metre.

$$\underset{\text{Coke}}{2C} + \underset{\text{Air}}{(O_2 + N_2)} \rightarrow \underset{\text{Producer gas}}{2CO + N_2}$$

Water gas. Water gas is obtained by passing a blast of steam through a deep bed of red hot coke. Its main constituents are CO_2, CO and H_2.

$$\underset{\text{Coke}}{C} + \underset{\text{Steam}}{H_2O} \rightarrow \underset{\text{Water gas}}{CO + H_2}$$

Compressed Natural Gas (CNG) is now being considered as an alternative fuel for automobiles.

Liquid Petroleum Gas (LPG) refers to hydrocarbons, such as propane, propylene, butane, butylene etc. which are liquified under moderate pressure and at normal temperature. It is widely used as domestic fuel.

Gaseous fuels have all the advantages of liquid fuels except case of storage. The gas operated power plants are generally located near the natural gas field and cost of transportation play an important role in selecting the location and the type of fuel for the power plant.

11.7 SOURCES OF ELECTRIC ENERGY GENERATION

As discussed earlier, in the age of development, mainly through industrialisation and due to urbanisation sources of energy discussed above are mainly used in the form of electrical energy. The electricity is the only form of energy which is easy to produce, transport, use and control. As electricity is the main form of energy being used its consumption per capita is considerd as the index of living standard of people of a place.

Classification of Sources for Electric Generation

Various sources of electrical generation can be classified in two ways, as discussed hereunder:

1. **Conventional and Non-Conventional.** The most important and common sources which are used for mass generation of electric power are known as **Conventional Sources**. These are:
   1. Nuclear Power
   2. Hydro Power
   3. Thermal Power

   The other sources of electric power generation are known as **Non-Conventional Sources** of power. The quantum of power production of these sources are limited. These sources are :
   1. Solar Power
   2. Wind Power
   3. Ocean Power
   4. Ocean Thermal Energy Conversion (OTEC)
   5. Tidal Power
   6. Geothermal Power
2. **Renewable and Non-renewable Sources.** As mentioned earlier in this chapter, the **renewable sources** are those which are perpetually available in nature or are generated continuously in nature. Examples of renewable energy are :
   1. Solar Power
   2. Wind Power
   3. Thermal Energy of Sea
   4. Tidal Power
   5. Geothermal Power
   6. Biomass Energy

The **Non-renewable sources of power** are those energy sources which have accumulated in nature over a long span of time (millions of years) and cannot be replenished even in hundreds of years. Therefore, these are the sources which are finite in quantity and cannot be reproduced. These are sure to be exhausted after some years. Examples of non-renewable sources of power are: Nuclear Power, Hydro Power and Thermal Power.

Installed Capacity of Power in India as on 31-3-2018	
Type of Power	**Installed Capacity in MW**
1. Thermal Power	
(a) From Coal	197,171
(b) From Gas	24,897
(c) Other Sources	838
Total Thermal Power	**2,22,906**
2. Nuclear Power	6,780
(A) Sub-Total Non-Renewable (1+2)	2,29,686
3. Hydro Power	45,293
Sub-total conventional (1+2+3)	2,74,979
4. Other Renewable sources*	69,022
(B) Sub-total Renewable source (3+4)	1,14,315
Total Installed Power (A+B)	**3,44,002**
Non-Utilised Capacity= 13% of total installed capacity	

* Details of other Renewable Power Sources :

Solar	21,651
Wind	34,047
Biomass	8,700
Waste to Power	138
Small Hydro	4,486
Total	**69,022 MW**

11.8 NUCLEAR ENERGY

As we have discussed earlier, cheap and abundant power is essential to modern world. The rapid increase in industry and living standard of the people creates the pressure on conventional sources of power (coal, oil and gas), but these sources are rapidly depleting and will soon be unable to meet the rapidly increasing demand of power.

The discovery that energy can be liberated by the nuclear fission of materials, like, uranium (U), plutonium (Pu), has opened up a new sources of power of great importance.

It has been found that complete fission of one kilogram of U can produce as much energy as can be produced by burning about 4000 tonnes of coal or 2,000 tonnes of oil. This shows that nuclear energy can be successfully employed for producing low cost energy in abundance as required by the expanding and industrialising population. Therefore, nuclear energy should be used to as much extent as possible to solve power problem.

In a nuclear power station instead of a furnace, there is a nuclear reactor, in which heat is generated by splitting atoms of radioactive material under suitable conditions. The heat produced to fission of U or Pu is used to heat water to generate steam which is used for running turbo-generators. A nuclear power plant differs from a conventional power plant only in the steam generating part. The main parts of a nuclear power station are the nuclear reactor and a heat exchanger, together with the steam turbine, condenser and a generator similar to that of thermal power station. Heat is provided by fissioning or splitting of uranium atoms, in a reactor. A cooling medium takes up this heat and delivers it to the heat exchanger, where steam for the turbine is raised. The reactor and heat exchanger are equivalent to the furnace and boiler in a conventional steam plant.

The reasonably assured **uranium** reserves in our country are placed at about 34,000 tonnes of uranium oxide of which about 15,000 tonnes are considered to be economically exploitable. The established uranium resources are capable of supporting natural uranium reactors of about 8,000 MW of installed capacity. This shows that the share of nuclear power generation in the total power programme will remain modest. However, when the fast breeder reactor technology is fully developed and U-233 cycle used, then there will be a large increase in nuclear power, as there exist an estimated deposits of about 3,63,000 tonnes of thorium.

India has the largest thorium reserves in the world. It would help the development of nuclear power. The other countries having significant thorium reserves are USA and Brazil. Although conversion of thorium into uranium 233, a fissionable form of uranium on commercial scale has not been achieved but it is hoped that it would be possible to do so and thus, thorium would be successfully used for power generation. This will solve India's power problems to a great extent. The nuclear reactors using thorium-uranium cycle are economical and have longer life. The utilization of thorium in fast breeder reactors is quite successful.

Homi Jehangir Bhabha, the father of Indian nuclear energy programme has given a "three phased Indian Nuclear Fission Power Programme". In the first phase, the available uranium resource of about 70,000 tonnes is to be used in thermal reactors of pressurised heavy water type with a total generating capacity of 10,000 MW. In the second phase, the plutonium obtained as a by-product from the first phase will be used to fuel fast breeders to sustain a generating capacity of 3,50,000 MW and at the same time irradiate naturally occurring thorium, U^{232} to convert to fissionable U^{233} variety, in the third phase, breeder reactors using U^{233} as fuel will sustain a total generating capacity of 7,00,000. But to use this resource (thorium), it is necessary to go through the first two phases, as described above. The two PHWR (Pressurised Heavy Water Reactors) of 235 MW capacity and the FBR (Fast Breeder Reactor) at Kalapakkam demonstrate the capacity of Indian nuclear engineers to build and operate reactors safety.

The major centre for research and development work in atomic energy in our country is the Bhabha Atomic Research Centre (BARC) at Trombay. The centre is the largest single scientific establishment in India. Besides BARC, three other national institutions associated with important aspects of atomic energy programme are as follows:

(i) Tata Institute of Fundamental Research, Bombay

(ii) Tata Memorial Centre, Bombay

(iii) Saha Institute of Nuclear Physics, Kolkata.

11.8.1 Advantages of Nuclear Energy

Nuclear energy is the most useful power available to mankind today. In large parts of the world, it is becoming a predominant source of electrical power and a versatile tool for use in many areas of human endeavour. In India too, atomic energy is being used to generate electricity and to bring about improvement in industry, agriculture, medicine and other fields through its varied applications.

The main advantages of nuclear power station compared to the conventional thermal power stations are:

1. It reduces demand for coal, oil and gas, the costs of which are tending to rise as their stocks are depleting. Greater nuclear power production leads to conservation of coal, oil etc., which can be used in other areas of utility.
2. The amount of nuclear fuel used in the plant is small. Therefore, no problems of fuel transportation, storage, etc. As stated earlier, one kg of uranium can produce, as much energy as can be produced by burning about 4,000 tonnes of high grade coal or about 2000 tonnes of oil.
3. Nuclear power plant requires less space compared to any other plant of the equivalent size.
4. Besides producing large amounts of power, the nuclear power plant can produce valuable fissile material, which is extracted when the fuel has to be renewed.
5. Nuclear plants create no air pollution, and are unaffected by adverse weather conditions.
6. Despite higher initial capital cost of a nuclear power plant, the lower fuel cost permits economical power production.
7. It minimises the ecological affects of power generation.
8. Siting of nuclear power station, generally in remote areas, would open a new field of general development in the area.
9. It does not have problems associated with mines safety.

11.8.2 Limitations

1. High initial capital required.
2. Difficulty in getting trained technologically competent man-power.
3. Danger of radioactivity, detrimental working conditions to health of workers, problem of disposal of radioactive waste etc.
4. These can be used as base load plants only. They are not suitable for variable load operations, as the reactors cannot be easily controlled to respond quickly to load changes.

Because of the above advantages, bulk of electricity is produced in nuclear plants in many countries, like France, Japan, UK, Russia, US, Canada, etc. In recent years, concerns are being raised in some quarters against the use of nuclear energy for power generation due to the problems related to nuclear safety, radioactive waste disposal and nuclear weapons proliferation. Despite these difficulties, the future of large capacity electricity generation includes nuclear energy as one of the main sources.

The biggest drawback of a nuclear power plant is that the uranium rods need to be changed periodically. This produces a lot of high level nuclear waste, which poses serious problems of disposal. When not disposed off properly, the radioactive nuclear waste produce a very adverse impact on the environment and human health. When human body is subjected to the radioactive radiations, large scale health hazards, like cancers, shortening of life spans, deformations, and genetic changes (long term effects which take several years and even generations to cure) may occur. The safe disposal of nuclear wastes is an extremely difficult and costly exercise. Besides the continuous production of nuclear wastes from nuclear power plants, which need safe disposal, enormous problems to arise in decommissioning of old plants.

In addition to the above, another problem is for the disposal of hot waste water, which damage aquatic ecosystems. Possibility of a nuclear accident is most threatening and dangerous. As a single nuclear accident can kill thousands of people, make many others seriously ill with incurable diseases, and destroy a wide area for decades by its radioactivity. Land, water and vegetables are also destroyed for long periods of time.

11.9 HYDROPOWER

The word "Hydro" is a Greek word meaning water. The water is being used since long as the source of energy. The first designed water wheels were used in nineteenth century. Later on water wheels were further developed as water turbine. Afterwards, several types of turbines were build up, some of them are: Jonval turbine (in 1837), Girard turbine (in 1850), Francis turbine (in 1865), Kaplan turbine (in 1916).

When water strikes the blades, fixed on the periphery of the turbine wheel, wheel starts rotating due to the force of the water. Thus, hydraulic turbines are the devices which convert hydraulic energy into mechanical energy. Mechanical energy, thus, produced is used for running the electric generator, which are coupled to the turbine shaft. The electric power, thus, produced is known as hydro-electricpower or water power. As water power is very cheap, it is now-a-days gaining the importance over thermal power.

Projects designed to produce electric power from water are known as "hydroelectric projects". Sometimes, these projects are made for controlling flood, drinking water supply, and irrigation purposes in addition to the electric generation, in which case these are known as "multi-purpose projects". Examples of such projects are Bhakra-Nangal Project, Sardar Sarovar Project, Hirakund, Nagarjun Sagar, Koyna, Rihand, Tehri Project etc.

For selecting the site for hydroelectric plant, a large quantity of water at suitable head should be available. Regarding the quantity of water which could be available for the plant, a survey of past-history of rainfall is done. From this, it can be known that during past years, what was the minimum and maximum rainfall. Maximum capacity of the plant should be decided on the basis of minimum quantity of water available, and the storage reservoir should be constructed for the maximum quantity of water, so that it can store the water even in the peak periods.

Initial cost of the hydroelectric plant is very high, but operating costs are too low. Thus, if the water is available in large quantity at a suitable head, hydroelectric plants are very advantageous.

The energy of water utilized for hydropower generation may be kinetic or potential. The kinetic energy of water is its energy in motion and is a function of mass and velocity, while the potential energy is a function of the difference in level of water between the two points, called the head. For the purpose of continuous availability of water, it is collected in natural lakes and reservoirs at high altitude may be utilized or water may be artificially stored by constructing dams across the flowing streams.

11.9.1 Advantages of Hydropower

Advantages of hydroelectric power generation over thermal power generation are:

(i) Besides power generation, hydropower plants are quite useful for irrigation, flood control, and for supply of drinking water, afforestation, navigation and aqua-culture.

(ii) Hydropower plants can be used both, as base load plant and peak load plant.

(iii) Running cost of hydro plant is low as compared to steam power plant or nuclear power plant of same capacity.

(iv) The hydropower plant system reliability is greater than that of other power plants.

(v) The hydraulic turbines can be put off and on in matter of minutes. Nuclear power plants and steam power plants lack this facility.

(vi) Modern hydropower plant equipment has a greater life expectancy which is about 50 years or more, whereas nuclear power plant has an effective life about 30 years.

(vii) Steam power plants have problem of ash disposal, whereas hydropower plants have no comparable problem.

(viii) Modern hydropower plants give very high efficiency over a considerable range of load.

(ix) Maintenance requirements are very less as compared to fuel burning plant. Thus, in addition to lesser maintenance expenditure, there is lesser requirement of reserve capacity, leading to economy of investment cost.

(x) Ability of varying, their outputs is much higher and faster in hydropower plants.

(xi) Unforeseen outages are less frequent.

(xii) The problem of emission of polluting gases and particulates to the atmosphere also does not exist. Therefore, it does not cause any acid rain and emit obnoxious oxides of nitrogen.

(xiii) Being simple in design and operation, the hydropower plants do not require highly skilled workmen. Manpower requirement is also low.

(xiv) The water stored in the reservoir also helps in developing tourism and recreation spots.

(xv) The efficiency of the hydropower plant does not change with age, whereas there is considerable reduction in efficiency of thermal as well as nuclear power plant with age.

(xvi) The hydropower plant and its associated civil structures have a long life.

(xvii) It operates on water, which is a renewable energy source.

(xviii) The cost of generation of energy varies little with the time.

(xix) The machine used in hydel plants are more robust and generally run at low speeds (300 to 400 rpm), where the machines used in thermal plants run at a speed of 3,000 to 4,000 rpm. Due to low speed and low temperature, there are no complications of special alloys required for construction.

Due to above mentioned reasons, hydel plant provides a stable and reliable operation of the system by taking fluctuations of all kinds, whether arising from system load or from forced outage of units. The hydel plants are best suited for peaking and is more economical in taking up variations of load generation that may occur through various causes and not merely covering the top of the load curve.

11.9.2 Disadvantages of Hydropower

The hydroelectric power plant have the following disadvantages:

(i) Initial cost of the plant including the cost of dam is high.

(ii) Hydro-station has special requirement of site which usually is an isolated area with difficult access.

(iii) Power generation by the hydel plant is dependent on the quantity of water available, which in turn depends on rain and vary from season to season.

(iv) The transmission of power from hydel stations to the load centre generally requires long transmission lines. Thus, the cost of transmission lines and losses in them are more.

(v) Hydel plants take long time for its construction as compared to thermal plants.

(vi) Large hydro power plants disturb the ecology of the area, by way of deforestation, destroying vegetation and uprooting people. Strong public opinion against erection of reservoir type hydropower plants is a deterrent factor. The emphasis is now more on small and run-of-river hydel stations.

Although there are some negative impacts of dams, but they also help in improving the general environment of the area.

Following are the some important **positive environmental impacts of dams:**

(a) Overall increase in wood production due to flourishing growth of trees and crops in the irrigated command area of the project.

(b) Excellent habitat for fisheries and water liking birds.

(c) Net improvement in public health due to availability of ample domestic water supplies leading to overall sanitation, cleanliness and better living conditions.

(d) Improvement in micro-climate caused in the adjoining areas due to evaporation from the open water surface of the reservior and from the irrigated command area of the project. This also increases overall improved oxygen production.

(e) Development of sanctuaries and wildlife becomes feasible, when adequate sweet water become available round the year by the implementation of the dam project.

11.10 BIOMASS ENERGY

Biomass is the organic matter produced by the plant or animals, and include wood; cattle dung; agriculture wastes, such as crop residues; organic municipal solid wastes like paper, food wastes, and other non-fossil-fuel derived material like textiles, natural rubber, etc. Biomass is used on a large scale to obtain energy, either by directly burning the same or by converting it into more convenient form of: (a) *solid fuels* like manure, dung cakes, etc. or (b) *into liquid fuels* like ethanol, biodiesel, etc., or (c) into *gas fuels* like biogas etc.

Biomass has provided a major source of energy to human beings, since the early days of human civilization. While fossil fuels take millions of years to form, the biomass is an alternate fuel, the source of which may be considered renewable, since plant life renews and adds to itself every year. It is the solar energy stored by the way of photosynthesis. The products of photosynthesis (or biomass); also called biofuels, are bulky and contain large amounts of water. Therefore, the stored energy in them must be utilized by direct burning in regions close to the source. The biomass can also be converted into variety of solid, liquid or gaseous fuels.

The fibrous waste of sugar industry is the world's largest potential source of biomass energy. Ethanol produced from sugarcane molasses is a good automobile fuel, and is now used in a third of the vehicles in Brazil.

Biogas is produced from animal dung, kitchen wastes, plants like hyacinth etc., and organic wastes from households and industries like fish processing, dairies etc. Since, in India, the biogas is primarily being produced from cattle dung (gobar), the biogas is called the "*gobar gas*". Biogas primarily contains, methane (60 to 70%) and carbon dioxide (30 to 35%), alongwith small quantities of nitrogen (1 to 5%), and oxygen (0-0.1%). The use of biogas not only gives a cleaner fuel for generation of energy, but also provides an efficient method of disposal of society's wastes. Biogas plants are increasingly becoming popular in India, particularly in the rural sector and is used as a fuel for cooking, heating and lighting. The reduction in kitchen smoke by using the gobar gas has given a great relief to the ladies cooking food in houses. It also helps in reducing indoor air pollution to a great extent. Once the biogas is produced from the waste biomass, some residue is left, which can be used as an agricultural fertilizer (manure).

The biogas generation is a chemical process of decomposition of organic matter. This gas is highly combustible, and therefore, can be used by burning for various applications.

As a result of energy shortages in the years to come, interest in the alternate fuel sources has increased considerably. Biomass is one of the such sources being considered.

As discussed above, biomass as a source of energy has following advantages over dwindling fossil resources:

(i) It is renewable,

(ii) It is environmentally clean, and

(iii) It is easily adaptable.

Biomass is produced in nature through photosynthesis achieved by solar energy conversion.

Solar energy → Photosynthesis → Biomass → Energy generation.

Biomass resources are as follows:

(a) Concentrated wastes like; industrial wastes, municipal wastes and manure at large lots

(b) Dispersed waste residue like crop residue, disposed manure

(c) Harvested biomass, standing biomass.

Energy from biomass is obtained in following three ways:

(i) Biomass in its traditional solid mass (wood and agricultural residue). This type of biomass is burnt directly to obtain the energy.

(ii) The biomass in non-traditional form (converted into liquid fuels). In this case, the biomass is converted into ethanol and methanol to be used as liquid fuels in engines.

(iii) The biomass is fermented aerobically to obtain a gaseous fuel called bio-gas.

11.11 NON-CONVENTIONAL ENERGY SOURCES

As already discussed, conventional sources of energy *viz.*, fossil fuels (coal, oil and natural gas), nuclear fuels and hydel power has very limited scope. The fossil fuels being the main fuels for thermal power, are likely to get exhausted in the current century. Therefore, other systems based on non-conventional and renewable sources are being tried by many countries. These sources are solar, wind, sea, geothermal, biomass, etc. However, efforts are being made to find a suitable generic term to describe the whole range of technologies designed to tap the earth's natural energy flows. These technologies have variously been called 'alternative', 'appropriate', 'natural', 'new' sources of energy. However, the term 'non-conventional' sources have gained the most widespread acceptance.

Renewable energy sources include both, 'direct' solar radiation intercepted by collector and 'indirect' solar energy, such as wind, hydropower, ocean energy (which includes tidal, seawaves, ocean thermal electric conversion, *i.e.*, OTEC), biomass resources. Biomass being the traditional source have been discussed in separate section. Similarly, hydropower have also been discussed separately. Geothermal fields tapped with present drilling technologies have a finite life, but being a new field, have been discussed in this Chapter.

Major obstacle to the implementation of non-conventional energy systems is the technical and economic uncertainties in comparison with conventional supply options and energy efficiency measures. In addition, there is a shortage of qualified personal to design, manufacture, market, operate and maintain such systems.

11.12 SOLAR ENERGY

Utilization of solar energy is of great importance to India, since it lies in a tropical climate region of the world, where sunlight is abundant for a major part of the year. Solar energy has the greatest potential of all the sources of renewable energy and if only a small amount of this form of energy could be used, it will be one of the most important supplies of energy especially when other sources in the country have depleted. The Sun is regarded as an inexhaustible source of useful energy.

The 'solar power' (solar power is referred to the electricity generated by solar energy) would eliminate most of the serious environmental problems associated with fossil fuel and nuclear power.

There are certain major limitations to the extensive application of solar energy, such as:

(i) The intermittent and variable manner in which it arrives at the earth's surface, and

(ii) The large area required to collect the energy.

The solar energy is effective only during the daytime, and if the power supply is to be made during night, then some reservoir of energy, such as storage battery or heat accumulator should be used. Solar energy cannot be used during cloudy weather and rainy season. Its intensity during winter is very less. At present harnessing, storage and use of solar energy is much more expensive than using fossil fuel. It is believed that solar power can become economically feasible with the following aims, achieved:

(a) Availability of better heat collectors

(b) Availability of improved materials and manufacturing techniques

(c) Better techniques for storage and cheap distribution of solar power.

India has the potentialities of solar and wind energy, but it needs the advanced technology to tap this potential.

11.12.1 Applications of Solar Energy

Traditionally, we have been using solar energy for drying clothes and food grains, for preservation of eatables, and for obtaining salt from seawater. However, in this modern age, we have already invented several techniques for harnessing the solar energy. Some of the important solar energy applications are:

1. Solar water heating
2. Heating and cooling of residential buildings
3. Solar drying of agricultural and animal products
4. Solar distillation on a small community scale
5. Salt manufacturing by evaporation of sea-water or inland brines
6. Solar cooking
7. Solar pumping of water
8. Food refrigerator
9. Solar furnaces
10. Solar power generation
11. Solar photovoltaic (PV) cells
12. Bioconversion, wind energy, ocean thermal energy conversion, tidal energy, sea waves etc., which are indirect sources of solar energy.

Electricity can be produced from the solar energy by photovoltaic solar cells, which convert the solar energy directly to electricity. This photovoltaic technology is also used for the energisation of pump sets for irrigation, drinking water supply and rural electrification covering street lights, community TV sets, medical refrigerators, and other small power loads.

11.13 WIND ENERGY

Winds arise primarily from temperature difference of the earth's surface resulting from unequal exposure to (or absorption of) solar radiation. Air heated by contact with a warmer surface tends to rise, and cooler air from a less heated surface flows in, to take this place; the resulting air flow constituted wind. Wind energy is, thus, a form of solar energy. Like direct solar energy, wind energy is also highly variable.

It is the conversion of kinetic energy of the wind into mechanical energy that can be utilized to perform useful work or to generate electricity. Wind energy is a renewable source of energy. India with its long coastline, blessed with steady winds almost throughout the year has tremendous potential of wind power.

Wind energy, which is an indirect source of solar energy, can be utilised to run wind mills which in turn drives a generator to produce electricity. Wind mills may be either horizontal axis wind mills or vertical axis wind mills.

Some **characteristics of wind energy** are as under:

1. It is a free and inexhaustible renewable source of energy.
2. Wind power systems are non-polluting, and has no adverse influence on the environment.
3. It avoids fuel provision and transport.
4. These are economical in the rural areas which are at far remote places from existing grids.

However, following **problems** are associated with wind power:

1. Wind energy available in unpredictable and fluctuating in nature.
2. Wind energy systems are noisy in operation.
3. Large areas are needed to install windfarms of electrical power generators.
4. Windmills usually kill the birds.

Wind being an intermittent source, cannot be used as the sole source of electricity, and requires some other back-up.

Wind power is a function of wind speed and, therefore, the average wind speed of an area is an important factor for economically feasible power. As wind speed increases with height, therefore, power available 30 m above ground is 60% greater than at 10 m heights. Therefore, wind mills need to be located at enough heights. As already discussed in the previous chapter, in India, high wind speeds are obtainable in coastal areas of Saurashtra, Maharashtra, Tamil Nadu, Kerela, Karnataka, Odisha, Andhra Pradesh, Western Rajasthan and some parts of central India. In these areas, there could be possibility of using medium and large sized wind mills for generation of electricity and feeding the same into the grid.

As a rough estimate, India has a potential of 20,000 MW of wind power. Traditionally, windmills were used extensively in the middle ages to mill grain and lift water for land drainage and watering cattle. Wind energy converters are still used for these purposes today in some parts of the world, but the main focus of attention now lies with their use to generate electricity.

11.14 OCEAN ENERGY

Solar radiation is absorbed by the seas and oceans, which causes, like the wind, ocean currents and moderate temperature gradients from the water surface downward, especially in tropical waters. The oceans and seas constitute about 70 per cent of the earth's surface area, hence, they represent a large storage reservoir of the solar input. The conversion of solar energy stored as heat in the ocean into electrical energy by making use of temperature difference between the warm surface water and the colder deep water. The facilities proposed for achieving this conversion are commonly referred to as **OTEC** (Ocean Thermal Energy Conversion) plants.

Periodic rise and fall of the water level of the sea which are carried by the action of the sun and moon on the water of the earth is known as **tide**. The tidal range varies from few centimeters to about 8 to 10 m in some parts of the world. The potential energy of the tides can be trapped to generate power.

Ocean and sea waves are caused indirectly by solar energy, like wind and OTEC. Wave energy drives from wind energy, which drives in turn solar energy. Devices that convert energy from waves can, therefore, produce much higher power densities than solar devices. But the wave energy development is not nearly as far along as wind and tidal energy.

The hydrological cycle results in rain-fall, which causes river flows, that can be trapped behind dams to even out the variations in river flows and, thus, become the source of **hydroelectric energy**.

Thus, the Ocean sources of energy are :

(i) Ocean Thermal Energy Conversion (OTEC)

(ii) Tidal energy

(iii) Wave energy, and

(iv) Hydrological cycle, which emanates from the sun—ocean system through surface water evaporation by solar heating.

11.15 OCEAN THERMAL ENERGY CONVERSION (OTEC)

Ocean cover a little more than 70% of the earth's surface. This makes them the largest solar energy collector and energy storage system. On an average daily basis, 60 million km^2 of tropical seas absorb an amount of solar radiation equal in heat content of about 250 billion barrels of crude oil. The surface of the water acts as the collector for solar heat, while the upper layer of the sea constitutes infinite heat storage reservoir. Thus, in this indirect form of solar energy at sea, collection and storage are free.

The technology which is used to convert this heat energy contained by the upper layers of the ocean into the electrical energy is known as **OTEC** (Ocean Thermal Energy Conversion). This technology makes use of the ocean's natural *thermal gradient*, caused due to warm upper layers and cold lower layers of water. An OTEC system can produce significant amount of renewable power. This potential is estimated to be of the order of 10,000 GW of baseload power generation.

The heat contained in the oceans, which is solar in original could be converted into electricity by utilizing the fact that the temperature difference between the warm surface water of the tropical oceans and the colder waters in depth (1000 – 1500 m) is about 20 – 25°C. Warm surface water could be used, to heat some low boiling organic fluid (such as ammonia), the vapour of which would run a heat engine. The exit vapour would be condensed by pumping cold water from the deeper regions. The process of OTEC requires that the warm surface water and cold water from depth, be brought into proximity, so they act as the heat source and heat sink, respectively for a heat engine, without the requirement of any depletable fuel. After providing the cooling effect in the condenser, a part of the cold sea-water is diverted for the development of aqua culture.

Althrough there is some seasonal variation in the ocean thermal resource at a given OTEC power plant location, there is little variation on day-to-day basis. OTEC systems can be used to generate electricity, desalinate sea-water, provide refrigeration and air-conditioning and may help in mineral extraction from sea-water. These complementary products may make the OTEC systems quite attractive to the industry and island communities.

11.16 TIDAL ENERGY

The water level in a sea rises and falls during tides, caused mainly by gravitational attraction of the moon and the sun on the water of solid earth and the oceans. About 70 per cent of the tide producing force is due to the moon and 30 per cent due to the sun. The moon, therefore, is the major factor in the tide formation.

Surface water is pulled away from the earth on the side facing the moon, and at the same time, the solid earth is pulled away from the water on the opposite side. Thus, high tides occur in these two areas (directly under the moon and diametrically opposite the earth's surface), with low tides on intermediate points. As the earth rotates, the position of a given area relative to the moon change, and so also do the tides. Thus, there are periodic succession of high and low tides. Therefore, the time between high tide and low tide at any given location is a little over 6 hours.

Thus, a full moon as well as no moon produce a high tide. In a period of 24 hours 50 minutes, there are, therefore, two high tides and two low tides. The difference between high and low water levels is called the **range of the tide**. The periodic rise and fall in levels is in rhythms with daily cycle of rising and setting of sun and moon.

The potential energy developed by the water head created during high tide can be used to run the turbines, to generate electricity. Electricity produced by these tides is called *tidal power*. If the differential head between high tide and low tide could be utilized in operating a low head reversible hydraulic turbine, the tidal energy could be covered into electical energy by means of an attached generator. During high tide, the sea-water flows towards the land reservoir through the openings in the dam or the barrage, and turns the turbines, generating electricity. During the low tide, as the seawater levels fall, the water from the upstream reservoir flows towards the downstream in seaward direction, again turning the turbine blades to generate electricity.

The main problem in harnessing the tidal power is that it is not possible to harness the tidal power along most of the coasts of the world. If the tidal range is less than about 8 metres or if narrow and enclosed bays are not available, then tidal power development becomes very uneconomical. Therefore, very few sites are available where tidal energy can be economically harnessed.

Following **characteristics** are to be noted regarding the tidal phenomenon:

1. The tidal are periodical phenomenon and two tides in any cycle are alike.
2. In any month, the tides on the full moon and no moon days are particularly higher than the rest.
3. The mean tidal range varies from place to place.
4. It is free from pollution.
5. It is inexhaustible and independent of the rains.
6. Tidal power plants require less area of valuable land and are located on seashore.
7. The tidal range is highly variable and, therefore, the turbines have to work on a wide range of head variation. This affects the efficiency of the plant.
8. High chances of machinery being get corroded due to sea-water.

11.17 GEOTHERMAL ENERGY

Geothermal energy is the energy which lies embedded within the earth as heat (*i.e.* thermal energy) in the earth's crust. The earth's crust constitutes a potentially useful and almost inexhaustible source of energy. This heat is apparent from the increase in temperature of the earth with increasing depth below the surface. The eruption of hot lava from volcanoes at the surface at about 1200°C temperature, clearly shows that the deep interior parts of the earth are very very hot, and may be in liquid or semi-liquid state.

From the extremely hot interior of the earth, a continuous upward current of heat flows towards the upper crust of the earth. Due to such currents, the upper 3 km portion of the earth's crust, also contains a huge amount of heat energy, which if removed, is liable to be replaced again by the new heat coming from the heat currents from the interior portion of the earth. This source of heat energy is, therefore, considered as a renewable energy source.

In the natural geothermal areas, water circulating systems are continuously transporting the heat energy to the surface, and this is observed as natural hot springs, called *natural geysers*. These natural eruptions of hot waters with steam with temperatures reaching upto about 350°C, may occur through the cracks existing in the crystal rocks.

The utilization of geothermal energy can be made by drilling holes or wells into these natural circulation systems, and to top the geothermal fluids through these wells. The naturally erupting geyser can also be utilised for this purpose. These hot fluids (water + steam) can either be used for generation of electricity, or for direct heating purposes. For large scale use bore holes are normally sink with depth upto 1000 m, releasing steam and water at temperatures upto 200 or 300°C and pressures upto 30 kgf/cm^2.

Geothermal energy is advantageous both, from the point of view of conserving fossil fuel and of pollution control. The hot water coming out of the ground with steam (which has about 1/3 of the available thermal energy) is discarded because it contains about 30% dissolved salts and minerals, which can cause serious rust damage to the turbine.

QUESTIONS

1. Name and explain various sources of energy.
2. What are the various types of fossil fuels? Explain them.
3. What are the advantages of liquid fuels and gaseous fuels over solid fuels?
4. Write short notes on:
   (a) Liquid fuel
   (b) Types of gaseous fuels
   (c) Wind energy
   (d) Geothermal energy
   (e) Wind mill
5. (a) What are the non-conventional sources of energy?
   (b) What are the renewal sources of energy?
6. Write a detailed note on wind energy, covering its characteristics, and factors to be considered for selecting the site.
7. Explain the following in brief:
   (a) Bio mass (b) Biogas
   (c) Gobar gas (d) Tidal power
8. What are the primary and secondary sources of energy?
9. Describe solar power. What are its advantages and limitations?
10. Discuss main applications of: (a) solar energy (b) wind energy.
11. What are the principal types of fuels used in power generation?
12. What are the main characteristics of a good hydel plant site?
13. Discuss the advantages and disadvantages of hydel power over thermal power generation.
14. Write a note on nuclear power plant.
15. What factors are considered for selecting a suitable site for tidal power plants?

12

Chapter

Food Resources

12.1 INTRODUCTION

As per estimates of biologists, there are about 2,70,000 different plant species, out of this just 12 species of plants provide the bulk of food for humans.

Out of these plants, just three—rice, wheat and corn are most common. These 12 common plant species are: Sugarcane, corn (maize), rice (paddy), wheat, white potato, cassava (sabudana), soyabean, barley, sweat potato, sorghum (jowar), peanuts and oats.

Animals provide foods that are protein rich. These foods include meat, eggs, milk, cheese and fishes. Sheep, chickens, goats, ducks etc. are the most common types of species of livestock.

Agriculture is the practice through which specific plant species are cared managed so as to obtain maximum yield of consumable parts of the plant. Land, soil and water are the basic resources needed for the agriculture. River valleys and coastal plains have relatively level land, fertile alluvial soils and abundant water supply which favour cultivation.

Availability of suitable land for cultivation is limited and due to increase in population, new areas are being cultivated by clearing forest land or grassland, terracing of hill slopes. Tractors and other machineries are also being increasingly used to cultivate large areas with minimum human labour. In recent decades, productivity in agriculture has been increased by using fertilizers, pesticides, and high yielding varieties of seeds.

Agriculture comprises the cultivation of variety of crops besides foodgrains. Other forms of crops are: fibre crops, like cotton, jute, commercial crops, like sugarcane, tobacco and variety of spices, plantation yielding tea, coffee and rubber. A variety of fruit trees are also grown. There are certain crops which provide raw materials for a variety of industries, which encourages the growth of agro-based industries.

Definition

Food is anything eaten to satisfy appetite and to meet physiological needs for growth to maintain all body processes and to supply energy to maintain body temperature and activity.

Needs for Food

Man needs food as a source of energy and for tissue replacements. Intake can be divided into metabolic needs necessary for the maintenance of the organisms, and the cultural and work behavioural needs.

Food is required for adequate energy intake together with sufficient protein, vitamins and minerals for continuous replacement of tissues.

12.2 EVOLUTION OF FOOD PRODUCTION

Primitive societies obtained their food through hunting and food gathering. Some of the tribes even today depend on hunting and food gathering.

The first major result in terms of food supply was due to the agricultural revolution which took place approximately 10,000 years ago. At that time many small social groups shifted from hunter-gatherers to producers of food. In this revolution, only some selected plants was produced and single-crop production was adopted. At that time, the pastoral grazing of domestic animals was also taken up.

The second major impact was due to scientific knowledge of food production resulting from scientific and industrial revolution that began about 400 years ago. This has also resulted in increase of output per unit of land or per animal.

Still the efforts are going onto increase the output of food per unit of land all over the world and substantial progress has already been achieved. Besides, efforts have been made for increasing the food-producing area.

12.3 TYPES OF FOOD PRODUCTS

As we know that vegetarians eat plant foods or the protein rich dairy foods, like milk, ghee, cheese etc. Produced by the animals. The non-vegetarians, besides consuming the plant food and the vegetarian dairy products, also consume the direct animal foods, like fish, meat, eggs etc.

Different types of food are:

1. *Plant Food.* As we have already discussed that only 12 plant species provide the bulk of food consumed by human beings. These are grown from *agriculture.* The plant species when grown through agriculture are called crops.

   Besides human beings, the plant food is also consumed by the livestock through whom they get their *animal husbandry food.*
2. *Animal Food.* These foods are protein rich. The animal food may include vegetarian dairy products, as well as the non-vegetarian food. Broadly, the animal food may be divided into the following two types:
   (a) Food from animal husbandry, and
   (b) Seafood coming from fishing.

12.4 WORLD FOOD PROBLEMS

Modern scientific and technological advancement have resulted in high increase in food production efficiency, nutritional quality, plant and animal genetics, diseases and pest management. During the last about 50 to 60 years, world food production has increased about three times. But at the same time high population growth has nullified it.

Following are the major food related problems:

1. *Starvation, Undernourishment and Malnutrition.* A person is at risk of malnutrition if the amount of energy and/or nutrients in the diet does not meet his nutritional needs. This leads to too weak body to function properly or to fight infection. Deficiency diseases are generally associated with lack of vitamins or minerals.
2. *Food Production not able to keep pace with growing demand.* Food production 64 of the 105 developing countries is lagging behind their population growth levels. Although India is able to produce enough food by cultivating a large proportion of land through irrigation.
3. Fertile soils are exploited faster than they can recuperate.
4. Forests, grasslands and wetlands are being put to agricultural use, which has resulted to serious ecological in balance.
5. Due to various types of pollutions, our fish resources, both marine and inland, show evidence of exhaustion.
6. In many countries, there is a shortage of cultivable productive land.
7. The supply of food from fisheries has increased manifold between 1950 and 1990, in several parts of the world. The FAO (Food and Agriculture Organisation) in 1995, reported that 44% of world's fisheries are fully or heavily or heavily exploited. This situation is not sustainable.

12.5 GREEN REVOLUTION CROPS

In order to keep pace with the growing human population, new methods and new varieties of crops are found and practice around the world. This results in dramatic increase in the food supplies, and this process is known as *'green revolution'*. The green revolution replaced (in past few decades) traditional crop varieties and growing methods throughout the developed world, and nearly half of all farmers in the developing world were using green revolution seeds, fertilizers and pesticides.

Most green revolution breeds yield more than other varieties, if given optimal revels of fertilizer, water and pest control. The new methods and inputs are expensive. Poor farmers who cannot afford hybrid seeds, fertilizers, machinery, fuel and irrigation are put at a disadvantage, compared to rich farmers.

Experiments have now shown that special programmes aimed at finding less expensive methods for low-income farmers have also found good results. Some of these suggest: diversified low-input farming, soil can be enriched with nitrogen-fixing plants and manure; rotating crops and inter-planting crops; better soil management. This will enable to exceed conventional production by improving fertility and reduces pest dispersal. These measures also reduce costly inputs.

Because of the 'green Revolution in India' the grain production has increased from 108.4 million tonnes in 1970 to 181 million tonnes in 1993 from almost same area.

12.6 AGRICULTURE RESOURCES

Agriculture is a process under which specific plant species are cared and managed by man to obtain the maximum yield of their consumable parts. The plant species when grown through agriculture are called *crops*.

We have already discussed that there are 12 crop species, which provide the bulk of plant food to the human beings, either for their own consumption or for the consumption of their livestock.

Land, soil and water are basic resources needed for agriculture. River valleys and coastal plains have relatively level land, fertile alluvial soils and sufficient water supply and, thus, favour cultivation. Rugged topography is a limiting factor in mountaneous and plateau regions. Thus, the extent of land suitable for cultivation is limited.

Use of fertilizers, pesticides, high yielding varieties of seeds have resulted in increased yields per hectare.

Agriculture is of different types, such as shifting agriculture, subsistence agriculture, commercial agriculture, and each type of agriculture is typical of certain regions and the type of people who live in such regions. Economic factors like cost of cultivation, changing prices of agricultural commodities are other factors, influencing the crops cultivated and the area under each crop.

Agriculture started with cultivation of foodgrains needed by the people. In course of time, other crops were also cultivated. These were:

1. *Fibre crops* like cotton, jute.
2. *Commercial crops* like sugarcane, tobacco and a variety of spices.
3. *Plantation agriculture yields* tea, coffee, and rubber.
4. *Fruit Trees.* A variety of fruit trees are also grown. In the tropical regions mangoes, bananas and citrus fruits like lime, orange etc. are cultivated. In the middle latitudes apples, olives, grapes etc. are suitable for cultivation.

   Each crop require specific environmental conditions for its cultivation.

These are discussed below:

1. **Agriculture** : The term agriculture is derived from two Latin words **agri** meaning soil and **culture** meaning cultivation. Agriculure is a broad term encompassing all aspects of crop production, livestock farming, fisheries, forestry etc. It includes :

(a) *Silviculture* (art of cultivating forest trees).

(b) *Sericulture* (rearing of silkworms for the production of raw silk).

(c) *Apiculture* (maintenance of honey bee colonies).

(d) *Olericulture* (science of vegetable growing).

(e) *Viticulture* (science for production and study of grapes).

(f) *Floriculture* (horticulture concerned with the cultivation of flowering and ornamental plants).

(g) *Pomology* (branch of horticulture for cultivation, production, harvesting and storage of fruits etc.)

Revolutions in Agriculture

Revolution	*Related with*
Green	Foodgrain production
Golden	Fruit production
Blue	Fish production
Round	Potato production
Red	Meat/tomato production
Silver	Egg/poultry production
White	Milk production

2. ***Tillage*** : It is the mechanical manipulation of soil with tools and implements for obtaining conditions ideal for seed germination, seeding establishment and growth of crops.

3. **Farm** : Farm is a piece of land with specific boundaries where crop and livestock enterprises are taken up under common management.

4. **Farming** : Farming is the process of harnessing solar energy in the form of economic plant and animal products.

5. **Organic farming** : It is a production system when all kinds of agricultural products are produced organically including grains, dairy, fibers such as cotton, flowers, and processed food products.

6. **Sprinkler irrigation** : In this method of irrigation, water is sprayed into the air and allowed to fall on the ground surface somewhat resembling rainfall. The spray is developed by the flow of water under pressure through such orifices or nozzles.

7. **Drip irrigation** : Drip (or trickle) irrigation involves dripping water onto the soil at very low rates from a system of small diameter plastic pipes fitted with outlets called emitters. In this system, water is applied close to plants so that only part of the soil in which the roots grow is welted.

12.6.1 Crops

Agronomy is a Greek word derived from *agros* meaning "field" and *nomos* meaning "management". It is a specialized branch in agriculture dealing with crop production and soil management. Crops refer to plants that are grown on a large scale for food, clothing, and other human needs.

Crop Classifications

Crops can be classified in following ways:

1. According to Climate

(a) **Tropical crops** : These grow well in warm and hot climate, *e.g.* rice, sugarcane, jowar, bajra etc.

(b) **Temperate crops** : These grow well in cold climate, *e.g.* wheat, oats, gram, potato etc.

2. **According to Growing Season**
   (a) kharif (rainy/monsoon crops), *e.g.*, cotton, rice, jowar, bajra.
   (b) Rabi (winter crops), *e.g.*, wheat, gram, sunflower.
   (c) Summer crops *e.g.*, groundnuts, watermelon, gourds.
3. **According to Agronomics**
   (a) Cereals
   (b) Millets
   (c) Pulses
   (d) Oilseed crops
   (e) Sugar crops
   (f) Starch crops
   (g) Fibre crops
   (h) Cotton
   (i) Narcotics
   (j) Fodder
   (k) Grasses
   (l) Plantation crops
   (m) Spices and condiments
4. **According to the Duration of Crops**
   (a) **Seasonal crops** : These complete in one season, *e.g.* rice, wheat etc.
   (b) **Two season crops** : These complete life cycle in two crops, *e.g.* cotton, turmeric, ginger.
   (c) **Annual crops** : These require one full year, *e.g.* sugarcane.
   (d) **Biennial crops** : These require two years to complete its life cycle, *e.g.* banana, papaya.
   (e) **Perennial crops** : These crops live for several years, *e.g.* fruit crops, mango, guava etc.
5. **According to Cultural Method**
   (a) **Rain fed crops** : Cultivation of these crops mainly depend on the availability of rain water, *e.g.* jowar, bajra, moong, etc.
   (b) **Irrigation crops** : These are cultivated with the help of irrigation water *e.g.* sugarcane, banana, papaya, chili etc.
6. **According to Economic Importance**
   (a) **Cash crop** : These are grown for earning money *e.g.* sugarcane, cotton, dry fruits etc.
   (b) **Food crops** : These are grown for raising foodgrain for the human beings and fodder for cattle.

12.6.2 Crops Rotation

Growing of different crops on a piece of land is a rotation is a pre-planned succession is known as 'crop rotation'. Its basic principle is to utilise the available resources so as to harvest maximum in a unit land without adversely affecting the soil health.

The **principles of crop rotation** are :

(1) Crops with deep rooted like cotton should be followed by those which have shallow rooted crops like sorghum or maize. This helps uniform use of nutrients from the soil.

(2) More exhaustive crops should be followed by less exhaustive crops because crops like potato, sugarcane, maize etc. require more inputs such as more fertilizers, better tillage, more number of irrigation etc.

(3) Selection of crop should be based on demand.

(4) Crops of same family should not be grown in succession.

(5) Farmers should select the crops considering their financial conditions.

(6) The crops should be selected so as to suit the soil and climate.

12.6.3. Sustainable Agriculture

This type of agriculture aimed at meeting the needs of present generation without endangering the resource base of the future generations. Sustainable agriculture is a system of cultivation with the use of manure, crop rotation and minimal to tillage and with minimum dependence on synthetic fertilizers, pesticides and antibiotics.

Sustainable agriculture is a balanced management system of renewable resources including soil wildlife, forests, crops, fisheries, livestock, plant resources and ecosystems without degradation and to provide food and the livelihood for current and future generations maintaining or improving productivity and ecosystem services.

It prevent land degradation and soil erosion. It replenish the nutrients and control weeds, pests, and diseases through biological methods.

12.6.4. Manures and Fertilizers

Manures are the plants and animal wastes that are as a source of plant nutrients. These release nutrients after their decomposition. Organic manures binds the sandy soil and improve its water retention capacity. The manures also open the clayey soil and help in acration for better root growth. The organic manures should be added before the sowing or planting because of slow release of nutrients.

Fertilizers are industrially manufactured chemical containing plant nutrients. Fertilizers have higher nutrient content in fertilizers than organic manures and the nutrients are released almost immediately.

Integrated Nutrient Management. Integrated nutrient management system is a judicious combination of organic, inorganic and biofertilizers which replenishes the soil nutrients removed (consumed) by the crops.

Important Food Crops of the World (in order of their production)	
S.No.	*Name of crop*
1.	Sugercane
2.	Corn (Maize)
3.	Rice (Paddy)
4.	Wheat
5.	White Potato
6.	Cassava
7.	Soyabean
8.	Barley
9.	Sweet Potato
10.	Sorghom
11.	Oats

Origin of Food Plants	
Plant	*Place of Origin*
Potato	Asndes, South America
Wheat	Turkey and Afghanistan
Rice	India
Soya	China
Coffee	Ethiopia
Orange	China

12.7 ANIMAL RESOURCES

As discussed above, animals provide meat and milk products which supplement food obtained from agriculture. Animals are also a source of wool, which is needed by man. In developing countries, animals supplement human labour for work on farms. These may be used for ploughing, drawing water from wells and pulling carts.

Special breeds of animal have been developed for specific purposes, such as meat cattle, dairy cattle, sheep for mutton, and sheep for wool.

12.7.1. Animal Food

The animal food is rich in proteins. Animal foods are of following types :

(i) Vegetarian dairy products like milk, ghee, cheese etc.

(ii) Non-vegetarian food like eggs, meat, obtained from animal husbandry, and

(iii) Non-vegetarian sea-food coming from fishing like fish, shellfish, crabs etc.

About 80 species of livestock are used to produce animal husbandry food. The important ones include : Cows, Buffaloes, Sheep, Pigs, Chickens, Goats, Ducks etc.

Meat consumption is high in high societies, and hence, large portion of crops grown in highly developed countries are used to nourish the livestock animals required for human consumption. Thus, half the cereal grains produced in highly developed countries are used to feed their livestock. The breeding of livestock and their ultimate consumption as meat is thus, a highly energy-loss process. That is why, it is stated that vegetarians cause less burden on the society in consuming food energy.

Fisheries

Fishes may it be marine fish or frehswater fish, have an important source of protein. Latest surveys have indicated that about 44% of the world's fishers are heavily exploited, 16% are already overexploited. Modern fishing technology is responsible for overexploitation which is not sustainable.

About 55% of the fish comes from the ocean, 33% from aquaculture, and 12% from inland fresh water fishing in rivers, lakes, reservoirs and ponds. A third of the world's fish harvest is used as animal feed, fishmeal and oils. It is incorrect to say that the vast open sea holds unlimited amount of fish. 80 to 90% global commercial fish catch comes from coastal waters within 300 km of the shorelines.

Freshwater fish species diversity is under threat. Over-harvesting, and pollution are the main problem in fisheries.

12.8 IMPACTS OF AGRICULTURE

Adverse impacts of both, the traditional agriculture as well as the modern methods of farming are discussed hereunder:

Impacts of Ancient and Traditional Agriculture

Ancient farming used *shifting cultivation.* In this method, the farmers grow crops on the same land for only a year or two, so as to keep the soil as fertile. The farmers, then, cultivate and grow their crops on some adjoining area or move to the grass and trees are cut down and burnt. Therefore, the shifting cultivation is also called *'slash-and burn Agriculture'*. The abandoned land returns to grassland or forest land over time. This method of agriculture is still being practised in many tribal areas.

The main **impacts of traditional agriculture** are:

1. *Deforestation.* Shifting cultivation involving cutting down and burning of trees and grasses and continuous shifting of locations of the cultivated area, results in large scale deforestation.
2. *Soil Erosion.* Clearing of forests exposes the soil bare, causing its erosion by water and wind, thereby resulting in loss of top fertile soil.
3. *Depletion of Nutrient.* The shifting cultivation results in destroying the organic matter in the soil. The land re-claimed thereafter, is usually weak in nutrients. The rich forest land is, thus, made a wasteland.

Impacts of Modern Agriculture

Modern agriculture use hybrid varieties of foodgrain seeds, and of the various chemicals (fertilizers, insecticides, herbicides *i.e.* plant killers and fungicides *i.e.* the chemicals to control plant as chemical diseases, etc.) supported with good irrigation. However, these modern advances have given rise to its own problems, which are proving harmful to the ecology and environment, as discussed below:

Major adverse impacts of modern agricultural practices are:

1. Fertilizer related Problems

1. Most of the chemical fertilizers used in modern agriculture have nitrogen (N), phosphorous (P), and potassium (K) in inorganic form. These cause micro-nutrient imbalance in the soil and reduce the productivity of soil. The nitrogen and phosphorous used in crop fields, either reaches down to the ground water or washed off and reach the neighbouring water bodies causing over-nourishment, which is toxic and adversely affects the water foodchain by depleting the dissolved oxygen.
2. Chemical fertilizers can also become air-pollutants, and also contribute to global warming and depleting ozone layer.
3. Excess use of fertilizers cause acidification and salinisation of soils.

2. Pesticide related Problems

Pesticide is a general term for any chemical that is used to kill unwanted organisms. It includes insecticides, herbicides, and other chemicals (that kill spiders, mites, worms, and rodents). Modern agriculture and sanitation heavily depend on pesticides. Pesticides have saved millions of lives by killing disease-carrying insects. The pesticides, by eliminating pests that attack plants, have increased our crop yields. These chemical pesticides remain in the environment for much longer than the natural ones. Most pesticides are toxic to humans.

According to WHO, there had been major loss in fishes due to oceanic pollution.

Pesticides are used for reduction of insect pests, plant pathogens and weed population to acceptable levels. Despite all the control efforts used, pests annually destroy about 35% of all crops. After considering the loss by birds, rodents etc, the total destruction is about 40-50%.

(i) Many of the pesticides not only kill the target species, but also non-target species which are useful for agriculture crops and other ecosystems.

(ii) Most pesticides now applied are long lasting synthetic compounds and are nondegradable and keep on accumulating in the food chain and are very harmful and lethal for the human beings and other animals.

(iii) Excessive use of pesticides poses a more direct problem for farm workers.

(iv) Pesticides do not stay where they are applied; rather they tend to move through the soil, water and air. When they enter the river water, they may damage to fish and also, cannot be easily removed through ordinary water treatment process.

3. Hybrid seeds related Problems

(i) Hybrid seeds encourage monoculture, as farmers tend to grow only that crop, which is likely to give maximum yield. This may lead to rapidly spread of the crop diseases.

(ii) Hybrid seeds generally need the support of lot of fertilizers, which have their own environmental impacts.

4. Water-logging Problems

Modern agriculture encourages production of only those crops which need a lot of irrigation water. The anonymous use of canal irrigation water may sometimes cause water-logging, thereby reducing the fertility of soil. In this excess water gets accumulated and gradually forms a continuous column in these fields due to inadequate drainage facilities and results in poor aeration to the crop roots, and reduces crop yield.

5. Salination Problem

Over-irrigation of salt-rich arid and semi-arid lands raises the ground-water table which brings with it dissolved salts from the lower soil profile. This initially reduces crop yield and after water evaporation, it leaves salt cakes on the surface making the field unfit for cultivation.

6. Other Problems

(i) Modern farming machines, like tractors, diesel pumps, harvesters, threshers etc. run on fossil fuels, which cause air pollution.

(ii) Heavy farm machines when used in the fields causes compression of soil, leading to decline in soil fertility.

(iii) Large scale development of agriculture has resulted in the conversion of valuable and biologically rich forests into agricultural land, thereby causing shrinkage of forest cover. This has resulted into increase in air pollution, and also habitat loss to various animal species.

(iv) The large scale use of land for agriculture has also resulted in destruction of wetlands, as several wetland areas have been filled up and re-claimed for agriculture.

QUESTIONS

1. What do you understand by food resources? Discuss the different types of food resources in brief.
2. Explain various food problems of the world and of India.
3. Describe environmental and other impacts of traditional as well as modern agriculture.
4. Describe in details as to how the use of pesticides do cause harmful effects on living organisms.
5. How does over-grazing impacts the grassland ecosystems?
6. What effects are caused on the availability of marine food due to over-fishing by man?
7. What is agriculture? What is its importance?

8. Define food. What are the needs for food?
9. Explain as to how the evolution of food production was done.
10. Explain the different types of food products.
11. What is the 'green revolution'? Explain it in detail.
12. (a) What are the basic resources for agriculture?
    (b) Explain the different types of agriculture.
13. How will you define a crop? Enumerate the different types of crops.
14. Discuss impacts of modern agriculture in detail.
15. Write short notes on:
    (a) Animal resources.
    (b) Fertilizer related problems.
    (c) Pesticide related problems.
    (d) Hybrid seeds related problems.
    (e) Water-logging and salivation problems.
16. Explain the following in brief:
    (a) Crop rotation
    (b) Sustainable agriculture
    (c) Manures and fertilizers
    (d) Animal foods
    (e) Fisheries.

PART IV : ENVIRONMENTAL POLLUTION

13

Chapter

Environmental Pollution (Hazards and Control)

13.1 INTRODUCTION

Air, water and surroundings are polluted by various activities of man, like industries, energy conversion plants, infrastructural projects etc. We need clean air, water atmosphere and surroundings for good health and better quality of living. Various process without pollution control provisions affect the environment adversely. These have certain degrading effect on the environment.

The large-scale emission of CO, CO_2, NO_2, SO_2, particulate matter, heat etc. into the atmosphere results in local pollution and global environmental degradation effects, such as:

- Adverse effects on human health, as discussed in the earlier chapters.
- Acid rains harming the agriculture, forests, animals, human heath, aquatic life, drinking water in the lakes etc.
- Global warming, melting of polar snow caps, increased level of ocean, environmental disturbance etc.

The emissions are influenced by power plants, industries, automotive vehicles etc.

The *environment is constituted by surroundings and includes:*

- Air, water, land, atmosphere.
- Land, mountains, deserts.
- Water, lakes, rivers, oceans.
- Trees, forests, pastures, agriculture.
- Biological organism in the surroundings.
- Dwelling houses, buildings and roads.
- Infrastructure projects including highways, bridge, power houses, ports, railways etc.
- Industries.

Normally, the nature has self-cleaning processes, like photosynthesis, water cycle, seasons, winds etc. resulting in clean environment and ecological balance. The extraction, processing, conversion and use of energy are accompanied by degradation of the environment.

Due to high tolerancing power of nature, the disturbances upto a certain level can be absorbed. But with the rise in agricultural production, urbanization, modern technology, high, standards of living, human interference with environment clubbed with rapid rise in human population, a stage has reached where nature is unable to sustain this ever growing pressure through its capacity for self-stabilisation. This indiscriminate neglect and disregard of environment may result in:

1. Large scale land degradation
2. Deforestation
3. Extensive pollution of rivers and other water bodies
4. Depletion of ground water level and deterioration of groundwater quality
5. Large scale atmospheric pollution in cities and industrial locations
6. Extinction of great number of plants and animal species
7. Rapid depletion of natural resources
8. Climate change
9. Acid rain
10. Global warming
11. Rise in sea-water level
12. Ozone depletion
13. Smog

Various *reasons of concern for degradation* of environment are given hereunder:

(i) The economic and population growth has put great pressure on consumption of water, energy, biological resources.

(ii) Ignoring the long term consequences of development activities, and not adopting the appropriate technologies for the purpose of sustainable development.

(iii) Failure to recognise that our economic system, our health and our lifestyles depend on the continued adequate functioning of ecological systems.

It should be understood clearly that the availability of natural goods and services is controlled by the hydrologic cycle, biogeochemical cycles, the climate system, and the maintenance of biological diversity and functioning of ecosystems.

13.2 ENVIRONMENTAL POLLUTION (DEGRADATION)

Environmental pollution can be defined in many ways, some of them are:

(i) Environmental pollution is a reduction in the quality of environment by the introduction of impurities.

(ii) Environmental Pollution means the act of destroying the purity or sanctity or making foul or filthy.

(iii) Environmental Pollution may also be defined as the introduction by man into the environment of substance or energy liable to cause hazards to human health, harm to living resources and ecological system, damage to structure or amenities or interference with legitimate uses of the environment.

(iv) Environmental pollution may be described as the unfavourable alterations of our surroundings because of the acts and omissions of the man. In our large and industrial towns, we are already experiencing the effects of air, water, soil and noise pollution which has already crossed human tolerance.

Therefore, in environmental pollution, we should study the causes and acts of destroying the purity of air, water and land and their effect on ecology *i.e.*, plants, animals, and human beings, and measures to control their effects. Air pollution is mounted in cities due to higher growth of populations and industrialisation. Water pollution has increased with the growth of population and industries. Disposal of industrial, commercial and household waste is becoming a problem. Noise has also become acute due to tremendous increase in motor vehicles.

Pollution has reached such magnitude that it not only threatens the health and well-being of the population in a particular area, but also produces effect on global scale. Man dumps tonnes of chemical and nuclear waste into the rivers and sea. The food eaten by him, is spread with lethal pesticides, which on long run can have serious side-effects, like paralysis, kidney trouble, cancer etc.

The environmental deterioration by man is attributable to three **major causative factors**:

1. Over-population,
2. Urbanisation and Higher living standards, and
3. Industrialisaton.

The increasing amount of wastes generated by these three phenomena, degrade the quality of land, water, air and food. Increase in population demands more foods, water and land—three major items that are limited on the earth. To increase food production, either more efficient method should be adopted, exerting more pressure on the limited cultivable area, or more land should be devoted for agricultural purposes diverted from forestland, thereby reducing carbon dioxide sinks which protect the world from a 'global-warming' effect. Man counts on the synthetic biocides to overcome crop losses due to pests, and relies on chemical fertilizers to boost the crop yield. Farmers who used to grow one crop a year, now have opted to cultivate as many as three. These practices have resulted in increased yields, soil degradation, nutrient depletion, water pollution by plant nutrients and contamination of the total environment with pesticides. Pesticidal chemicals are potential contaminants of air, water and food.

Every industry is associated with an emission of one type or other pollutants directly or indirectly. Industries are not only responsible for the pollution of air but also contamination of water. The untreated effluents by industrial facilities are primarily responsible for the water problem. Pollutants are chemical, physical or biological agents that exert undesirable effects on human health and the environment. The air we breathe, the water we drink, the food we consume and the land we live on, are all contaminated with one or the other from of pollutants, as mentioned hereunder:

Types of Environmental Pollutants

1. *Chemical Agents*. Gases and particulates, heavy metals, pesticides, petroleum, solid and liquid wastes.
2. *Physical Agents*. Heat, odour, noise, radiation.
3. *Biological Agents*. Micro-organisms, population.

13.2.1 Environmental Pollution and Health

One of the report of World Health Organisation (WHO) reveals that 24% of global diseases and 23% of all deaths are caused by environmental exposures, many of which can be prevented by better environmental management. The four diseases most influenced by poor environment were diarrhoea, lower respiratory infections, unintentional injuries and malaria:

1. *The diseases caused by air pollution*. Some of the major air-borne diseases are: common cold, cold influenza, chicken-pox, tuberculosis, silicosis, black lung cancer, bronchitis, asthma etc.
2. *The diseases caused by water pollution*. Some of the common water-borne diseases are: cholera, typhoid, dysentery, hepatitis, intestinal disorders etc.
3. *Food-borne diseases*. Contaminated food leads to various diseases, which includes: bolutism, viron food poisoning, fungal food poisoning, staphy lococcal poisoning.
4. *Deficiency diseases*. These may be: avitaminosis, anaemia, rickets and osteomalacia, teeth and gum diseases.

Between 60 to 90 percent of all cancers are directly or indirectly related to the environment that is related to factors: water, air, solar radiation, living and working environments and personal choices of diet and ways of life, such as tobacco smoking and alcohol consumption.

13.3 AIR POLLUTION

Air pollution may be defined as the "presence in the outdoor atmosphere of one or more contaminants or combination thereof in such quantities and of such duration as may be, or may tend to be injurious to human, plant or animal life, or property, or which unreasonably interfere with the comfortable enjoyment of life or property, or the conduct of business."

Air pollution can also be defined as "the presence of any solid, liquid, or gaseous substance (including noise) present in the atmosphere in such concentrations that may or tend to be injurious to human beings, or other living creatures, or plants, or property, or enjoyment.

Atmosphere is a dynamic system, and absorbs various pollutants from natural as well as man-made sources, thus, acting as a natural sink. Gases, like CO, CO_2, H_2S, SO_2 and NO_2

as well as particulate matter, such as sand and dust, are continuously released into the atmosphere through natural activities, such as forest fires, decaying of vegetation, winds and sand or dust storms. Man-made pollutants *e.g.*, CO, NO_2, CO_2, hydrocarbons, particulates are also released into atmosphere. Now-a-days the problem of air-pollution has alarmingly increased due to population and other human activities. The pollutants are now-a-days enter the atmosphere at a faster rate than absorbed by natural sink, thus, accumulating in the air, and disturbing the equilibrium in the atmosphere. Green house effect on global warming is outcome of excessive pollution.

All the pollutants, which are emitted directly from the identifiable sources, either from the natural hazardous events, like duststorms, volcanoes, etc. or from human activities, like burning of wood, coal, oil in homes or industries or automobiles etc. are called the *primary pollutants.* The following five **primary pollutants** contribute to about 90% of the global air pollution:

1. **Oxides of sulphur,** like sulphur dioxide (SO_2);
2. **Oxides of carbon,** like carbon monoxide (CO) and carbon dioxide (CO_2);
3. **Oxides of nitrogen,** like NO, NO_2, NO_3;
4. **Volatile organic compounds,** mostly hydrocarbon; and
5. **Suspended particulate matter** (SPM).

(i) **Sulphur oxides** are produced when sulphur containing fossil fuels are burnt.

(ii) **Carbon monoxide** is a colourless, odourless toxic gas produced when organic materials like natural gas, coal or wood are incompletely burnt. Vehicular exhausts are the largest single source of carbon monoxide.

(ii) **Nitrogen oxides** are found in vehicular exhausts. These are significant pollutants as they produce secondary air pollutants, such as ozone, which is harmful for the atmosphere.

(iv)**Hydrocarbons** are a group of compounds of carbon and hydrogen atoms. These either evaporate from fuel or are remains of fuels that did not burn completely. These hydrocarbons are washed out of the air during rains. These create problems when they react to form secondary pollutants.

(v) **Suspended particulate matter** are small pieces of solid material dispersed into the atmosphere, for example, smoke particles from fires, ash from industries and dust particles. Liquid particles include mist and fog, while solid particles are dust, smoke and fumes.

Indoor Air Pollution

The lack of ventilation indoors concentrates air pollution where people have the greatest exposure. Pollutants are also found indoors from the infiltration of polluted air from outside and from various chemicals used or produced inside buildings. Building materials, plywood, paint and solvents emit gases.

The air pollution is intentionally introduced with the use of air fresheners, incense, and other scented items. Fires in stoves, controlled woodfires in fireplaces also add significant amount of smoke paticulates into the air. Indoor pollution fatalities may be caused by using pesticides and other chemical sprays indoors without proper ventilation.

The most important source of indoor air pollution in rural villages is the combustion of domestic **biomass fuel,** such as cowdung cakes, wood, charcoal and crop residues, which are used for cooking, particularly by the poor rural and tribal families. The incomplete combustion of such unprocessed solid biomass fuels increases the level of indoor air pollutants, such as particulate matter, and noxious gases like carbon monoxide. Other major by-product of such burning of biomass fuel is the fine soot particles that can penetrate deep into the lungs causing acute respiratory diseases and lung cancer.

The indoor air pollution is much more worse than outdoor air pollution since a pollutant released indoor is more likely to reach the lungs than a pollutant released outdoors.

Top priority should be given to reduce air pollution by promoting the use of cleaner fuel such as biogas, LPG, solar cookers, improved stoves and better ventilation of homes and kitchens.

Smog

Smog is a combination of two words **smoke** and **fog.** Smog refers to hazy air that causes difficult breathing conditions. Smog is a combination of various gases with water vapour and dust. A large part of the gases that form smog is produced when fuels are burnt. Smog forms when heat and sunlight react with these gases and fine particles in the air.

The most harmful components of smog are ground level ozone and fine airborne particles. Ground level ozone forms when pollutants released from gasoline and diesel-powered vehicles and oil based solvents react with heat and sunlight. This ozone is harmful to humans, animals and plants.

Smog occurrences are often linked to traffic, high temperatures and calm winds. During the winter, wind speed are low and cause the smoke and fog to stagnate near the ground, hence pollution levels increase near ground level. Smoke particles trapped in the fog gives it a yellow/ black colour and this smog generally settled over cities for many days during winter.

Smog has following problems :

1. It harms the environment and creates visibility problems.
2. It creates respiratory problems.
3. It is responsible for bronchial diseases, which some times result into deaths.
4. Heavy continuous smog may result in the decrease of natural vitamin D production.

Bhopal Gas Tragedy

In the night of 2 December 1984, deadly methyl isocyanate (M/C) and other lethal gases escaped out of the Union Carbide's pesticide manufacturing plant located in Bhopal, Madhya Pradesh. Due to the accidental entry of water in the tank, the reaction mixture got over-heated and exploded, because its cooling system had failed. The M/C is a highly toxic gas, affecting eyes, lungs and skin. At higher concentrations, it removes oxygen from lungs and causes death. Due to accident, the gas spread in more than 40 sq km area.

The gases poisoned about 5 lakh people, leaving them with lifetime of ill health and mental trauma. Many of the victims were people, living in the slums surrounding the factory. Many of the victims after the gas effects, became too weak to do any work. About 8000 people died in a few days following the disaster. The death toll over the yeats has risen to about 20,000. Death toll is still rising due to exposure-related illnesses. More than 50,000 people have become handicapped.

The gas exposure effects have been passed onto generations. Today, atleast 1,50,000 people, including children born to the gas exposed parents continue to suffer exposure related health effects.

These primary pollutants often react with one another or with water vapour, aided and abetted by the sunlight, to form entirely a new set of pollutants, called the *Secondary Pollutants*. These new pollutants are often more harmful than the original primary pollutants that produce them. The important **secondary pollutants** are:

1. Sulphuric acid (H_2SO_4),
2. Ozone (O_3);
3. Formaldehydes; and
4. Peroxy-acyl-nitrate (PAN) etc.

13.3.1 Sources of Air Pollution

Major sources of air pollution are:

1. Burning of different kinds of fuels, including combustion-fired power plants.
2. Controlled burn practices used in agriculture and forestry management.
3. Motor vehicles generating air pollution emission.
4. Marine vessels.
5. Wood, coal, fuel oil or natural gas burning.
6. Oil refining, power plant operation and industrial activities.
7. Chemicals, dust, and crop waste burning in farming.
8. Fumes from paints, varnish, aerosol sprays and other solvents.
9. Waste deposition in landfills which generate methane.
10. Military uses, as nuclear weapons, toxic gases, germ warfare and rocketry.
11. Dust from natural sources usually large areas of land with little or no vegetation.
12. Pine trees, which emit volatile organic compounds.
13. Smoke and carbon monoxide from wildfires.
14. Volcanic activities, which produce sulphur, chlorine, and ash particulate.

Various sources of air pollution can be grouped into the following categories:

A. Fuel burning
B. Industrial processes
C. Automobile exhaust

Each of these sources are described further:

(A) Fuel Burning

These operations are performed for heat and power generation. The fuel burnt may be coal, gases, oil, refuse, wood etc. These are responsible for generating air pollution of the following types:

(i) **Smoke:** This is a product of incomplete combustion of coal, carbon monoxide, carbon dioxide etc.

(ii) **Dust:** It is finely divided powder and remains in atmosphere in floating condition. These are generated in large quantity in mining, quarrying, construction, cement industries etc.

(iii) **Fumes:** Solid particles generated by condensation from gaseous state and often accompanied by a chemical reaction, such as oxidation, are known as fumes.

(iv) **Gases:** These are formless fluids.

(v) **Smog:** This is a formation of heavy, cloudy, hazy, floating layer formed by the mixture of smoke, dust, fog, mist etc.

(B) Industrial Processes

As mentioned earlier, industries, like cement, chemical, textile, thermal power station, steel plants, oil refineries, fertilizers etc. are the main contributors to air pollution. These are responsible for emitting harmful gases, fly ash, dust, soot.

With the proliferation of lead, mercury, oxides of nitrogen, ozone, cyanides, sulphur compounds etc. are responsible for health hazards and are increasing regularly. The workers working in mines, construction and cement industries have to inhale lot of dust and suffer-from Asthama, Tuberculosis, Silicosia, Byssionosis etc.

(C) Automobile Exhaust

Most of the vehicles and construction equipment are operated by diesel engines and their exhaust is responsible for major air-pollution. The construction equipment may be either stationary equipment without prime mover, like air compressors, diesel generating sets, concrete mixers, dewatering pumps or may be with prime movers *e.g.*, dumpers, transit mixers, compactors, shovels, excavators, graders, loaders, pavers etc.

Automobile exhausts has following main constituents:

1. **Gases:** Carbon monoxide, nitric oxides, unburnt hydrocarbons, partial oxidation products.
2. **Particulates:** These include lead compounds, carbon and various other inorganic or organic materials.

Furthermore, additional hydrocarbons and fuel additives are contributed by the evaporation of gasoline from automobile tanks.

13.3.2 Effects of Air Pollution

Air pollution has many adverse effects. It damages the building structures, materials, vegetations, causes health hazards and obstructs visions. The effect of air pollution can be categorised as:

(a) Effects on human health
(b) Effects on animals
(c) Effects on plants
(d) Effects on materials
(e) Effects on atmosphere

(A) Effects on Human Health. Air pollution seriously affects the respiratory system. Carbon monoxide when absorbed into the lungs and reduces the haemoglobin available to carry oxygen to the body. This may cause injury to vital organs. Sulphur dioxide also causes cough, shortens breath and can cause irritation to the eyes. It then converted to sulphuric acid, damage the lungs. It also helps in severe infections and allergic response. Even very low concentration of ozone can cause damage of the lungs. Hydrogen sulphide and other organic oxides can create odour problems. Hydrocarbons are believed to cause cancer in lungs, whereas, lead emitted by automobile exhausts is a cumulative poison and can cause brain damages.

Some of the chronic diseases due to exposure to polluted air are lungs cancer, bronchitis, emphysema and asthama.

(B) Effects on Animals. The effects of air pollution due to its inhaling on animals are similar to those on human. Air pollutants also enter the animal body through ingestion of contaminated plant materials. Among the metallic contaminants, arsenic, lead and molybdenum are important. Air contaminated with ozone, flouride effects the forage crops, which when eated continuously by the animals, affects the animal adversely.

(C) Effects on Plants. As we know that leaf veins function as the transport system for water, minerals and food. The leaf tissues form cells in between them. Air pollution damages these leaf veins and tissues. The damage can be in the form of Necrosis (refers to dead areas on a leaf structure), Chlorosis (resulting in yellowing of the leaf), Abscission (dropping of leaves). When the absorption of sulphur dioxide exceeds a particular level, the cells become inactive and are killed, resulting in tissue collapse and drying of plants adversely. Chlorine, ammonia, hydrogen sulphide and ethylene also affects the vegetations adversely. Smog also causes injuries to the plants, and certain species are injured even with very small concentration of smog.

(D) Effects on Materials. Air pollution also adversely affects building structures and materials. Smoke, dust and oxides of sulphur are most destructive pollutants. Sulphur dioxide changes with moisture and air to sulphuric acid which is corrosive. Moisture, temperature, sunlight and air movement decide the rate of attack of pollutants. Aluminium, copper, iron and steel are corroded, when exposed to contaminated air. Building structures are damaged/disfigured due to polluted air, sooty deposits, normal weathering.

(E) Effects on Atmosphere. Increased combustion of fossil fuel and oil, enchances the carbon dioxide concentration of the atmosphere. Carbon dioxide absorb heat and the radiative cooling effect of the earth is, thus, decreased. CO_2 is the single culprit responsible for half of global warming. This is known as "greenhouse effect" and is expected to raise the planet's temperature by 1.5–4.5°C in about 60 years.

'Green house' is a shield that keeps our planet warm enough to sustain life, but is now becoming a heat trap threatening to disrupt the global environment. A thin layer of certain gases, about 25 km like a green house glass letting heat through but stopping enough radiation back into space to warm out world. But now "Green house" layer is becoming denser, keeping in more heat.

13.3.3 Prevention of Air Pollution

In order to prevent air pollution following steps must be taken:

1. Everyone of us must feel responsibility and take initiative immediately to prevent air pollution.
2. We should start planting trees, as they purify the air, and thus, keep the sky clean. Trees breathe carbon dioxide and release oxygen. Forests act as vast sinks for carbon dioxide and tend to reduce global warming.
3. Curtail vehicular emission by using them to minimum possible, using fuel efficient engines, keeping vehicles in maintained condition. Conserve fuel, encourage mass transport, use cars in pools, monitor the fuel emission.
4. Polythene, plastic containers, polyurethane foams and P.V.C. are eco-non-friendly items, therefore, their use should be avoided, disposal of these plastics is not an easy task, as burning produces soot, corrosive gases and highly offensive odour.
5. Maximise use of solar energy—use of solar heaters, solar energy, electric plants be encouraged.
6. Air pollution monitoring should be carried out regularly.
7. While preparing a project, environment management plan should be prepared for formation, implementation and monitoring of environmental protection measures during and after commissioning of projects.
8. Emission levels of pollutants from chimneys should conform to the pollution control standards prescribed by the central or state pollution control boards, for which necessary arrangements must be made. Generally, used air cleaning devices are:

   (a) Electrostatic precipitators (b) Cyclone separators
   (c) Wet collectors (d) Gravity settlers
   (e) Bag filters (f) Cyclone filters
   (g) Wet scrubbers (h) Absorbers or Adsorbers.
9. For indoor pollution, use proper ventilation so that used air or vitiated air gets removed by air.

13.3.4 Control of Air Pollution

A. Control of Pollutants

For the purpose of air pollution control, the pollution can be divided into following categories:

1. Particulate control in chimneys.
2. Particulate control in atmosphere.
3. Dust control.

1. Particulate Control in Chimneys. Plants used in construction, cement industries, thermal power plants etc. emit particulates through chimneys. These emit dusts of coal, stone, cement, lime, sand etc. and combustion gases. The persons engaged on these plants and in nearby areas are badly affected by this type of pollution. These also adversely affect the nearby residential areas, and cultivated farm etc. Dust collectors are required to be installed with such plants to remove suspended particles from the chimney emission. The Dust collectors are of several types, and their suitability is required to be decided depending upon the characteristics of the particulate.

2. Particulate Control in Atmosphere. In construction, cement and mining industries, thermal plant areas, coal, stone aggregates, ore, sand etc. are stored at site or plant areas. These should be maintained in such a condition so as to prevent dust from becoming wind borne while in storage or being handled, loaded or transported. Following steps may help in this respect:

(i) Wetting of material to prevent dust, becoming wind borne.
(ii) Location of stock piles in an area, protected from prevailing winds.
(iii) Installation of wind breaks *i.e.*, trees, wall etc., to protect stock piles from direct exposure to the prevailing winds.
(iv) Storage of materials in bins, silos or any other type of enclosures.
(v) Plant and plant areas should be cleaned regularly of accumulated dust, dirt etc.

3. Dust Control. Dust emitted due to haulage, working of heavy earth moving equipment etc. are needed to be controlled by adopting following measures:

(i) Maintaining haul roads, moving the vehicles at reasonable speed, sprinkling water on unpaved haul roads.
(ii) Heavy earth moving equipments should be provided with cabins so as to protect the operators.

B. Site Selection

For installing an industry, it is necessary to select the site which will have minimum adverse effects on people, animals, agriculture etc.

C. Control of Motor Vehicles Emission

In order to control emission from motor vehicles proper maintenance including engine tuning, use of catalytic reactors etc. are necessary.

D. Use of Latest Technology

Equipments with latest technology, fuel efficient engines, plants with dust collectors etc., should be used so as to have better control over air pollution.

13.3.5 Ambient Air Quality Standards

Ambient air quality standards prescribed in India are given in the following table:

S.No.	*Pollutant*	*Time weighted average*	*Concentration in Ambient Air in $\mu g/m^3$*		
			Industrial Areas	*Residential Areas*	*Sensitive Areas*
1.	Sulphur Dioxide (SO_2)	–Annual Average –Daily Average	80 120	60 80	15 30
2.	Oxides of Nitrogen (NO_2)	–Annual Average –Daily Average	80 120	60 80	15 30
3(A).	Total suspended Particulate Matter (TSPM)	–Annual Average –Daily Average	360 500	140 200	70 100
3(B).	Respirable Particulate Matter (Size ≤ 10 μm) (RSPM)	–Annual Average –Daily Average	120 150	60 100	50 75
4.	Lead (Q_b)	–Annual Average –Daily Average	1.0 1.5	0.75 1.00	0.50 0.75
5.	Carbon Monoxide (CO)	–8 hour Average –1 hour Average	5,000 10,000	2,000 4,000	1,000 2,000

13.3.6 National Air Quality Index

National Air Quality Index (AQI) was launched in April, 2015 starting with 14 cities to disseminate air quality information. The AQI has six categories of air quality, *viz.,* Good (0-50), Satisfactory (51-100), Moderately polluted (101-200), Poor (201-300), Very poor (301-400), and severe (401-500). All these categories have been allotted distinct colour scheme. AQI considers eight pollutants, namely PM 10, PM 2.5, NO_2, SO_2, CO, O_3, NH_3 and Pb for which National Ambient Air Quality Standards are prescribed.

13.3.7 Air Pollution in India

India is placed as the "bottom performer" on several indicators such as environmental health impact, air quality, water and sanitation.

13.8.8 Steps for Controlling Air Pollution

1. Formulation of a comprehensive policy for Abatement of Pollution.
2. Supply of improved auto-fuel.
3. Formation of improved vehicular and industrial emission norms.
4. Mandatory environmental clearance for specified industries and projects.

5. Management of municipal, hazardous, and bio-medical wastes.
6. Promotion of cleaner technologies.
7. Strengthening the network of air quality monitoring stations.
8. Public awareness.

13.4 WATER POLLUTION

Water is essential to all forms of life. 20% of the surface of the earth is land and the rest is all covered by water. Nothing could survive on the earth without water. Conservation and keeping up of the good quality of water are of prime importance, since water as a natural resource is 'limited' in a sense, because 97% of the water is contained in oceans as salt water and additional about 0.3% of salt water is buried under ground. Of the remaining 2.5% about 2.2% is frozen in the polar ice caps and in glaciers. The remaining small quantity of about 0.3% is available for our use.

Water pollution can be defined as the adverse change in composition or condition of water so that it becomes less suitable for the purpose for which it would be suitable in its natural state. These changes may be chemical, physical, biological or radioactive. Thus, water pollution means addition of any thing to water which changes the natural quality of water so that people residing in downstream does not receive the natural water of the stream.

Man has polluted much of this limited quantity of water with sewage, industrial wastes and synthetic chemicals. We are harming our planet to the point where organisms are dying at a very alarming rate, in addition, our drinking water has become greatly affected. Water pollution occurs, when a body of water is adversely affected due to the addition of large amounts of undesired materials to the water. When it is unfit for its intended use, water is considered as polluted.

Types of Water Pollutants

1. Oxygen-demanding wastes.
2. Disease-causing agents.
3. Synthetic organic compounds.
4. Plant nutrients.
5. Inorganic chemicals and mineral substances.
6. Suspended solids.
7. Radioactive wastes.
8. Thermal discharges.

13.4.1 Sources of Water Pollution

1. *Point Source Pollution.* These occur when harmful substances are emitted directly into a body of water, like industrial discharge and discharge of untreated or poorly-treated sewage/waste water.
2. *Non-point or diffused pollution.* Non-point or diffused pollution can arise from many sources. These are mainly related to the way we use and manage land and soil. Groundwater are vulnerable from and affect by, leaching of pollutants from the land

surface and from areas of contaminated land, while surface water are affected by rainfall that washes over and off the land (run-off). Run-off has increased as agricultural activities have intensified and as we have built more roads and houses, particularly where we have degraded the natural permeability of the landscape and reduced its capacity to retain water.

Water Contaminants

Contaminants may include organic and inorganic substances:

1. ***Organic, water pollutants.***
   - Insecticides and herbicides, and other chemicals;
   - Bacteria from sewage or live-stocks operations;
   - Food processing waste;
   - Tree and brush debris from logging operations.
2. ***Inorganic water pollutants.***
   - Heavy metals including acid mine drainage;
   - Acidity caused by industrial discharges;
   - Chemical waste as industrial by-products;
   - Fertilizers in run-off from agriculture;
   - Silt in surface run-off from construction sites, logging, burn practices or land clearing.
3. ***Additional pollutants.***
   - Petroleum;
   - Radioactive pollutants;
   - Heat, it increases temperature resulting in the death of many aquatic organism.

13.4.2 Effects of Water Pollution

1. ***Ecological Impacts***
   - Rise in Biochemical Oxygen Demand (BOD).
   - Lowering of dissolved oxygen (DO).
   - *Algal boom.* Chemical run-off from farmers' fields bring extra nutrients into a lake, allowing increasing growth of algae in the water.
   - *Bio-accumulation and Bio-magnification.* Inside a biological system, non-essential and harmful substances may either be stored or are reacted upon by the enzyme system and excreted.
2. ***Chemicals Affecting the Human Health***

A large number of chemicals that either exists naturally in the land or are added due to human activity dissolved in the water, thereby contaminating it and leading to various diseases.

- *Fluoride.* Though it protects the dental caries, but higher levels can have an adverse effect on health. Excess fluorides can cause yellowing of the teeth and damage to the spinal cord.

- *Arsenic.* High concentration of arsenic in water can have an adverse effect on health.
- *Lead.* Higher levels of lead in the water may affect the central nervous system. Children and pregnant women are most at risk.
- *Petrochemicals.* These may cause cancer.
- *Other heavy metals.* These contaminants come from mining/quarrying waste, landfills, or hazardous waste dumps.
- *Chlorinated solvents.* Metals and plastic effluents, fabric cleaning, etc., are often discharged and contaminate ground water.
- *Pesticides.*
- *Nitrates.* It restricts the amount of oxygen that reaches the brain.

3. ***Water-Borne Diseases***

These are the infectious diseases which spread primarily through contaminated water. These may be:

(a) *Bacterial infections.* Typhoid, Cholera, Paratyphoid, Bacillary dysentery.

(b) *Viral infections.* Infectious Hepatitis (Jaundice), Poliomyelitis.

(c) *Protozoan infections.* Amoebic dysentery.

13.4.3 Control of Water Pollution

1. **Water Pollution Treatment**

   (a) Treatment of municipal waste water.

   (b) Treatment of industrial discharge.

2. **Preventive Measures**

In order to prevent the spread of water-borne infectious diseases, people should take adequate precautions. Water pipes of city water supply should be regularly checked for leaks and cracks. The water at home should be boiled or filtered to ensure that it is free from infection.

We should learn ways for disposing harmful household wastes so that they don't end up in sewage treatment plants that cannot handle them or landfills not designed to receive hazardous materials.

State of Rivers in India

Major rivers in India though retain their quality at the originating stage in hilly areas, but they become polluted once they reach the plains. Agricultural run off, industrial effluents, and domestic sewage find their way into most rivers. In the middle stretches of the rivers, both the extraction rates as well as pollution levels are high.

Tests of the river water near the cities show that in most cases the water is unfit for drinking purposes.

Urbanisation, industrialisation, excessive withdrawal of water, agriculural run-off, improper agricultural practices and various religious and social practices all put together contribute to river pollution in India. The water flows in the river now is water from small *nalhas* and streams that carry with them sewage and industrial effluents. The residual fresh water is unable to dilute the pollutants, and the rivers, thus turn into stinking sewers.

Ganga Action Plan

In 1985, Ganga Action Plan (GAP) was launched (the largest ever river clean up operation in the country), though the plan was criticised for overspending and slow progress. GAP phase II in 1991 included cleaning operations for the tributaries of Ganga—the Yamuna, Gomati and Damodar. In 1995, the National River Conservation Plan (N.R.C.P.) was launched. In this, all the rivers in India were taken up for clean-up operations. Under this plan, the main activities include treatment of the pollution load from sewer system of towns and cities, setting up sewage treatment plants, electric cremation, and low cost sanitation facilities, river front development, afforestation, and solid waste management. But, due to large number of practical difficulties, lack of clear-cut defined responsibilities, no provision for problem of pollution from agricultural run-off, this programme was not very successful.

Namami Gange Project

The Ganga, largest river in india, due to pollution poses significant threats to human health and the larger environment. Severely polluted with human waste and Industrial contaminants, the river provides water to about 40% of India's population. Today, the Ganges is considered to be the 6th most polluted river in the world.

The main causes of water pollution in the river is the high population density, various human activities such as bathing, washing clothes, the bathing of animals and dumping of various human and industrial wastes. The river flows through large number of cities with high population density. A large proportion of the sewage water with higher organic load is also discharged into the river.

A number of initiatives were undertaken in the past to clean the river, but failed to deliver the desired results. In July 2014, *"Namami Gange Project"* was announced as an integrated Ganga Development Project.

The objectives of this project were : effective abatement of pollution, conservation and rejuvenation of the Ganga. The project has a budget outlay of Rs. 20,000 crores for the next 5 years.

The *Namami Gange* will focus on pollution abatement, interventions namely interception, diversion and treatment of waste water flowing through the open drains through bio-remediation, appropriate *in-situ* treatment, sewage treatment plants (STPS). Effluent treatment plants, rehabilitation and augmentation of existing STPs, and immediate short term measures for arresting pollution at exit points on river front to prevent inflow of sewage etc.

The pillars of *"Namami Gange Programme"* are :

1. Sewerage treatment infrastructure.
2. River front development.

3. River surface cleaning.
4. Bio-diversity.
5. Afforestation.
6. Public awareness.
7. Industrial effluent monitoring.
8. Ganga grams.
9. Development of public amenities.

13.4.4. Water Quality and Characteristics

For the purpose of safety to public health, economy and utility it becomes imperative to thoroughly check, analyse, and treat the raw water to safe and permissible limits, before supplying to the public. The quality standards may vary depending upon the requirement of each category of use *e.g.*, agriculture, type of industry, drinking, bathing, washing etc.

The water for drinking purposes should be : hygienic, aesthetically acceptable, non-corrosive to the distribution system, free from excessive amount of minerals and organic matters, should not be hard, and free from radioactive substances. The general criteria for water quality are discussed in following paragraphs.

1. **Physical Characteristics** : These characteristics relate to the quality of water for domestic use and are usually associated with the appearance of water, its colour, temperature, taste, odour, and turbidity.

(i) **Colour** : The fresh and clean water should be colourless. The dissolved organic matter from decaying vegetation or some inorganic materials may impact colour to the water. The presence of colour in water may or may not be objectionable from health point of view but is objectionable from aesthetic and psychology point of view.

(ii) **Tastes and Odours** : The dissolved materials, salts or gases may impart tastes and odours to the water. Taste and odour may be caused by the dissolved gases such as, H_2S, CH_4, CO_2 etc., combined with organic matter, mineral substances like NaCl, iron compounds, carbonates and sulphates of other elements, and phenols and other tarry or oily matter.

The odour is generally determined when water is cold, as it may change with the change in temperature.

(iii) **Turbidity** : The turbidity of water is due to the presence of suspended matter that interferes with the passage of light through the water, clay, mud, algae, silica, rust, bacteria and calcium carbonate are common substances found suspended in water.

2. **Chemical Characteristics** : Chemical analysis of water is carried out to determine the chemical characteristics of water.

These characteristics are as under :

(i) Total solids include both the suspended and dissolved solids. Water with a high solid content is inferior and may be polluted. Suspended solids give the water a dirty muddy appearance and is unfit for drinking. Dissolved solids may not produce any colour. The upper limit for dissolve solids (DS) for drinking water is 500 mg/lit.

(ii) **Hardness** : Hardness in water prevents the formation of sufficient leather or foam, when such hard waters are mixed with soap. It is usually caused by the presence of calcium and magnesium salts present in water. Hard waters are undesirable because they may lead to greater soap consumption, scaling of boilers, causing corrosion and incrustation of pipes etc.

(iii) **Acidity, Alkalinity and pH value**

Acidity : Natural water may contain both dissolved carbon dioxide (CO_2) and mineral acids. Acidity contributes to corrosiveness.

Alkalinity : Alkalinity is due to carbonates and hydroxides. It is the capacity to neutralise acid. Ground water often contains carbonates, bicarbonates and hydroxides, due to its interaction with limestone. Besides these, phosphates, silicates ions are also present giving water an alkaline character.

pH Value : These pH value denotes the nature and intensity of water towards acidity or alkalinity. The pH value of 7 indicates neutrality, 0-7 acidity and 7-14 alkalinity. pH value of 6-8 is desirable for drinking. However, it is preferred that the water should be supplied slightly alkaline (pH 7-8) and with no free carbon dioxide.

(iv) **Dissolved Oxygen (DO)** : The amount of oxygen dissolved in water is called the dissolved oxygen (DO). It is expressed in milligram per litre (mg/l). The solution of atmospheric oxygen in fresh water depends on the atmospheric temperature and pressure.

When an aqueous solution is saturated with air at 1 atmospheric pressure and 25°C, the amount of oxygen present in the solution is 8.32 mg/litre. It may be in the range of 7-15 mg/litre. The solubility of oxygen is less in salt water (sea water) than in fresh water. In polluted water, also the solubility of oxygen decreases. DO is necessary for the life of fish and other aquatic organisms. When its concentration is less than 4 ppm, the water system is unsuitable for aquatic life.

Low concentration or absence of oxygen is an indicator of pollution in water. Its determination is, therefore, essential for assessing the water quality.

The saturation of dissolved oxygen (DO) is 14 to 15 ppm at 0°C in fresh, clean water. With increase in temperature the capacity of water to dissolve oxygen is reduced and becomes 0 ppm at 100°C.

(v) **Biological Oxygen Demand (BOD)** : Bio-chemical oxygen demand (BOD) is defined as the amount of oxygen required by bacteria for stabilizing decomoposable organic matter in water under aerobic conditions, BOD is calculated by measuring the oxygen consumed (decrease in DO) by bacterial and chemical action in a closed sample of water maintained at 20°C for five days. BOD is expressed in mg/l.

(vi) **Chemical Oxygen Demand (COD)** : The chemical oxygen demand (COD) represents the amount of oxygen in milligrams required to oxidise all the organic pollutants present in water to carbon dioxide and water by a strong oxidising agent. COD is a more complete and accurate measurement of total depletion of dissolve oxygen in water. Water is considered seriously polluted when its BO or COD causes the DO content to fall below 4.5 ppm and gravely polluted if the DO level falls below 4 ppm.

13.4.5 BIS (ISI) Water Quality Standards

The surface water has been classified into following categories:

1. *Class A:* Drinking water sources without treatment but after disinfection.
2. *Class B:* For outdoor bathing, swimming etc.
3. *Class C:* Drinking water sources with conventional treatment followed by disinfection.
4. *Class D:* Waters for fish culture and wildlife.
5. *Class E:* Waters for irrigation, industrial cooling.

S.No.	*Main characteristics*	*Tolerance limit*				
		Class A	*Class B*	*Class C*	*Class D*	*Class E*
1.	pH value	6.5 – 8.5	6.5 – 8.5	6.5 – 8.5	6.5 – 8.5	6.5 – 8.5
2.	Dissolved O_2, mg/l, min	6	5	4	4	–
3.	BOD, mg/l, min	2	3	3	–	–
4.	Taste	Tasteless	–	–	–	–
5.	Total dissolved solids, mg/l, max	500	–	1500	–	2100
6.	Total Hardness, mg/l, max	300	–	–	–	–
7.	Copper, mg/l, max	1.5	–	1.5	–	–
8.	Iron, mg/l, max	0.3	–	0.5	–	–
9.	Manganese, mg/l, max	0.5	–	–	–	–
10.	Surphase, mg/l, max	400	–	400	–	1000
11.	Nitrate, mg/l, max	20	–	50	–	–
12.	Floride, mg/l, max	1.5	1.5	1.5	–	–
13.	Mercury, mg/l, max	0.001	–	–	–	–
14.	Arsenic, mg/l, max	0.05	0.20	0.20	–	–
15.	Cyanide, mg/l, max	0.05	0.05	0.05	–	–
16.	Lead, mg/l, max	0.10	–	0.10	–	–
17.	Zinc, mg/l, max	15	–	15	–	–
18.	Pesticides	Absent	–	Absent	–	–

13.5 NOISE POLLUTION

Sound in the environment is caused by vibrations in the air (or some other medium) that reach human ears and stimulate a sensation of hearing. When the sound becomes loud, or disagreeable, or unwanted, it becomes noise. Since noise produces several undesirable effects on our body health, it can be termed as an environmental pollutant. The Air (Prevention and Control of Pollution) Act, 1981, includes noise as one of the air-pollutant.

Noise may be defined as unwanted sound at a wrong time and at wrong place. Generally prolonged and loud sound is considered as noise. Meaningful sound is generally meant for inviting attention or expecting a consequent response, such as the cry of a baby. Excessive noise can result in loss of hearing. Today, noise pollution has become a serious health hazard in the big cities, as it disturbs the sleep, increases heart beat, effects respiratory system, dizziness, loss of physical control and other changes due to stress, nausea and vomiting may be caused, deterioration in concentration and even mental disorientation have been noticed temporarily.

Loud continuous noise reduces the working efficiency, interferes with communication increase the frequency of errors in working which may even cause accidents.

Noise Control Methods

Following are some of the common methods used for **noise control:**

1. Replacing noise producing machinery with quiet alternatives.
2. Interrupting the path of sound with insulating materials.
3. Use of baffles, sound proof materials.
4. Mounting the noise and vibration source on flexible supports.
5. Providing conveyors to avoid fall of material on metals from height.
6. Doing proper maintenance and lubrication.
7. Protecting the workers by providing earplug, earmuff, helmets, control cabins etc. Thus, shielding the receiver.
8. Modify the present practices and procedures so as to minimise noise level.
9. Trees are supposed to be the best for noise control, therefore, more and more trees should be planted along the roads, rail tracks and around the aerodromes. These should be planted in plenty in colonies, hospitals, school campuses etc. to protect the people from noise pollution from outside.

Sources of Noise

Motor vehicle noise, aircraft noise and rail noise, industrial noise, high volume sound from loudspeakers, noise due to poor urban planning for the residential accoustic environment. Besides appliances, power tools, audio entertainment systems are the common noise sources.

Measuring Sound

Sound is measured in decibels (dB). It is not a linear, but a logarithmic scale. For example, a change from 30 dB to 60 dB represents a 10,000 fold increase in loudness. Apart from loudness, the frequency or pitch of the noise also determines whether it is harmful or not.

Impact of Noise Pollution

1. Noise can do physiological or psychological damage if the volume is high or if the exposure is prolonged. A loud, high pitched noise damages the fine hair cells in the Cochlea of the ear. Since the body does not replace damaged hair cells, permanent hearing impairment is caused by prolonged exposure to loud noise.
2. Noise can also produce other effects like heart pulpitation, pupil dilation, or blood pressure. Migraine headaches, nausea, dizziness, gastric ulcers, and constriction of blood vessels are some of the other possible impacts.
3. Noise can cause serious damage to wildlife.

4. It creates annoyance to the receptors due to sound level fluctuations.
5. The working performance of workers/human is affected as it distracts the concentration.
6. It effects the nervous system as it causes pain ringing in the ears, feeling of tiredness, thereby effecting the functioning of human system.
7. It effects the sleeping thereby inducing people to become restless and loose concentration during their activities.

Ambient Air Quality Standards in Respect of Noise

S.No.	*Category of Area / Zone*	*Limits in dB*	
		Day Time	*Night Time*
1.	Industrial Area	75	70
2.	Commercial Area	65	55
3.	Residential Area	55	45
4.	Silence Zone	50	40

Notes: 1. Daytime means from 6.00 a.m. to 10.00 p.m.
2. Nighttime means from 10.00 p.m. to 6.00 a.m.

Acceptable Noise Levels for Residential Areas in India

S.No.	*Location / Areas*	*Noise Levels in dB*
1.	Rural Areas	25 – 35
2.	Sub-urban Areas	30 – 40
3.	Urban Residential Areas	35 – 45
4.	Residential & Business Urban Areas	40 – 45
5.	City Areas	45 – 55
6.	Hospitals	35 – 40
7.	Apartments, hotels, homes	35 – 40
8.	Restaurants	50 – 55

Noise Pollution in Construction Industry

High level noises are created due to collision or friction of metal against metal, against aggregate, and metal against aggregate; movement of compressed air into the atmosphere, and the combustion of fuel in the burner. Areas of high level noises are: crushers, burners, blowers, exhaust fans, pug mills, jack hammers, etc. Operators in most plants are, therefore, housed in cabins, where exposure level to noise is reduced. If an employee is required to work near plant area outside cabin, it should be for short durations, as prolonged exposure in excessive noise is irritating, produce mental fatigue and reduce the efficiency of persons around the plants, especially asphalt plants and crushing plants where noise level is too high.

13.6 SOIL POLLUTION

Formation of a soil basically depends upon the mechanical weathering of rocks caused by the temperature changes and abrasion. Climate and time are, thus, the two most important factors responsible for the development of soils. The soil layer, primarily consists of mineral particles, which are formed by disintegration (weathering) of rocks. These mineral particles, after disintegration from rocks, are transported by water or wind or glaciers and deposited over the land surface, to form a soil layer. In normal cases, 1 cm of top soil formation may take 200-1000 years.

Soil is treated as nature's dispose-all, it is sewage treatment plant, it is water purifier and at times it is also a pollutant. The soil acts as a physical (sieving action), chemical (absorption and precipitation) and biological filter (decomposition of organic waste materials) as well as the receptacle for all materials buried and disposed of beneath and on the surface.

Soil pollution usually originates from the development of industry, infrastructure, intensive agriculture associated with modern systems of cultivation using chemical fertilizers, pesticides, sewage, sludge, city compost etc.

Gaseous emissions from industrial areas, thermal power plants, burning fossil fuels, automobiles, mining etc., frequently leads to soil pollution through entry into terrestrial ecosystem as acid rain or directly as particulate matters on the soil surface.

Major Sources of Pollution of Soil (land)

1. Application of fertilizers to agricultural land.
2. Application of pesticides to agricultural land and forests.
3. Application of insecticides to control vector borne diseases, like malaria.
4. Use of salty groundwater for irrigation; and excess use of irrigation water resulting into water-logging.
5. Acidic precipitation from air as dust, or droplets, or rain, or snow.
6. Dumping of solid wastes on land, such as at landfill sites/industrial dumps etc.

Land Degradation

The degradation of land, through soil erosion, alkali-salinization, water logging, pollution, and reduction in organic matter content has several proximate and underlying causes. The proximate causes include loss of forest and tree cover (leading erosion by surface water run-off and winds), excessive use of irrigation (in many cases without proper drainage), improper use of agricultural chemicals (leading to accumulation of toxic chemicals in the soil), diversion of animal wastes for domestic fuel (leading to reduction in soil nitrogen and organic matter), and disposal of industrial and domestic wastes on productive land.

Soil pollution and land degradation have been discussed in detail in the chapter on "Land (soil) Resources."

Effects of Soil Pollutions

1. **On Agriculture** : Soil pollution effects agriculture due to reduced soil fertility, increased erosion, loss of soil and its nutrients, reduced crop yield, increased salinity, deposition of silt in irrigation tanks and reservoirs.

2. **On Health** : Soil pollution impacts health, because dangerous chemicals may enter underground water, release of pollutant gases, and release of radiative rays.
3. **On Environment** : Soil pollution effects the environment due to reduced vegetation, ecological imbalance, and imbalance in soil fauna and flora.
4. **On Urban Areas** : It effects urban areas due to clogging of drains, inundation of areas, foul smell and release of gases, and waste management problems.

13.7. RADIO- ACTIVE POLLUTION

Radioactivity is a phenomenon of spontaneous emission of proton, electrons, and gamma rays due to disintegration of atomic nuclei of some elements. These cause radioactive pollution.

Radioactivity can also be defined as a property of certain elements (radium, thorium, uranium etc.) to spontaneously emit protons, electrons, and gamma rays by disintegration of their atomic nuclei.

Sources of Radioactive Pollution

There are two types of sources of radioactive pollution, namely:

1. **Natural Sources** : These include cosmic rays from space and terrestrial radiations from radio-nuclides present in earth's crust such as radium-224, uraninum-238, thorium-232, potassium-140, carbon-14 etc.
2. **Man-made Sources** : These are :
   (a) Nuclear power plants
   (b) Nuclear weapons
   (c) Transportation of nuclear material
   (d) Disposal of nuclear waste
   (e) Uranium mining
   (f) Radiation therapy

Radioactive Pollution Control Measures

Since there is no cure available for radiation damage, prevention is the best control measure.

(a) All safety measures should be strictly enforced. Leakage of radioactive elements should be strictly checked.
(b) Safe disposal of radioactive waste.
(c) Safety measures against nuclear accidents.
(d) Regular monitoring through frequent sampling analysis.
(e) Nuclear explosions and use of nuclear weapons should be completely banned.
(f) Steps should be taken to protect from accidental exposure.

13.8 MARINE POLLUTION

A sea may be defined as an enormously large continuous body of water, formed over the depressed parts of the earth's surface. Exceptionally deep seas, generally with an average depth of more than 4 km are known as oceans. Since all the rivers flowing through the land surface, drain into this body of oceans, huge amounts of salts are continuously added to this body, which make the ocean's water (termed as marine water) highly saltish. The present salt concentration of marine water has been estimated to be 3.5% by weight.

Unfortunately, due to this huge volume of water being present in a single body man has treated it to be a source for dumping several kind of liquid and solid wastes generated by the humanity. This has caused pollution of the seas and other marine water bodies. The virtually uncontrolled dumping of domestic and industrial waste-waters, solid waste, and radioactive and other hazardous wastes into the ocean body, particularly by the developed countries, have made this valuable body of water as a dumping yard for all wastes.

The marine water bodies provide us food, energy, minerals and a medium to carry cargo. This also help to keep the earth's climate healthy by regulating the air temperature and by supplying the moisture for rainfall. Considering these important uses of seas, it becomes our duty to protect the seas (including oceans) from pollution, and to make use of their resources in a sustainable manner.

Sources of Marine Pollution

Major sources of the pollution of a sea are:

1. Polluted discharge of rivers outfalling into a sea.
2. Discharge of sewage and sludge into sea.
3. Discharge of wastes from beach houses and hotels into a sea.
4. Disposal of solid refuse by barging it out into a sea.
5. Dumping of industrial wastes into a sea.
6. Dumping of dredged material from works associated with shipping activities into a sea.
7. Dumping of radioactive wastes into a sea.
8. Oil spills and other oil discharges into a sea.

Impact of Marine Pollution

Impact of marine pollution on marine bio-diversity can be summarised into following categories:

1. *Habitat loss.* Conversion of mangroves and other wetlands due to urbanization and agricultural expansion, oil drilling, dredging and filling etc., contributes to habitat loss of marine life.

   Damming of rivers and water diversion projects lead to changes in downstream marine communities. Intense exploitation of marine resources can also lead to habitat loss. Fishing with dynamine and harvesting of corals are major threats to coral reef areas.

2. *Intense Overexploitation.* Much of the global fishing effort is targeted at a few species. Overexploitation of these species has three effects:

   (i) loss of genetic diversity as fish populations decline,

   (ii) mix of different species within an ecosystem, and

   (iii) depleted fisheries have direct economic impacts.

3. *Contamination and Sedimentation.* Run-off and point-discharges from the land (and rivers) account for some three-fourths of the pollutants entering marine ecosystems.
4. *Species Introduction.* Ships carry an enormous variety of exotic species within their ballast water. Accidental introduction of exotic species may be one factor in the apartment spread of toxic blooms; it is also the suspected cause of a disease affecting corals.
5. *Oil Pollution.* The most serious types of oil pollution occur, when an oil tanker goes ashore or hits a reef and spills its contents. As the oil drifts ashore, great damage is done to beaches, rocky shores, salt marshes, or mangrove forests. Another reason for oil pollution is accidental spills when loading or unloading, or from washing out oil tanks, are widespread and of significant concern.

13.9 THERMAL POLLUTION

The hot water released by power plants and industries, using large volumes of water for cooling processes, may result in a sharp and sudden rise in the temperature of the local water bodies, causing what is known as the *Thermal pollution.* The release of large quantities of heat into a water body may kill heat-intolerant plants and water animals. Elimination of heat intolerant species would lead to flourishing of heat tolerant species, which are usually less desirable species.

Thermal pollution can be controlled by constructing 'cooling ponds' for collecting hot water for cooling, before its release into the nearby water body. 'Cooling towers' are also constructed and used in modern days to dissipate heat energy before releasing the waste-water to avoid its hot discharge into the water body.

Environmental Effects

Common environmental effects of thermal pollution are as under:

1. Direct thermal shock, which changes in dissolved oxygen, and the re-distribution of organisms in the local community. Hot water has decreased oxygen carrying capacity.
2. It can also adversely affect the reproductive systems of these organisms and also make them more susceptible to disease.
3. Changes in the environment (cool to hot) may also result in a migration of organisms to another, more suitable environment. As a result, one has the problem of compromising food chains of the old and new environments. Biodiversity can be decreased as a result.

13.10 ROLE OF INDIVIDUAL IN POLLUTION PREVENTION

There are large number of problems caused by human actions on the environment. When we respond to these problems, we recognise that each one of us is individually responsible for the quality of the environment in which we live. The actions of individual can worsen or improve the quality of our environment. This necessitates that individuals should not only be aware of various environmental issues and the consequences of their actions on the environment.

Some concepts that help individuals contribute towards a better quality of environment and human life are:

1. Conserving energy: (a) minimizing electricity consumption, (b) regular checking of car pollution and maintenance of car.
2. Develop respect for all forms of life.
3. Plant trees and look after them. Preserve existing trees. Also, plant shrubs to help prevent soil erosion and promote infiltration of water into the soil.
4. Encourage 're-cycle and re-use' of items whose production creates pollution.
5. Avoid putting toxic substances litter into drains, avoid dropping, minimizing the use of pesticides and fertilizers.
6. Make it a habit to walk or ride a bicycle for small distances. This will help in his contribution to hydrocarbon emissions to the atmosphere.
7. Reduce the use of wood and paper products wherever possible. As these items lead to loss of forests which release oxygen and absorb carbon dioxide. Try to recycle paper products and use re-cycled paper, wherever possible.
8. Do not purchase furniture, doors or window frames made from hardwoods.
9. Help in restoring a degraded area near your locality or join on afforestation programme.
10. Advocate organic farming, this will help in reducing the use of pesticides.
11. Shut off the lights, fans and A.C., when not needed.
12. Use rechargeable batteries.
13. Buy consumer goods in refillable glass containers instead of cans or throwaway bottles.
14. Avoid asking for plastic carry bags when you buy groceries or vegetables or any other items. Carry your own cloth bag.
15. Use washable cloth napkins instead of paper ones.
16. Do not use disposable paper and plastic plates, cups and glasses, when reusable versions are available.
17. Set up a compost bin in the garden to produce manure for your plants.
18. Try to lobby for setting up grabage separation and re-cycling programme in your locality.
19. Remember environment protection begins with you.

QUESTIONS

1. Define and explain 'Environmental pollution'.
2. Define and explain 'Water pollution'.
3. Enumerate various types of pollutants which cause pollution of water bodies.
4. Explain the following terms:

   (a) DO (b) BOD
5. Discuss the parameters of water pollution and ecological balance of large streams.

6. How do pesticides and fertilizers cause water pollution of ground water sources as well as surface sources?
7. Briefly discuss marine pollution and its prevention from disposal of waste waters of a coastal city.
8. Write an essay bringing out strategies for controlling water pollution.
9. Mention water quality standards for classifying surface water.
10. Define air pollution, and differentiate between natural and man-made air pollutions.
11. (a) Discuss three major causative factors for environmental degradation.
    (b) Discuss about environmental pollution and health.
12. Write short notes on:
    (a) Sources of Air pollution
    (b) Effects of Air pollution
    (c) Prevention and Control of Air pollution
    (d) Indoor Air pollution
    (e) Smog
13. Explain the following in brief:
    (a) National Air Quality Index
    (b) Namami Gange Project
    (c) Water quality characteristics
14. Discuss about 'Noise pollution'. Also discuss about impact of 'noise pollution'.
15. Explain the following:
    (a) Radioactive pollution
    (b) Effects of soil pollution
16. Describe the cause of air pollutions.
17. Differentiate between the primary and secondary air pollutants.
18. How does air pollution affect human health?
19. Enumerate the different major air pollutants.
20. Describe the important local, regional and global impacts of air pollution.
21. Describe the harmful effects of important air pollutants on plants, animal life, aquatic life and materials.
22. Mention the national ambient air quality standards of important air pollutants in India.
23. Write short notes on:
    (a) Air pollution caused by automobiles.
    (b) Indoor air pollution.
24. Define 'noise'. Explain as to how and why it should be regarded as an environmental pollutant?

25. Mention typical values of accepted sound levels as per I.S. Code for different categories of areas.
26. Differentiate between continuous noise, intermittent noise, and impulse noise.
27. Discuss various sources of noise, and their typical noise levels.
28. Write a note on noise pollution, its causes, effects and remedies.
29. What is a soil and how is it formed?
30. How are soils polluted? Explain the various sources which pollute the soils.
31. Explain as to how and in what manner, the modern agricultural practices have polluted our soils?
32. How are soils affected (polluted) due to water-logging?
33. "The use of synthetic fertilizers and pesticides pollute our soils and environment beyond retrieval." Discuss the statement.
34. How are the "acid-rains' responsible for damaging?

14

Chapter

WASTE MANAGEMENT

14.1 INTRODUCTION

Waste is defined as every substance or object arising from human or animal activities, that has to be discarded as useless or unwanted. Waste can be any trash, rubbish (combustible or non-combustible, garbage, refuse, filth, bottles, glass, crockery, cans, cartons, scrap metal, junked vehicles, appliances or hardware, waste soil, rock, construction material, animal carcasses, sludge from a waste treatment plant or discarded matter, or parts or portions thereof, or any industrial, commercial, mining, agricultural or other waste.

Different waste management and disposal methods and strategies are followed, depending on the nature and volume of waste. Waste matter is classified as solid, liquid or sludge based on its consistency and hazardous or non-hazardous based on whether or not they have properties, such as ignitability, reactivity and leachable toxic components are bio-hazards.

Waste management is the collection, transportation, processing and disposal of waste materials in an effort to reduce their effect on human health or local aesthetics. A sub-focus in recent decades has been to reduce waste materials' effect on the natural world and the environment.

Types and Sources of Waste

Category	*Source of generation*
1. Municipal solid waste	- *Residential:* Houses, apartments etc.
	- *Commercial:* Hotels, restaurants, market, shopping complexes etc.
	- *Open areas:* Recreational areas, parts, play□grounds etc.
2. Industrial waste	- Chemicals, textile by-products etc.
3. Agricultural waste	- Agricultural by-products and residues etc.

4. Hazardous waste - Chemicals and oil from refineries, waste from ordinance and fire work factories etc.
- *Hospitals:* bio-medical waste

The term solid waste includes all those solid and semi-solid materials that are discarded by a community. The solid waste generated through domestic and commercial activities is classified as *Municipal Solid Waste,* and is also called as *'rufuse'.* The solid waste generated by industries is known as *Industrial Solid Waste.* The solid waste generated by agricultural activities may be known as agricultural waste and may also become a part of the 'municipal solid waste.'

14.2 MUNICIPAL SOLID WASTES (MSW) OR REFUSE

These are the all sorts of solid wastes from community. The Municipal Solid Waste (MSW) is a heterogeneous mixture of various kinds of solid wastes, which are not transported with water as sewage and may include: (*i*) biodegradable (putrescible) food wastes called *garbage*, and (*ii*) non-putrescible solid wastes, called *rubbish.*

1. *Garbage:* It includes all sorts of bio-degradable organic wastes obtained from kitchen, hotels, restaurants etc. Since all these wastes are organic in nature, and thus, likely to decompose quickly, producing foul odours and health hazards, fly and insect problems when it is scientifically processed and are composted, then it is possible to obtain valuable products like fertilizer.
2. *Rubbish:* It includes a variety of materials, which may either be combustible (such as plastic, paper, textile etc.) or incombustible (such as broken glass, crockery, metal, masonry etc.).

Besides garbage and rubbish, the municipal solid waste may include ashes, fine dust, silt and sand obtained from street sweepings. These also include certain special wastes, such as construction debris, abondoned appliances, automobile parts etc.

The quantity of MSW produced by a society depends upon the living standards of its residents. The degree of commercialisation, urbanisation, and industrialisation has resulted in vast increase in the amount of refuse generated per person. With the modernisation and industrialisation, new types of solid wastes have come in abundance. Examples of such wastes are plastic bottles, cans, packages, containers, plastic toys electronic appliances, etc. All these factors have increased the MSW by a large extent.

Collection and Transportation of Refuse

Refuse collection is generally divided into two categories, namely: primary collection and secondary collection.

Primary collection involves delivery of the wastes by the house-holders to community storage. Dustbins kept at each house are unloaded to wheel borrows or rickshaws, from where it is taken to the community storage of large capacity (1 to 2 m^3) at an intervals of 50 to 100 metres.

Secondary collection and transportation to the disposal sites is often done by heavy motor vehicles. The refuse should be dumped into sanitary refills to avoid making it a breeding ground of mosquito, fly etc. These refuse are placed in trenches or lowlying areas in layers of about 1.5 meters, covering each layer by about 20 cms of the earth. The area till covered, should be sprayed with solution of creosote or paraffin.

Average Composition of MSW (Refuse)		
S.No	*Item*	*Approx. Average Composition (% by weight) in Indian Cities*
1.	Garbage	45
2.	Rubbish (paper, glass, rags, etc.)	15
3.	Ashes	15
4.	Fine dust, silt and sand	25
	Density of MSW	400 to 600 kg/m^3
	Calorific Value	5000-6500 kJ/kg

14.3 WASTE MANAGEMENT TECHNIQUES

An integrated waste management strategy includes following three main components:

1. Source reduction
2. Re-cycling (re-using)
3. Disposal
   - (i) Sanitary landfills
   - (ii) Shredding
   - (iii) Incineration
   - (iv) Composting

These common techniques (methods) for waste management are discussed hereunder:

1. Source Reduction: Source reduction means making the waste fit into less space and easy to handle. This can be achieved by compaction or fragmentation by shearing or grinding. This can also be achieved on individual level by reducing the use of unnecessary items while shopping, buy items with minimal packaging and avoiding purchase of disposable items, and avoid asking for plastic carry bags.

2. Re-cycling or Reusing: Re-cycling means to reuse a material that would otherwise be considered as waste. Some components of waste that may have some life may be re-used. This conserve resources, reduce energy used during manufacture and also reduce pollution levels. Some materials, such as steel and aluminium can be recycled many times.

We should select re-usable products, such as sturdy and washable utensils, table ware, cloth napkins etc. Re-use newspapers, boxes, packaging materials etc. The items can be re-sold or donated after repairing as charity.

Papers, plastics, glasses, rubbers, ferrous and non-ferrous metals, and other materials that can be re-cycled are salvaged from the municipal solid wastes to produce low-cost products extensively used for lower income groups of society.

3. Disposal: In general, following methods are used for disposal of municipal solid waste (MSW) or called as refuse:

(i) Sanitary land filling

(ii) Shredding and pulverisation

(iii) Composting

(iv) Incineration

Out of these above methods, only the method given at serial 1 is complete, while the other require further disposal of some left out residues or by-products.

(i) *Sanitary Landfilling*: Sanitary landfills differ markedly from the open dumps. Whereas, open dumps are simply dumping places, sanitary landfills are engineered operations, designed and operated according to acceptable standards.

The basic principle of a landfill operation is to deposit the refuse, compact it with bulldozers and cover the material with atleast 15 cm of the earth at the conclusion of each days' operation and a final cover of 60 cm, when the area is full.

The engineering aspects include:

(a) *drainage* for rapid runoff to lessen mosquito problems

(b) wind blown should not affect the community

(c) size should be sufficiently large

(d) ultimate use—generally these are best suited for playgrounds or gardens

(e) many landfills have a gas extraction system after they are closed, to extract the gas generated by the decomposing waste materials.

(ii) *Shredding and Pulverisation:* The size and volume reduction of MSW is accomplished by the physical processes of shredding and pulverisation. **Shredding** refers to action of cutting and tearing; whereas, **pulverisation** refers to action of crushing and grinding. These actions reduce the overall volume of the original MSW, by as much as 40%. This process also changes the physical character of the waste, which becomes practically odourless and unattractive to the insects.

In order to further reduce the waste volume, and facilitate handling of the pulverised MSW, it is usually compacted, under high pressure, to form *rectangular blocks* or *bails.*

(iii) *Incineration and Thermal Pyrolysis:* Burning of refuse at high temperature in furnaces, called **incinerators,** is quite a sanitary method of refuse disposal and is widely adopted in developed countries, where the collected refuse is of high calorific value and is quite suitable to burning.

Normally, only the combustible matter, such as garbage, rubbish and dead animals are burnt. The incombustible matter, like broken glass, chinaware, metals etc., are either left unburnt or are separated out for re-cycling and re-use, before burning the solid wastes. Prior separation of such materials will reduce the load on furnaces, and shall more than compensate the cost of this separation. The left out ashes and clinkers from the incinerators measures only 10 to 25% of the original MSW, which shall have to be disposed off by sanitary landfilling.

The heat produced during burning of the refuse is used in the form of steam generation for power plant.

(iv) *Composting*: Composting of refuse is the process of controlled decomposition of biodegradable organic wastes, such as food and yard wastes, into humus. It is nature's way of re-cycling organic waste into new soil composting. Keep organic wastes out of landfills and instead, put it to use in gardens, where it provides nutrients and increases beneficial soil organisms.

The decomposition can be affected either under aerobic conditions, or anaerobic conditions, or both. The final end product, is a manure, called the compost or humus as fertilizer for farms.

Composting is considered to be aerobic process, because it involves piling up of refuse and its regular turning, either manually or by mechanical devices, so as to ensure sufficient supply of air and oxygen during its decomposition by bacteria, fungi and other micro-organisms.

Moisture content of the compost mass should be controlled to ensure optimum aerobic decomposition, because excessive moisture will make it difficult to maintain aerobic conditions, while deficient moisture inhibits biological life. A moisture content of about 55% should be established, so that aerobic biological activity may proceed at an optimum rate.

14.4 INDUSTRIAL SOLID WASTES

The solid wastes produced by industries can broadly be divided into the following two categories:

1. Non-hazardous Solid Wastes
   (a) Biodegradable wastes
   (b) Non-biodegradable wastes
2. Hazardous Solid Wastes

Biodegradable Wastes

The major industries in urban areas that generate substantial amount of biodegradable non-hazardous solid wastes are: fruit processing, cotton mills, paper mills, sugar mills etc.

Non-biodegradable Wastes

These solid wastes of industries are usually referred to as the industrial solid wastes. These are usually produced by industries, like integrated iron and steel plants produce blast furnace slag and steel melting slag; non-ferrous industries produce, red mud and tailings; fertilizer and allied industries produce phospho-gypsum, and thermal power plants produce coal ash (also known as fly ash).

Hazardous Solid Wastes

Some of the industrial wastes are 'hazardous wastes', because they contain substances that are **toxic to plants, animals, and human beings, or are flammable, corrosive, explosive, or highly reactive chemically.** The major industries that produce hazardous wastes include metals, chemicals, drugs and pharmaceuticals, leather, pulp and paper, electroplating, refining, pesticides, dyes, rubber goods etc.

Hazardous waste is that which can cause harm to humans or the environment. Waste is classified as hazardous, when it causes or significantly contributes to an increase in mortality, or an increase in serious irreversible or incapacitating reversible illness or poses substantial present or potential hazard to human health or the environment, when improperly treated, stored, transported or disposed off.

Characteristics of Hazardous Wastes

(i) **Toxic wastes** are substances that are poisonous even in very small or trace amounts.

(ii) **Flammable (ignitable) waste** is that which burns at relatively low temperature and is capable of spontaneous combustion during storage, transport or disposal, such as gasoline, paint, thinners, alcohol etc.

(iii) **Corrosive waste** is that which destroys materials and living tissue by chemical reaction, such as acids.

(iv) **Reactive waste** has a tendency to react vigorously with air or water, is unstable to shock or heat, generates toxic gases, or explodes during routine management, such as gunpowder and nitroglycerine.

(v) **Infectious waste** includes human tissues from surgery, used bandages, hypodermic needles, and microbiological materials.

(vi) **Radioactive waste** is the output from nuclear power plants that can persist in the environment for thousands of years before it decays appreciably.

Impacts of Hazardous Wastes

1. When disposed of on or in the land its serious impact is contaminated ground water, which is not possible to reverse the damage.
2. Lead, mercury and arsenic (referred to as heavy metals) are hazardous substances.
3. Exposure of chemicals used in industries expose many problems to both, humans and wildlife.

Health Impacts of Solid Wastes

Improper handling and disposal of solid wastes can lead to infections of nose, throat and lungs, cause allergy or asthma.

Poorly maintained landfill sites may lead to groundwater contamination because of leachate generation.

Open dumping of garbage can facilitate breeding of disease vectors, such as flies, mosquitoes, cockroaches, rats and other pests.

14.5 AGRICULTURAL SOLID WASTE MANAGEMENT

Agricultural wastes are defined as the residues from the growing and processing of raw agricultural products, such as fruits, vegetables, meat, poultry, dairy products and crops. They are the non-product outputs of production and processing of agricultural products that may contain material that can benefit man but whose economic values are less than the cost of collection, transportation and processing for beneficial use.

The agricultural wastes can be in the form of liquid, slurries or solids. The waste generated is dependent on the type of agricultural activities carried out.

The first goal of any waste management system is to maximise the economic benefit from the waste resource and maintain acceptable environmental standards. The management system must be affordable and suitable to the operation. If wastes are not properly handled they can pollute surface and groundwaters and also contribute to air pollution.

Management that puts into practice the principles of the 4 Rs. of Reduce, Reuse, Recycle and Recover is the best option. The meaning of 4 Rs. is explained hereunder:

1. **Reduce** the amount of waste generation.
2. **Reuse** the waste product on the farm or provide it to others for use.
3. **Recycle** the product either on-farm or off-farm, after reducing and reusing as much of the product as possible.
4. **Recover** methane gas from manure waste. This aspect have been discussed in detail under the heading 'Biomass energy' and 'Biogas' in the chapter on "Energy Resources".

The farm waste (agricultural waste) is disposed of only after considering the 4 Rs.

Important Note: Never burn garbage or farm waste, as it seriously affects the quality of air.

14.6. HOSPITAL WASTE

The hospital waste is generated during the diagnosis, treatment, or immunization of human beings or animals or in research activities. This waste include chemicals and mercury, which is used in thermometers or equipment that measure blood pressure. It may also include disposables, anatomical waste, cultures, discarded medicines, body fluids, disposable syringes, swabs, bandages, human excreta.

These wastes are highly infectious and can be a serious threat to human health if not managed in a scientific manner. It has been observed that the health care establishments in India are not giving due attention to their waste management. However, after the notification of the Bio-medical Waste (Handling and Management) Rules, 1998, these establishments are now slowly streamlining the process of waste segregation, collection, treatment and disposal.

14.7 LEGISLATION ON SOLID WASTE MANAGEMENT

The Municipal Solid Waste (Management and Handling) Rules, 2000 framed under the Environment (Protection) Act, 1986 by the Ministry of Environment and Forests, Govt. of India provides requirements for solid waste management.

These rules do not cover hazardous wastes (generated from industries) and Bio-medical wastes (generated from Hospitals), as they are covered under the Hazardous Wastes (Management and Handling) Rules, 2003 and the Biomedical Wastes (Management and Handling) Rules, 2001 respectively.

Requirements of Municipal Solid Wastes (MSW) Rules

Every municipal authority is responsible for collection, segregation, storage, transportation, processing/treatment and final disposal of solid wastes and for development of related infra□structure.

1. **Collection**
   - Organize house-to-house collection of solid wastes through community bin collection, using bell ringing-vehicles on pre-informed timings
   - Collection of waste from slums, hotels, offices and commercial areas
   - Separately managing biodegradable wastes (meat, fish, fruit, vegetables)
   - Ensure that industrial and biomedical wastes are not mixed with solid wastes
2. **Transportation**
   - Use of closed containers for transportation of collected wastes
   - Transportation vehicles shall be so designed that multiple handling of wastes, prior to final disposal, is avoided.
3. **Segregation and Storage**
   - Organize awareness programmes for communities to promote segregation of wastes and re-cycling or reuse of segregated materials
   - Segregate dry waste from wet waste
   - Segregate and separately store re-cyclable waste
   - Storage facilities or 'bins' shall have 'easy to operate' design for handling, transfer and transportation of waste
   - Bins for storage of bio-degradable wastes shall be painted green, those for storage of re-cyclable wastes shall be painted white and those for storage of other wastes shall be painted black
   - Ensure that stray animals are not allowed near waste storage facilities
   - The storage facilities shall be daily attended for clearing wastes
4. **Processing/Treatment**
   - Adopt suitable technology or combination of such technologies to make use of wastes so as to minimize burden on landfill.
   - The biodegradable wastes to be composting, vermin-composting, anaerobic digestion or any other appropriate biological processing for stabilization of wastes.

- Recoverable resources to follow the route of re-cycling.
- Ensuring that the wastes are not burnt in open.
- Incineration with or without energy recovery including palletisation can also be used for processing wastes.

5. **Disposal**
   - Landfilling should be restricted to only non-biodegradable inert and other wastes that are not suitable either for re-cycling or for biological processing.
   - Landfilling should also be carried out for residues of waste processing facilities as well as pre-processing rejects from waste processing facilities.
   - Landfilling of mixed waste shall be avoided unless the same is found unsuitable for waste processing.

QUESTIONS

1. Define the 'municipal solid waste'. Explain the necessity of its safe disposal.
2. Explain the importance of 're-cycling and re-use' in disposal of solid waste of a society.
3. Enumerate the various methods which can be used for disposal of municipal solid waste.
4. Explain the impacts of improper disposal of solid wastes on human health and the environment in general.
5. Explain the term 'refuse', and give its composition and classification.
6. Describe various methods employed for the collection and disposal of the refuse.
7. What do you understand by 'sanitary land filling' and how is it practised?
8. Explain the term 'composting'. How is it done?
9. Write short notes on:
   (a) Garbage and rubbish
   (b) Incineration
   (c) Composting
10. Differentiate between the 'hazardous' and 'non-hazardous' industrial solid wastes.
11. Explain the salient provisions of the Municipal solid wastes (Management and Handling) Rules, 2000.
12. Explain the impacts of improper disposal of solid wastes on human health and environment in general.
13. What is the role of rag-pickers in segregation of Muncipal Solid Waste?

15
Chapter
DISASTER MANAGEMENT

15.1 INTRODUCTION

Disaster means an event, which brings sudden great misfortune bringing disruption to normal life and causing colossal loss of life and property and disruption of human activities. Disasters are the situations of great calamity, like earthquake, floods etc.

Disaster can also be defined as a catastrophe, mishap, calamity or grave occurrence in any area arising from natural or man-made causes, or by accident or negligence which results in substantial loss of life or human suffering or damage to, and destruction of, property, or damage to, or degradation of environment, and is of such a nature or magnitude as to be beyond the coping capacity of the community of the affected areas.

Natural disasters are global phenomena, which may occur anywhere at any time. These adversely affect the lives of the people and cause considerable damage to property, environment and infrastructure worldwide. Although the basic cause of most of the natural disasters is linked with natural processes yet the degree of intensity of a disaster is also dependent on man-made changes caused in the environment. However, hazards and disasters are intimately related and are synonymous to each other. Virtually hazard is a threat, whereas disaster is an event.

India is one of the most disaster-prone countries of the world. The following data clearly show that India is a disaster-prone country:

- 65 per cent of the land area is vulnerable to earthquakes.
- 12 per cent of the land area is vulnerable to cyclones.
- 70 per cent of the land under cultivation is prone to drought.

Crisis is an event of acute danger, which can cause sudden disruption of work. The event may be caused due to natural cause, human error, equipment failure or sabotage by anti-social elements. The basic idea in crisis (emergency situation) is to avoid wastage of crucial time. Terrorist threats and attacks, bomb threats, hoax and bomb explosion, strike, accidents, fire accidents, etc., are the examples of crisis situations.

Disaster management is a comprehensive term which includes all aspects of preventive and protective measure, preparedness and relief operations for reducing the impact of disaster on humans.

Disaster management is a continuous and integrated process of planning, organizing, coordination and implementing measures, which are necessary or expedient for:

(i) Prevention of danger or threat of any disaster
(ii) Mitigation or reduction of risk of any disaster or its severity or consequences
(iii) Capacity building
(iv) Preparedness to deal with any disaster
(v) Prompt response to any threatening disaster situation or disaster
(vi) Assessing the severity or magnitude of the effect of any disaster
(vii) Evacuation, rescue and relief
(viii) Rehabilitation and re-construction.

Poverty and lagging development amplify the adverse effects or natural disasters. Therefore, developing countries (particularly then most densely populated regions) are more vulnerable to natural disasters because of their limited capacity to prevent and absorb these effects. Poor communities get forced to live in crowded, makeshift houses and in areas vulnerable to bad weather or seismic activity having low quality of infrastructure.

Mitigation efforts lessen the impact disasters have on people and property. Mitigation measures include keeping homes away from flood plains, designing structure to withstand earthquakes. The vulnerability to natural hazards can be reduced by altering, averting, avoiding and adapting to the hazard.

Besides the natural disasters, there are many man-made or technological oriented physical events, which results in a significant disruption in the normal working. The event which causes the sudden disruption of work, is known as crisis. The *crisis management* plan shall be able to respond quickly and effectively to such unexpected events and situation. The crisis management includes the preventive measures need to be adopted to avoid any crisis and the ways and means to tackle a crisis, if it occurs inspite of preventive measures. Thus, crisis management helps in avoiding (or atleast minimizing) wastage of crucial time and cost.

15.2 OBJECTIVES

Following are the major objectives of "disaster/crisis management":

1. To improve state of preparedness to meet any contingency.
2. To reduce response time in organizing the assistance.
3. To identify major resources, manpower and equipment needed to make the disaster management plan operational.
4. Making optimum use of the combined resources.
5. To identify mitigating measures and take timely action.
6. To avoid panic situations and rumours at the time of contingency.

The objective of preparing "*disaster management plan*" is to enable various departments/ agencies to plan for, quickly respond to and recover from unexpected events and situations. This plan shall provide necessary guidelines for assistance to various sections for ensuring safety of people, protection of property and restoration of normal life, including water and power supply, communication etc.

15.3 ORGANISATIONAL SET-UP FOR MANAGING/HANDLING DISASTERS

Planning and managing the disasters is very essential. It is generally seen that help arrives from several quarters but the same does not reach the affected persons in time and in adequate measures. Total preparedness and planning to handle such unforeseen situations play a key role in successful handling of disaster situations.

For managing disasters effectively and efficiently, close co-ordination with various government and non-government agencies is essential. We are having an integrated administrative machinery for management of disasters at the national, state, district and tehsil/block levels. The prime responsibility of undertaking rescue, relief and rehabilitation measures lie with the concerned State Government in the event of natural disasters. The Union Government adds to the efforts of the concerned State Government(s) by providing financial/logistic assistance.

15.3.1 Disaster Management at National Level

The instant response from the Union Government is based on the following factors:

(i) The gravity of the disaster

(ii) The overall scale of the relief operations

(iii) The requirements of the Central help for financial resources and logistics support.

Following table is showing various disasters and their nodal ministry:

S. No.	*Type of Disasters*	*Nodal Ministry*
1.	Natural disasters (except drought)	Ministry of Home Affairs
2.	Chemical disasters	Ministry of Home Affairs
3.	Biological disasters	Ministry of Home Affairs
4.	Nuclear disasters	Ministry of Home Affairs
5.	Epidemics	Ministry of Health and Family Welfare
6.	Drought	Ministry of Agriculture
7.	Railway accidents	Ministry of Railways
8.	Air accidents	Ministry of Civil Aviation

Following are the technical organizations:

1. Indian Meteorogical Department – Cyclone and earthquake
2. Central Water Commission (CWC) – Floods

3. Building and Material Promotion Council – Construction laws
4. Bureau of Indian Standards (BIS) – Norms
5. DRDO – Nuclear/Biological disasters

For co-ordination for disaster response and management functions the Director General of Civil Defence provides sufficient logistical assistance.

15.3.2 National Disaster Management Authority (NDMA)

National Disaster Management Authority has been established in accordance with the Disaster Management Act, 2005. This national authority consists of the Prime Minister as its chairperson and other members, not exceeding nine, to be nominated by the Chairperson of the national authority. The chairperson may designate one of the nominated member as Vice-chairperson of the National Authority. The officers, employees and consultants as considered necessary are provided by the Central Government.

The National Authority have the responsibility for laying down the policies, plans and guidelines for disaster management for ensuring timely and effective response to disaster. In addition, the national authority has following functions:

- Co-ordinate the enforcement and implementation of the policy and plan for disaster management.
- Recommend provision of funds for the purpose of mitigation.
- Take measures for the prevention of disaster, mitigation, preparedness and capacity building for dealing with the threatening disaster situation or disaster.

The national authority may constitute advisory committee consisting of experts in the field of disaster management to make recommendations on different aspects of disaster management. The Central Government has also constituted a **National Executive Committee** to assist the national authority in the performance of its functions. The National Executive committee shall consist of Secretary, Ministry of Home Affairs (nodal ministry for disaster management) as *ex official*-chairperson, and the Secretaries of Govt. of India in the Ministries of Agriculture, atomic energy, defence, drinking water supply, environment and foreset, finance (expenditure), health, power, rural development, science and technology, space, telecommunication, urban development, water resources and the chief of the Integrated Defence Staff of the chiefs of staff committee.

The National Executive Committee may, as and when required, it considers necessary, constitute one or more sub-committees, for the efficient discharge of its functions. The National Executive Committee shall assist the National Authority in the discharge of its functions and have the responsibility for implementing the policies and plans of the National Authority and ensure the compliance of directions issued by the Central Government for the purpose of disaster management in the country.

15.4 ELEMENTS OF DISASTER MANAGEMENT

The prime aim of the disaster management is to establish essential systems, programmes, structures, capabilities and guiding principles for reducing disaster risk and preparing for and responding to disasters and threats of disasters. The disaster management is essential so as to reduce the detrimental effects of disasters which cause extensive and tremendous damage of life, property and environment.

Following are the **four elements of disaster management** which need to be incorporated in the development plans at community, regional, state and national levels:

1. Preparedness
2. Mitigation
3. Prevention
4. Response.

The main purpose of these four elements PMPR is to:

(i) Understand about hazards and disasters, their behaviour and the amount of risk they pose to the community.

(ii) Reduce and avoid physical and economical losses.

(iii) Reduce personnel sufferings by removing affected persons to safer place and providing them food, shelter and medicines.

(iv) Fasten recovery process.

(v) Provide emergency supports.

These elements are based on the fact that disaster management is a positive approach to solve the problem by enhancing the resources and capabilities. It is absolutely imperative that we (as community members) prepare ourselves properly and quickly to prevent and respond to disasters. If the community members are adequately trained and prepared to face any disaster, they would prove life-saving assets in such critical situations. This help communities to perform a positive role in risk reduction and minimizing damage to life and property. It is quite evident that where communities are well prepared to face disasters, the damage to life, property and environment can be minimum.

The *pre-disaster phase* consists of preparedness, mitigation and prevention. Following are the initial activities that will be carried out in this phase:

(i) Comprehensive data collection.

(ii) To maintain an extensive list of resources.

(iii) To develop action plans.

(iv) Capacity building.

(v) Proper training.

(vi) Community awareness.

15.5 DISASTER PREPAREDNESS

Disaster preparedness includes preparation of a contingency plan which consists of institutional arrangements, rescue and relief operations and an action plan. Usually non-government organizations and trusts come forward with various kinds of help in case of disaster, whose participation in relief supply and operation should be dovetailed in the disaster preparedness planning.

It is essential to develop an evacuation plan for which sites should be identified and reserved. Together with evacuation plan, it is necessary to provide transportation facilities for which an inventory may be prepared for procuring vehicles in case of emergency.

Disaster preparedness plan should include the elements as discussed hereunder:

15.5.1 Pre-disaster Preparedness

The rescue operation for any disaster has to start right from the warning received from the intelligence, meteorological organization or any other State/Central Government agencies. The rescue operation should concentrate on life safety as the prime objective followed by attending the injured and stopping the disaster from spreading further.

Pre-disaster preparedness and mitigation measure are for facing disaster in a better and planned manner.

For effective preparedness to face the disaster and to avoid last minute arrangements in panic condition, the following aspects shall be covered:

(i) Well-documented emergency plan.

(ii) Data on availability of resources and buffer stocks of restoration materials.

(iii) Identification of key personnel with their skills and experience on the disaster management.

(iv) Allocation of budget for emergencies.

(v) "Delegation of Power" at various levels for disaster condition.

15.5.2 Disaster Preparedness

- Good knowledge about vulnerability of target area
- Clear roles and responsibilities of Governmental and non-governmental agencies
- Appropriate arrangement to ensure security and Law and Order
- Training of disaster managers and relief workers and volunteers
- Adequate rescue equipment and relief items
- Adequate warehousing and transportation facilities
- Appropriate mechanism for receipt and distribution of internal/external assistance
- Identification of safe houses, shelters including schools, other government, panchayat and community building for persons and livestock, safe evacuation routes etc.
- Stock water, food, clothes, kitchen utensils, candles, petromax lights, medicines, etc. for emergency use, procure rescue materials, water testing and cleaning equipment, kerosene relocation plan for vulnerable habitations, install forecasting and warning systems.

Various phases of disaster management for which full preparedness and action teams required are:

1. Prediction/forecasting and warning in the form of precautionary measures
2. Threat, emergency action and survival methods

3. Environmental, social and economical impact
4. Taking stock of situation, resources, decisions for actions
5. Rescue operation for survival of maximum victims
6. Ensuring timely relief, medical and security measures to victims
7. Rehabilitation, recovery, relocation and reconstruction.

15.5.3 Pre-disaster Actions

The stage occurs when, the prior information is available about a situation that may lead to disaster in near future. The emergency management group, depending upon the nature of emergency should be put on high alert and following actions are taken:

Pre-alert Notification. This type of notification is mainly used for disseminating an important piece of information and is not intended to convey whether or not an emergency situation can arise. For example, notification that dam attained full reservoir level is a pre-alert notification. This notification may include, slowly developing dam. Site emergencies which can develop immediately, or even after 10 to 15 days or may not develop at all.

Alert Notification. An alert notification shall be used to provide notice that although a crisis/disaster is not imminent, a more serious situation could occur unless conditions improve. Local officials and residents in affected areas should be alerted that an unsafe situation could be developing. Examples of such situations could be full reservoir with forecasts for heavy storms in the catchment, which could lead to release of water through spillways that can submerge downstream areas.

Warning Notification. A warning notification implies that a crisis/disaster is imminent, an advance action may be initiated for minimizing the damages/rescue operations. The warning notification, indicating the magnitude of crisis/disaster should be communicated to all concerned in the region.

15.5.4 Mapping

The main aim of the mapping exercise is to provide a pictorial base to planning process. The mapping exercise is done by drawing the maps on the ground using stone pillars and other locally available resources. Following are the types of maps:

1. Social Mapping. These maps show:

(i) Exact location of the habitat and natural topography.
(ii) Number of kuchcha and pucca dwellings.
(iii) Other common infrastructures of the locality, such as religious place, school, health center/hospital, safe shelter, etc.

2. Resource Mapping. Resource mapping mainly centers on the locally available resources and other assets which can be utilized for building the capacity of the community members during and after disasters. It includes:

(i) Identification of individual skills, *e.g.*, doctors, community leaders, divers etc.

(ii) Available resources around the region, *e.g.*, food stock, boats, etc.

(iii) Important locations/places in the area, like open land, elevated areas, etc.

(iv) Drainage facilities in the area.

(v) Protection bund in the area.

(vi) Shelter belts, mangrove plantations, agricultural zone, forest area, etc.

3. Vulnerability Mapping. It helps to identify the hazards and the possible regions which are likely to get affected, such as:

(i) Number of vulnerable houses and out of use infrastructure.

(ii) Lowlying regions prone to floods.

(iii) Landslide prone regions.

(iv) Exact location of the dangerous industries, electrical installation, tall weak structures and buildings, etc.

4. Safe and Alternate Route Mapping. The members of the community are interested to know about the places/locations that are safe. In areas-that are generally affected by floods, we need to identify the buildings and houses that are situated on raised platforms. This map also shows the alternate safest approach route to the area. It can be either roadway or water way.

15.5.5 Disaster Zoning

Natural calamities, like earthquake, cyclones, floods, etc., have been studied based on data collection and from them, it has become possible to predict their intensities with certain degree of confidence so that structures/plants, etc., can be designed to withstand the effect of these forces. Seismic effects and wind loads have been measured by Indian Meteorological Department, Geological Survey of India, etc. with regard to floods, Central Water Commission is doing the measurement of discharge data of major rivers.

The concept of disaster zoning identifies the areas, having similar parameters on the average. Suitable margins and factor of safety are considered in design of structures to sustain the severity according to the zone characteristics. However, the intensity suddenness and extent of any natural calamity are beyond any perfect assessment and have to be effectively managed in the event of its occurrence:

1. Zoning for Earthquakes. Our country has been demarcated from the point of view of intensity of seismic loads in five zones *i.e.*, Zone-1 to Zone-5. Zone-1 indicates that part of the country where seismic risk is minimum and Zone-5 covers that portion of the country where seismic risk is the highest. Bureau of Indian Standard (BIS) has published in code namely, IS: 1893 for the earthquake resistance designs of various structures. In case of large dams, site specific studies of the project area need to be carried out and the design of structures are finalized according to the Maximum Credible Earthquake (MCE) and geophysical conditions of the area.

2. Zoning for Wind Load. IS: 875 (Part III) indicates the wind speeds and wind load intensities in various parts of the country, based on measured and collected data on wind speeds. The said code specifies detailed procedures for working out the design wind speed and design wind pressure for the design of structures depending on the life, tallness and importance of the structure.

3. Flood Plain Zoning. Flood plain zoning is used to regulate landuse in the flood plains in order to restrict the damage by floods, which are likely to occur from time to time. This aims at determining the locations and the extent of areas likely to be affected by floods for different magnitudes/frequency and to develop these areas in such a way that the resulting damage is minimum.

As per flood plain zoning, the flood plains have been classified as follows:

(i) *Prohibitive zones:* In this no building activity should be carried out, as this area is highly susceptible to flooding.

(ii) *Restrictive zones:* These are moderately flood prone areas.

(iii) *Warning zones:* These areas would be safe to floods upto the large extent and are very less liable to flooding.

For the purpose of regulating landuse in different flood zones, the different types of buildings and utility services have been grouped under following priorities:

(a) **Priority 1.** Defence installations, aerodromes, railway stations, industries, public utilities like hospitals, electric installations, water supply, telephone exchange, etc.

(b) **Priority 2.** Public institutions, offices, residential area, etc.

(c) **Priority 3.** Parks and playgrounds.

15.6 MITIGATION AND PREVENTION

Disaster Mitigation

Mitigation efforts lessen the impact disasters have on people and property. Mitigation measures include keeping homes away from flood plains, designing structure to withstand earthquakes, protecting buildings from hurricane, earthquake. The vulnerability to natural hazards can be reduced by altering, averting, avoiding, and adopting to the hazard.

Upto some time ago, it was generally noticed that so far, the emphasis of disaster management had been on post-disaster response involving rescue, evacuation, sheltering, relief and rehabilitation. Now, it has been felt that there is need that the elements of disaster mitigation and prevention be included in the strategies for urban and regional planning and therefore, there is a need for shifting the emphasis from post-disaster reactive approach to pre-disaster proactive approach.

In view of this, following mitigation strategies should be adopted:

- Identification of major hazards and dissemination of the information.
- Developing a framework for landuse on the basis of an analytical study of potential hazard areas. Many of the disasters can be avoided by proper planning and avoiding habitations in the areas prone to natural hazards.

For the purpose of mitigation, it is necessary to empower the concerned authorities through appropriate legislation and trained personnel to implement the various legislations and by adopting the other measures, such as:

- Use of vulnerability atlas for planning appropriate disaster reduction measures.
- Creating awareness, education and training at various levels as implementation strategy.
- Implementation of the various building codes for earthquake, cyclone, flood resistant buildings and infrastructures.

It is not possible to avoid the natural hazards completely, but by adopting a plan of disaster mitigation, its ill-effects can be reduced to a large extent.

From the above, it is clear that disaster mitigation can be defined as "all actions to reduce the impact of a disaster that can be taken prior to its occurrence including risk reduction and man-made hazards reduction measures." Mitigation is important because basically it is to prevent or minimize loss of life, property and infrastructure from disasters.

From the point of view of sustainable development, it is necessary that the disaster mitigation is built into the development-process. Investment in mitigation measures is more cost effective and justifiable than expenditure on relief and rehabilitation. When the disaster strikes, a state of preparedness and ability to respond quickly to a natural calamity can considerably mitigate loss of life and property and the human suffering and re-store normalcy at the earliest. Therefore, an 'action plan' should be formulated and update periodically for dealing with contingencies that arise in the wake of natural calamities.

Disaster Prevention

Disaster preventive action refers to measure designed to prevent natural phenomena from causing or resulting in disaster or other related emergency situations. Preventive actions concern the formulations and implementation of long-range policies and programmes to prevent or eliminate the occurrence of disasters, on the basis of vulnerability analysis of all risks, preventive actions including legislation and regulatory measures.

Vulnerability represents the extent to which an individual, community, sub-group, structure, service, or geographical area is likely to be damaged or disrupted by the impact of a particular disaster hazard.

15.7 RESPONSE

Response is the first phase which occurs after the onset of an emergency and is intended to provide emergency assistance for disaster effects and casualties. This includes search rescue, evacuation, shelter, medical care, relief measure, clearance of debris, disposal of bodies, damage assessment and other efforts to reduce the probability or extent of secondary damage.

Emergency response plan is a detailed programme of action to control and/or minimize the effects of an emergency requiring prompt corrective measures to protect human life, minimize injury etc. The sufferings are not just due to disaster, but due to post-disaster. Many more people die and suffer because of:

- Chaos all round
- Lack of food, drinking water, shelter
- Lack of medical attention
- Hygiene issues, causing health hazards

Emergency Response Phase include the following actions:

1. Maintenance of law and order
2. Evacuation of people
3. Recovery of dead bodies and their disposal
4. Medical care for the injured
5. Supply of food and water
6. Restoration of electricity and water supply system
7. Temporary shelters
8. Restoring communication line and transport routes
9. Assessment of damage and demarcation of damage areas
10. Cordoning of severely damaged structures liable to collapse
11. Preparation of proposals about the re-construction requirements and strategy to be adopted.

Response Features for Efficient System

1. Mobilization of teams for establishment of base camps/infrastructure
2. Pre-defined staff for co-ordination with other agencies
3. Clear hierarchy of command system
4. Mobilization of damage assessment teams
5. Officer for communication with press/media etc.
6. Management of funds and other resources.

In this case, helps of pre-disaster planning/preparedness is taken. In general, response to a disaster is in terms of relief and rescue operations after the event.

Specialist Response Teams/National Emergency Response Force (NERF)

Following forces are designated as National Emergency Response Force (NERF) and specialist response teams are deployed by the Minister of Home Affairs being the nodal authority:

(a) Indian Army
(b) Border Security Force (BSF)
(c) Central Reserve Police Force (CRPF)
(d) Indo-Tibetan Border Police (ITBP)
(e) Central Industrial Security Force (CISF).

The main objective to introduce these forces as NERF is to make them more agile and vigilant, and can cope with the emergencies and disasters more skillfully. Specialist training center are running at state Police Training Colleges and State Fire Training Centres.

Search, Rescue and Relief

As we know that we cannot stop disasters from happening but adequate and timely preparedness can make all the difference between life and death. People residing in highly vulnerable regions, cope with frequent disasters on the basis of their ability to understand and judge things quickly and clearly. On the basis of their accumulated knowledge, skills and resources of the community, the affected people have managed to survive the fury of the nature for centuries. Whenever and whatever be disasters, search and rescue always starts at the local levels. First thing before taking part in 'search and rescue' operations is to make sure that you do not put yourself also at risk by unnecessarily exposing yourself to a hazard.

Search and rescue team, on the basis of its capability, far-sightedness, preparedness, planning, strength and effectiveness can save more lives. Search and rescue operations are organised in close co-operation with the community and in a team approach.

Objectives of Search and Rescue Team

1. To rescue the survivors trapped under the debris or from a flood.
2. To provide first-aid services instantly to the trapped survivors.
3. To provide temporary support and protection to endangered/collapsed buildings.
4. To create awareness amongst the community members regarding self rescue.

15.8. CLASSIFICATION OF DISASTERS

The disasters are generally classified in two major categories: natural and man-made disasters.

Natural Disasters

Natural disasters can further be divided as under :

(1) Hydro-meteorological Disasters

(a) Floods including flash floods and cloud bursts.

(b) Droughts

(c) Cyclones

(d) Forest fires

(2) Geological based Disasters

(a) *Tectonic Disasters*

- Earthquakes
- Tsunamis
- Volcanoes

(b) *Topographical Disasters*

- Landslides
- Avalanches

(3) Infestic Disasters

(a) Epidemics

(b) Pest attacks, locust invasion of crops.

Man-made Disasters

(1) Chemical and industrial hazards.

(2) Nuclear (radiation) hazards.

(3) Major power breakdown.

(4) Traffic accidents (road, rail, air accidents).
(5) Fire hazard
(6) Mine flooding
(7) Major building collapse
(8) Stampede

Natural disasters may be broadly grouped into major and minor types depending upon their potential to cause damage. For example :

(i) **Major disasters** : Earthquakes, floods, tsunamis, cyclones, drought. (Floods and droughts have also been discussed in chapter on "water resources").

(ii) **Minor disasters** : Landslides, avalanches, tornadoes, hailstorm, heat wave, cold wave, etc.

Common disasters are being discussed in this chapter.

15.9 EARTHQUAKES

Earthquakes consist of various forms of waves originating at the center of disturbance and causing horizontal and vertical ground movements or vibrations. The movements are complicated due to forced and superimposed vibrations.

Earthquakes are one of the worst natural hazards that generally turn into disaster causing immense destruction and loss to human lives. Earthquakes are the sudden tremors caused by displacement of a part of the earth's crust due to tectonic forces. Such earthquakes of tectonic origin are the most devastating and are associated with large scale strain at the crust of the earth. Earthquakes of non-tectonic origin are caused by volcanic eruptions, subsidence in mines, rock bursts and impounding of reservoirs. However, damages caused by non-tectonic earthquakes are not widespread.

A major part of India is prone to earthquakes. As per IS: 1893 code, the country has been demarcated from the point of view of intensity of the seismic loads in five zones, *i.e.*, Zone-1 to Zone-5. In Zone-1 seismic risk is minimum, while in zone-5 seismic risk is highest.

The earthquakes are considered a sudden on-set hazard. They may occur at any time in any region of the world. They may also occur in the oceans and generate underwater quakes known as Tsunamis. Earthquakes generally occur along geological fault (fractures in the earth's crust). Buildings with thick walls and heavy roofs, and settlement on steep hillsides and in river flood-plains are vulnerable to earthquakes. Weak structures with high rate of occupancy increase vulnerability. Congested cities in poor countries are virtually rubble in waiting.

Damage or collapse of buildings, cracks in ground, bridges, roads, water tanks, walls, bund and floods due to dam failures or landslides are the result of earthquakes. The number of casualties are high, if the people reside near to the epicenter and in places where the density of population is high and earthquake-resistant buildings are not made. Breakdown of sewerage, collapse of water supply system, major fires caused by earthquakes, collapse of transport network, electricity and communication network causes the whole situation very serious.

Intensity and Magnitude of Earthquakes

Intensity of an eathquake is the rating of an earthquake based on the actual effects produced by the quake on the Earth. These effects may range from simple harmless vibration to complete overturning and collapse of buildings. The effects depend partly upon the number of jerks and tremors, but will mainly depend upon the maximum rate of change of the movement of the ground *i.e.* by its maximum acceleration (measured with the help of seismograph).

The intensity of the earthquake (represented by ground acceleration) varies from place to place during an earthquake, but **its magnitude (M)** remains the same, because the magnitude corresponds to the maximum intensity reached or the total energy released in an earthquake.

The earthquake magninute scale is represented by **"Richter's scale"**, as Dr. Charles F. Richter devised it. This scale classifies the various shocks in magnitude varying from 1 to 10, and its every succesive higher number represents a little over 30 fold increases in energy released. Thus an earthquake of magnitude 9 has over 30 times more energy than an earthquake of magnitude 8, and 900 times more energy than an earthquake of magnitude 7.

No shock smaller than 5 magnitude causes severe damage. Magnitude 2 represents the smallest tremor that can be felt. Many of the destructive earthquakes are more than 6 in magnitude. Value of the earthquake on the Richter scale and the level of the earthquake are as under :

Value on Richter Scale	*Earthquake Level*
< 2	Smallest tremor felt
4.0 - 4.9	Minor
5.0 - 5.9	Moderate
6.0 - 6.9	Severe
7.0 - 7.9	Very severe
8.0 - 8.9	Extremely severe

Bhuj (Gujarat) Earthquake 2001

On the Republic day, 2001, a severe earthquake with a magnitude of 6.9 on the Richter scale had hit Gujarat near the town of Bhuj in Kuchch. The shock was felt in most parts of the country. Due to the impact of this earthquake the districts of Kuchch, Bhavnagar, Surendra Nagar, Rajkot, and Ahmedabad were devastated.

This earthquake was the most severe earthquake in the last more than 50 years in India, with 20,000 to 30,000 dead, 150,000 injured and about 1.6 crore people affected. The total economic loss was estimated to be Rs. 225 billion.

The Bhuj was the worst affected town with about 10,000 people killed, and almost half of its structures had been levelled. The state government was grossly unprepared to meet such an unprecedented emergency.

15.9.1 Mitigation Strategies

Before erecting buildings and starting major construction work, proper and scientific test of soil should be made mandatory. Structures should be designed and built to withstand ground shaking. The Bureau of Indian Standards has made building bylaws and codes for safe constructions against earthquakes. Building plans should be duly checked by the Municipal Corporation or Municipal Committee in accordance with the laid down by-laws, regulations and designs.

Existing lifeline buildings, like educational institutions, hospitals, offices, malls, etc., must be build with earthquake safety measures. If the existing lifeline buildings are without proper safety measures, the same should be upgraded by retrofitting techniques.

To safeguard from the effects of earthquakes, all components of building should be firmly tied together and so stiffly braced that the building will tend to move as a unit. Floors and cross-walls should be continuous throughout the building and opening should be avoided near the outside corners. The different parts of the same building dissimilar in mass or stiffness should be separated by a space.

A continuous spread foundation of reinforced concrete increases the resistance of a building to the earthquake forces. Hence, it is essential requirement of design and construction of a building to provide a rigid foundation of reinforced concrete strongly tied together. A raft foundation is the best for earthquake resisting structures but is expensive and therefore, used for large/multi-storey buildings or in poor soils.

In earthquake zones, framed buildings should be constructed, in which all the loads is taken by the frame and no portion of the load should be considered transmitted to the wall. In reinforced concrete construction, wall panels should be sufficiently tied to the frame at points of supports. Now-a-days, for earthquake resistance, rubber bearings are placed on the foundations. These rubber bearings are designed to take the vertical load of the building and the calculated horizontal displacement due to earthquake.

Tips for keeping safe during earthquake

(a) Take cover. Go under a table or other study furniture, kneel, sit or stay close to the floor. Hold onto furniture legs for balance. Be prepared to move, if your cover moves. If no cover is nearby, kneel or sit close to the floor next to a structurally sound interior wall. Place your hands on the floor for balance.

(b) Do not stand in doorways. Violent motion could cause doors to slam and cause serious injuries.

(c) Move away from windows, mirrors, book cases and other unsecured heavy objects. If in bed, stay and cover yourself with pillows and blankets.

(d) Do not run outside, if you are inside. Never use the lift.

(e) If living in kutcha house, move to an open area, where there are no trees, electric or telephone wires.

(f) If outdoor, move into open, away from buildings, street lights and utility wires.

(g) If your home is badly damaged, you have to leave. Collect water, food, medicine, and other essential items and important documents with you.

(h) Avoid places, where there are loose electrical wires and never touch metal objects that are in touch with loose wires.

(i) Do not re-enter damaged buildings.

(j) If in a moving vehicle, move to a clear area away from buildings, trees, over passes and utility wires. Stop and stay in the vehicle.

15.9.2 Structural Damage

While there is too much loss instantaneously, but what makes the situation worse is:

(a) With roads, rails, airports damaged the total transportation system comes to a collapse. Hence, it takes several days for relief teams to arrive.

(b) When damage is everywhere and too many people are injured, medical facilities fall in major short-supply. This refers to both space in hospital, availability of trained medical personnel, medical equipments, medicines, etc.

(c) Collapse of electricity, water supply and no food.

(d) Communication infrastructure fails.

After the immediate impact, many more people could die during next few days due to:

- hunger,
- cold,
- epidemic,
- lack of medical attention,
- shock and trauma.

15.9.3 Action Plan

A detailed action plan for taking necessary actions before, during, and after the earthquake shall be prepared, which *inter-alia* should include the following:

Before the Earthquake

- Evacuation in less than one minute without pushing or failing
- Identify the open areas where people can assemble after evacuation in an emergency
- Make sure that necessary supplies are approachable
- Practice mock drill
- Do not hang heavy items or items made of glass
- Keep emergency items, flash lights, first-aid kits, etc., at defined places, which should be easily accessible in emergency situations.

During the Earthquake

- Don't create panic
- Do not use lifts
- Do not rush outside and stay where you are, keep away from walls and windows
- Keep away from buildings, high cracked walls, handling electric wires, power lines, etc.
- Do not use candles, match box, lighters, etc.
- At once leave the building, if a fire starts

After the Earthquake

- Stay from damaged buildings/structures.
- Listen to emergency announcements on your Radio/T.V.
- Have a close watch on other members and structures around you.
- Turn off gas connections and extinguish open flames.
- Always remain alert and be prepared for secondary tremors that follow the main tremor.
- Maintain calm and order. Provide immediate first-aid to the injured, and start rescue operation. Ask for help for fire fighting, police ambulance, trained rescue teams.

15.9.4 Role of Incidence Response Team

The rescue team, equipped with rescue related materials and equipments, like fire tender, ambulance, medical officer, paramedical staff, transportation team, reaches the spot and rescue the wounded and after preliminary treatment and dispensation of medical aid, send them to the nearest hospitals.

This team, called as Task Force. The fire tenders are engaged in dousing the fire in the buildings. The transportation branch immediately swings in the action and assists in moving the relief materials to the camps, relief camps, where the rescued beneficiaries are housed. In case of serious wounded patients, the helicopters are geared for their movement to the nearest safe hospitals for treatment. In case of serious damage of road and rail communication, the air operation is undertaken for transportation of men, materials and supply of relief goods to the relief camps.

The incident commander, who is overall in-charge of management of on-site response, oversee the relief operation by surveying situation status, like number of people and area affected, availability and procurement of resources requirement facilities. However, keys to establish immediate priorities, like search and rescue and relief distribution strategies. He is also required to maintain law and order, traffic management in the site and relief camps and help the local police.

In case of total destruction of buildings, he is to raise temporary shelters or pitch tents for housing the rescued population. Pie is expected to deploy the volunteers for search and rescue operation, food preparation and orderly distribution, giving first-aid, ensure proper sleeping and resting facilities, organize and provide toilet, bath and sanitation, make lighting arrangements, maintain general cleanliness in the incident base, relief camps and command post, provide clean drinking water and mobilize the medical team to treat the inmates in relief camps, if they fall ill due to various adverse situations, prevailing in the camp area, ensure gender sensitive needs and their safety and security of the camp. He is to effectively manage the dead bodies by deploying the dead body management group after their identification and disposal. He immediately takes steps in restoration of communication facilities, power and water supply (if any) and other essential services.

Treatment of the wounded and disables, shifted to the local hospitals is a major challenge. Hospital preparedness is very much crucial to any disaster response system. Each hospital should have an emergency preparedness plan to deal with mass casualty incidents and the hospital administration/doctors need to be suitably trained for this emergency. The local volunteer organization, if any, the NGOs and the civil defense force and homeguards be utilized in search and reuse, first-aid and relief operation on the very site of the disaster. The another most important step on the site to be performed by the commander, is mobilization of fire fighting.

Recovery: The aim of recovery phase is to restore the affected area to its previous state. It starts with rebuilding destroyed property, re-employment and repair of other infrastructure. Efforts should be made to build back better aiming to reduce pre-disaster risks inherent in the community and infrastructure. External assistances as discussed earlier, be procured after observation of set procedures, like UN, Red Cross, scores of NGOs, worldwide and foreign countries in the matters of search, evacuation, quarantine, rescue, first-aid, relief, recovery and rehabilitation work.

Major Earthquakes in India

1618 Bombay earthquake killing 200 people.

1737 Bengal earthquake killing 3 lakhs people.

1819 Kutch earthquake chief towns of Tera, Kathara and Mothala were devastated.

1869 Assam earthquake.

1897 Shillong earthquake.

1905 Kanga earthquake in Himachal Pradesh killing 20 thousand people.

1934 Bihar earthquake killing 6 thousand people.

1935 Quetta earthquake (now in Pakistan) killing 25 thousand people

1941 Andaman island earthquake.

1950 Assam earthquake killing 1,500 people.

1988 Assam earthquake affecting 20 million people.

1993 Latur earthquake killing around 8 thousand people and injuring around 30 thousand.

2001 Gujarat quake in the town of, Bhuj, Bacho, Anjur, Ahmedabad and Surat killed 25 thousand and affected 6.3 million people.

2004 Earthquake generated tsunami in Tamilnadu, Kerala, Andhra Pradesh and Pondicherry killed 10,749 people 5,640 were found missing and around 2.79 million were affected. Total loss of this tsunami in world was 2,26,00 people.

2005 Earthquake in Kashmir killed 1,309 people.

2009 Andaman Islands earthquake of 7.5 magnitude.

2015 Earthquake in Nepal and India of 7.8 magnitude which killed 8964 people and injured 21952 people.

15.10 FLOODS

Floods refer to huge amount of water reaching the land in a short span of time, causing land surface to be submerged under water at places, where land surface is usually not covered with water. Floods are caused by discharge of huge volume of water in a short span of time, at a rate, such that water cannot be carried away from the scene of discharge.

Floods are temporary inundation of land. Floods occur when there is an increase in reservoir level or of rivers thereby flooding their banks. This may happen due to heavy rains, cloudburst, cyclones, storm surge along coast, melting of snow, dam burst, release of water from spillways or tsunami. Floods are a serious cause of human misery in our country almost every year. India is one of the most flood affected countries in the world. It is a natural disaster which turns out to be a social disaster, since the people who are badly affected are generally the poor.

Floods can occur slowly and take hours or sometimes, can even occur all of a sudden due to heavy rains or cloud bursts. Some other **causes of floods** are:

(i) Breach of the water storage and control structures

(ii) Spillover

(iii) Storm surge.

The areas which are prone to flood risks are:

(a) Places which have a history of flooding

(b) Areas receiving heavy rainfall with not much naturally sloping landscape

(c) Areas of the lower levels, where the higher areas are receiving heavy rainfall

(d) Areas around river banks or seacoasts

(e) Areas downstream of dams, etc., where release of water becomes necessary to safeguard the dam, which might cause flooding of downstream areas.

The flood warning is issued by the Central Water Commissions (CWC), irrigation and Flood Control Department, and Water Resources Department. Everything in the flood plains get inundated, buildings and houses made on the earth, weak foundations and water soluble ingredients collapses, endangering human beings and their property. Houses built on silts at the edge of the rivers are at risk. Agricultural fields, livestock, food stock, water supply, communication lines, sewerage and powerlines are also under risk. Settlements in lowlying areas of flood plains are also at risk.

Floods also lead to outbreak of epidemics. In the rainy season diarrhoea, viral infections, malaria, typhoid, jaundice and other waterborne diseases are common. Due to encroachment of flood plain by the people, the flood affected areas, casualties and property damages have also gone up.

15.10.1 Risk Assessment

(i) Identify threats

(ii) Determining probability of occurrence

(iii) Estimating the impact of the hazard

(iv) Determining measures that can reduce the risk

(v) Taking action to reduce the risk.

15.10.2 Preparedness and Prevention

(i) Identify areas which are prone to high risk

(ii) Identity existing infrastructure, like roads, shelters, hospitals/dispensaries, search and rescue tools

(iii) Install warning system and conduct drills

(iv) Institute a hierarchical structure for preparedness down to village level. Encourage participation of community

(v) Train man-power in various tasks in emergency

(vi) Implement land zoning regulations in flood plains

(vii) Create necessary awareness leading to self-help

Sometimes, prevention is not possible, in such cases certain action should be taken to reduce the impact significantly:

1. Keep the drainage system clean so as to allow water to be carried down very fast.
2. Rain water harvesting. If there are several rainwater harvesting systems, the rainfall in the area would go the sub-soil of the region.
3. The drains should be desilted before the onset of rainy season.
4. Inspection and repairs of dams, canals and river embankments.
5. Afforestation helps in binding the loose soil. It also reduces the impact of flowing water and also reduces the structural damages.
6. Construct local embankments around low-lying houses, etc.

Long Term Preventive Actions

(i) Identify flood diversion area. Flood water are diverted to these unpopulated areas.

(ii) Construction of dams, etc., at strategic locations.

(iii) Construct sea walls in coastal areas.

(iv) Embankments are constructed around populated areas lying along rivers/sea coasts.

(v) Conversion of flood-prone areas into wetlands, where, settlements are not allowed.

15.10.3 Flood Mitigation

Disaster mitigation involves measures to reduce the effect of disaster. Disaster mitigation can be defined as "all actions to reduce the impact of a disaster that can be taken prior to its occurrence, including preparedness and long term risk reduction measures. It also includes the planning and implementation of measures to reduce the risks or man-made hazards."

Mitigation is important because basically, it is to prevent or minimize loss of life, property and infrastructure.

The main reasons for loss due to floods is because of rapid increase in population which has occupied the flood plains and deforestation, re-claiming areas for occupation and shifting cultivation.

As flood in a river is caused by excessive precipitation in the catchment area of the river. The methods adopted to manage such heavier river discharges are known as "flood control measures." Inspite of making best efforts, sometimes it is not possible to fully eliminate floods, and we can only check and reduce their devastating effects by pre-warning the area residents.

Various flood mitigating measures are discussed in following paragraphs:

1. Mapping of the Flood Plain. To reduce the risk of the area, an appropriate mapping of the flood plain is the primary or initial step.

2. Land use Control. When the flood plains and coastal areas are inundated by water, land use control would reduce danger of life and property. The areas that are subject to flooding should not be allowed to have residential, commercial and other development projects. The elevated land may be used for this purpose and water-holding regions can be made into lakes, or ponds.

3. Structural Measures. In the flood plains, the structure should be made and strengthened in such a way that they can withstand flood forces and seepage. The buildings should be made on an elevated area. Other structural measures are: Construction of dams, reservoirs, dykes, retaining walls, construction of alternate drainage courses.

4. Flood Control Measures. The main aim of flood control is to reduce or decrease the flood damages.

There are many ways to check and control floods; one among them is to have dams and reservoirs, which stores excess water during heavy rains, but when the reservoir is almost full, the flash floods may compel the opening of spillway gates. These gates should be managed properly, otherwise they may bring flood in the downstream. Flood control can also be made possible by flood reduction that reduces the amount of runoff by treatment, such as protection of vegetation, afforestation, clearing of deposits from water channels and other water holding areas.

Another measure to control floods is "Natural detention measure." In this measure flood is controlled by reducing the run-off by allowing the water to percolate into groundwater through increasing the permeability of soil. The simplest, cheapest, most effective and highly beneficial method of reducing water flow is artificially induce infiltration of water and recharging the groundwater so as to allow this water to infiltrate. The dug wells can be sunk along the beds of ephemeral streams.

5. Non-structural Measures. These are people's participation through education and awareness, sedimentation clearance from rivers, afforestation, effective warning systems, changing farming practices, flood evacuation preparedness, arrangement of boats and rescue equipments, etc.

Non-structural measures contrary to the structural measures for aiming to mitigate the flood damage by trying to keep the flood water away from the people, strive to keep the people away from the flood water, bearing in mind the fact that the flood plains belong to the river. It contemplates use of flood plains judiciously, simultaneously permitting vacating of the same for use of the river, whenever the situation calls for. This technique allows the use of flood plains reducing hazard, while retaining its beneficial effects.

The non-structural measures can be broadly grouped under two categories, *i.e.*:

(i) Flood forecasting; flood proofing, disaster preparedness; and flood plain zoning.

(ii) Disaster relief; flood fighting; and flood insurance.

(a) **Flood Forecasting:** It enables forewarning as to when the river is going to use its flood plain, to what extent and for how long. With reliable advance information/ warning about impending floods, loss of human lives and moveable properties can be reduced to a considerable extent. People and cattle, etc., can be shifted to safer places.

(b) **Flood Proofing:** These measures help greatly in the mitigation of distress and provide relief to the population in flood prone areas. These techniques include providing raised platforms for flood shelter and raising the public utility installations. These include drinking water and sanitary arrangements, human dwellings and animal shelter, storage facilities for food, fodder and other essential commodities, communication links-telephone/wireless/road/boat.

(c) **Flood Plain Zoning:** The rapid growth and spread of population in hazardous areas is a matter of increasing concern because it leads to mounting cost of disasters in terms of lives lost and damage to property and investment.

The basic concept of flood plain zoning is to regulate the landuse in the flood plains in order to restrict the damage by floods which are bound to occur from time to time. Flood plain zoning, therefore, aims at determining the locations and the extent of areas likely to be affected by floods of different magnitudes/frequencies and to develop those areas in such a fashion that the resulting damage is reduced to minimum. For the purpose of flood plain zoning, we can broadly classify the areas into three zones, *i.e.*:

(i) *Prohibitive zones.* In these zones no building activity is to be permitted as this area would be highly susceptible to flooding.

(ii) *Restrictive zones.* These are moderately flood prone areas.

(iii) *Warning zones.* These zones would be safe to floods upto the large extent. These are very less liable to flooding.

(d) **Flood Preparedness and Relief:** Flood disaster relief is rendered as soon as flood damage occurs. The flood preparedness and relief has, therefore, to be planned in advance to meet any disaster. Such preparedness generally include measures like:

(i) Flood forecasting
(ii) Flood warning
(iii) Flood fighting
(iv) Evacuation
(v) Community shelter for people, and
(vi) Relief camps, etc.

(e) **Flood Fighting:** Flood fighting covers construction of temporary dykes along the river, closing breaches immediately. Stagnant water becomes the breeding ground for mosquitoes affecting public health. Public health operations should ensure availability of supplies and equipment, co-operation with other organizations engaged in disaster relief, for immediate mobilization of personnel to health hazards.

Flood fighting measures normally involve:

- Strengthening of central, state and district flood control rooms

- Evacuation of flood victims
- Air dropping of food pockets
- Supply food and other rations including kerosene, etc.
- Supply of fodder
- First-aid and health operations
- Restoration of road and rail links
- Restoration of tubewells, etc.
- Pumping out water from ponds and lowlying areas in cities/villages
- Restoration of public assets and structures, power installations sewerages and water supply schemes, roads and drains, etc.

Most of the above activities are of immediate nature. Participation by voluntary organizations and local people should be encouraged for flood fighting.

(f) **Flood Insurance**: Flood insurance has several advantages as means of modifying loss of burden. The insurance does not reduce the flood loss potential directly, but it provides a mechanism for spreading the loss over large number of individuals.

15.10.4 Pre-Flood Measures

1. **Communication:** Communication plans need to be drawn on the assumption that normal communication system goes hay ware. Communication plan should be able to provide foolproof warning for impeding danger.

Considering the above, following actions shall be taken:

- Activate control room and teams
- Disseminate information/warnings/instructions
- Activate information centre
- Public awareness is essential for inducing a mindset of disaster prevention, mitigation and preparedness.
- Order evacuation, and arrange evacuation of vulnerable people.

Besides the telephone, wireless sets are to be established at important locations and also at central control room, at district headquarter and also at state headquarter.

A list of telephone numbers and the radio frequencies of wireless sets are to be maintained with each official involved with disaster management plan.

2. **Resource Mobilization** : Resource mobilization activities shall include :

- Identification of trained and motivated human resources
- Mobilization of life support systems
- Mobilization of rescue and search equipments, including transport arrangements
- Ensuring availability of human and material resources at vulnerable places
- Maintenance and constant updating of data, regarding availability of resources
- Arrangement for drinking water
- Assess hazard intensity and evacuation
- Move resources to nearest central location

- Co-ordinate resources coming in from various agencies, adjacent districts and from outside.

15.10.5 Measures during Flood

- Invoke medical aid
- Arrange temporary shelters
- Co-ordinate relief operations
- Arrange food relief operations and first-aid to victims
- Assist urgent repairs of buildings and other critical infrastructural facilities
- Co-ordinate relief operations between governmental and non-governmental relief agencies
- Use of Armed/Para-military forces
- Ensure adequate funds
- Distribute relief material/cash promptly
- Disposal of dead bodies/carcasses
- Preventive measures for epidemics by ensuring availability of medicines, vaccine and drugs.

Rescue Operations

For the purpose of rescue operations personal be drawn from :

- Police
- Fire service
- Army
- Paramilitary forces
- Medical services
- Critical engineering service
- Members of community
- Member of NGOs.

Support will be provided to the district administration of having an emergency rescue kit with some essential equipments, like a boat, portable power generator set, early warning equipments, tents, power saw, etc., to meet the emergency need at the time of natural disasters, like cyclones or flood or earthquakes. Each selected district will be provided the equipment kit as per their need. Equipments will be procured in consultation with district administration and the maintenance will be the responsibility of the district administration.

Relief

Relief measures are the immediate need in the post-disaster scenario, when an unknown number of victims are affected and of even their locations are not clearly known. In such circumstances search, rescue and evacuation processes are carried out immediately after the disaster has struck a certain area. These are the most important operations, which are usually performed by the local volunteers, voluntary organizations and the district and state agencies.

After the search, rescue and evacuation, some important steps required to be taken to provide relief to the evacuees are:

1. To provide **shelter** for victims whose houses have been destroyed/submerged.
2. To organize and distribute **food** to victims and volunteers.

3. To establish **communication** system.
4. To clear **key roads and airfields.**
5. To **re-store water and power supply** system or to make temporary arrangements for them.
6. To **maintain law and order**, especially to prevent looting and crime.
7. There is need to **protect livestock** in disaster situations, as in Indian families livestock has important role to play. It helps the community in the farm and provides extra income to poor people.

15.10.6 Post-Flood Measures

Planning and implementing human re-settlement to avoid recurrence of disasters:

- Identify shortcomings and prepare plans for future accordingly
- Disaster should be documented
- Assess losses/damages critically for preparing future contingency plans
- Co-ordinate and monitor post-disaster actions
- Ensure accessibility for saving valuable lives
- Training and awareness campaigns, based on lessons learnt
- Provide livelihood to the most needy
- Drinking water and sanitation restoration.

15.10.7 Training and Mock Drills

The Training of Trainers (ToT) to enhance the capacity of disaster management committees and prepare a core team of trainers and trainees. Training would be a continuous process on disaster risk management programme. The trained cadre will facilitate the process of contingency plan development at different levels.

Specialized training will be organized at different levels for the disaster management team members of enhancement of skills to effectively carry out their responsibilities, such as warning dissemination, search and rescue operation, shelter management, first-aid, trauma counselling and damages, assessment, etc. The DMT members will be provided a specific type of apron or jacket for easy identification after the training. Adequate training will be provided to the women DMTs to carry out activities during emergency situations.

Special groups, such as women, disabled persons, children, etc., are more vulnerable in an emergency situation and hence, require special attention. The programme will aim at strengthening capacities of these, groups to respond to disasters.

Special training shall be provided to women for enhancement of their capacities to carry out the activities effectively. Capacity building of women groups will include skill upgradation in use of the latest know-how for effective response and sustainable recovery in disaster situations.

An effective disaster risk management campaign strategy will be developed in consultation with all stakeholders of the selected areas for public education to take preventive measures in the wake of natural hazards to minimize the loss. The district administration, with the help of civil society response groups, would take up a massive awareness campaign through out the selected districts for preparedness, through mass meeting, drawing competitions, etc., among school students, poster, and leaflets. Similarly, wall paintings will be done in each village explaining Do's and Don'ts of various disasters, showing the safe shelters and safe routes for evacuation, etc.

Based on the disaster preparedness and response plan, there will be mock drills before disaster seasons to find out the feasibility of the plan and to ensure greater role clarity of the key players. It will also ensure the availability and functional condition of equipments and resources.

15.10.8 Combating the Flood Disasters

In combating the flood disasters, following steps should be taken:

(a) Storage of sufficient non-perishable food items and drinking water and essential medicines in advance

(b) Battery operated flash lights/inverters

(c) First-aid box

(d) Gas cylinders

(e) Sufficient petrol and lubricants for the vehicles.

(f) Since the toilets flush systems become inoperative due to break, damage to plumbing, pipe fittings and water-logging of safety tanks, outbreak of cholera, is a regular sight in the flood affected area due to unhygienic conditions. A simple technique can help toward-off this situation, several plastic polythene garbage bags be stored for use in excretion. After using these for several times, close their mouth tightly and keep it in a corner. On arrival of relief team and normalcy re-stored these bags could be dumped in garbage collecting trucks for disposal by the civic authorities. In rural set up, these can be conveniently unloaded in a large dumping yard and then covered with the earth for processing as compost.

(g) To avoid missing of kids or minors, or separated, select a pre-destinated place, where all of them would meet/send them to locations at the first available opportunity. Even if one cannot move to this place, just call up and leave a message about the location and well-being so that the friend of the pre-designated place owner would work towards this union with the missing/separated kith and kin.

(h) The administration's disaster management authority at various levels should swing to action in rescue and providing relief to the victims. The operation section chief, the designated officer, with its battalion headed by incident commander is to start rescue, search, transportation, and supply of food and water, medical aid, etc. as elaborately discussed earlier. The services of police/army/navy/air force are utilized for this operation. At times, some villages are marooned and surrounded by flood water. The inhabitants are trapped in that island for days together without food and water. People are also found taking shelter on rooftops, trees, etc. Hence, airlifting these unfortunate inhabitants become inevitable.

The services of navy/air force be called for immediately for rescue and air dropping of food, water and other essential items packaged at the base camps.

Flash Flood

Flash flood is a rapid flooding of geomographic low-lying areas. It may be caused by heavy rain, associated with a severe thunderstorm, hurricane, tropical storm, or melt water from ice or snow. Flash floods may occure after the collapse of a dam or cloud burst.

The flash flood occurred in the form of huge and deadly cloud bursts, causing flash floods in the River Ganga, Sudden heavy rains caused dangerous landslides which may kill hundred of people and cause miseries to the people.

Major Floods in India

Every year floods in India do occur. On an average, every year 1588 lives are lost, 7.5 million hectares of land are affected and crops damaged, including houses and public utilities to the tune of Rs. 18 billion. As per data collected, 84,207 lives were lost between 1,953 to 2,000.

Heavy flood damages have occurred during the monsoon years of 1955, 1971, 1973, 1978, 1980, 1984, 1988, 1989, 1998, 2004, 2005 and 2008. The infamous Bihar flood that occurred in 2008 (Koshi river flood) killed 1500 people.

Following are the major floods in recent pasts.

2014 Kashmir Floods. Death toll of more than 500 was caused by continuous torrential rainfall in Sept 2014 from massive floods in the areas of Srinagar, Bandipur, Rajouri etc. In this flood, around 500 bridges were damaged across the state of Jammu and Kashmir. The damage of properties was estimated over Rs 5000 crores.

2015 Tamil Nadu floods. Heavy rains in November and December 2015 resulted in flooding of Adyar river in Chennai, Tamil Nadu.

2016 Assam floods. Heavy rain in July and August 2016 resulted in floods affecting 1.8 million people and flooding the Kaziranga National Park killing around 200 wild animals.

2017 Gujarat flood. Following heavy rain in July 2017 Gujarat was affected by severe floods resulting in more than 200 deaths.

15.11 CYCLONES

Cyclones are violent storms in which strong winds move over the ground in a circle. A cyclone is a low pressure area in the atmosphere in which winds spiral (circular motion) inward, making up a large whirling mass of air, at the center of which the barometric pressure is the lowest. All cyclones, therefore, have two characteristics: (1) the atmospheric pressure is the lowest at the centre, and (2) the winds spiral in towards the center.

Cyclones, also known hurricanes or typhoons can be termed as the heat engine whose heater is the oceanic surface. The released heat after condensation changes it into kinetic energy for the cyclone. A tropical cyclone forms, when the horizontal temperature gradients are extremely high and are of low pressure. Thus, cyclones are caused by warm tropical moisture bearing clouds developing in open oceans or seas. It has a warm core. A cloudless spot in the centre of the cyclone is known as the *eye of the cyclone*. A cloud of great vertical extent encircles the eye.

A cyclone is formed in following stages:

- The ocean surface temperature rises over 26°C.
- A closed isobar is observed.
- Low pressure falling below 1000 mb.
- Regions of circular movement, initially to a radius of 30-50 km, then increasing to 100-200 km and even shooting upto 1000 km. Right outside the eye, very strong and ferocious winds, reaching to as much as 290 km/h exist. The wind speed gradually decreases towards the outer edge. The pressure also increases outwards.

These high speed wind storms, when moving towards the land at higher altitude may bring in terrific rains due to cooling and condensation of the water vapour contained in the air. Such rains are usually associated with violent thunder and lightening. These cyclones usually move to higher altitudes in an irregular path at a speed of about 10-30 km/h, deriving energy from the latent heat of condensation of ocean water vapour, and increase in size, as they move over the ocean. However, when they reach and move on land, this source of energy gets cut off, and the cyclone dissipates its energy fast by bringing heavy rains.

The satellite tracking can track the movement of cyclones since the build up and the expected path can be projected or planned. The dissemination of cyclone warning is done by radio or televisions, public announcements, etc. Everything in the path of cyclone is at great risk, particularly light weight structures build of mud, wood, old houses and buildings with weak walls without adequate anchorage to the foundation are always at a great risk. Settlements that are situated in low-lying coastal areas will be prone to the direct effect of the cyclones, like wind, stromsburg and rain. Due to heavy rains, settlements in the nearby areas are vulnerable to floods, landslide or mudslides.

Tornado

Tornado is a special kind of intense cyclone, and is a powerful, twisting windstorm. It has a diameter of generally 100 to 2000 m and has the most violent winds that occur on the earth. Most tornadoes travel a distance of about 30 km at a speed of 15 to 40 km/hr and last less than an hour. However, some tornadoes may last for several hours and may measure upto 2.4 km in diameter and may travel upto about 300 km at a speed of upto 100 km/hr.

15.11.1 Impacts of Cyclones

Physical damage: The severest damage is caused by the floods, wind forces and storm surges, landslides, weak and light weight structures are completely razed to the ground.

Casualties and public health: Cyclones cause great danger to health. Viral infections are caused by violent winds. Epidemic diseases, such as diarrhoea, malaria, cholera, typhoid, jaundice, etc., spread very fast.

Effect on water supplies: Flood water also causes havoc by contaminated ground and piped water supply, and results into acute shortage of drinking water.

Damage to crops and food supplies: Cyclones accompanied by violent winds and torrential rains cause damage to the standing crop and stocked foodgrains in the low-lying areas. Sea-water flooding shall enhance the salinity and make the soil unsuitable for cultivation purpose.

Damage to communication and transport: The cyclone winds completely smash the communication and transport system.

Damage to essential services: The cyclonic winds badly damage and disrupt the essential services, like electricity, sewage disposal and water supply, etc.

15.11.2 Mitigation Measures

1. *Multi-purpose cyclone shelters* of adequate designs and engineered structures should be constructed in vulnerable locations of coastal areas.

2. The *Engineered structures* that can withstand the forces of the winds and torrential rains need to be built. However, poverty or people living in coastal areas are a hurdle, as they cannot afford the cost of engineered structures. Moreover, most of the houses and other buildings in coastal areas are constructed with locally available building materials. Following precautionary steps should be taken:

- Thatched roofs should be avoided in coastal areas
- Wind and water-resistant buildings should be constructed
- Houses should be built on earth mounds, etc.
- Houses should be properly strengthened so as to resist wind and flood damage
- In order to resist the uplift or flying off of the objects, all elements holding the structures require to be adequately anchored
- Large overhangs should be avoided, as they get lifted and broken
- Projections should be properly tied down
- Rows of trees be planted in the direction facing the wind so as to decrease the wind energy.

15.11.3 Mitigation Strategies

- *Hazard mapping* is done to show the regions prone to the cyclone.
- *Land-use control* should be adequately designed so that least critical activities are included in vulnerable regions. Vulnerable sites should be reserved for parks, playgrounds or grazing lands.
- *Cloud seeding* can also be adopted by using latest technologies to break the cyclones.
- *Shelter belts* can be planted along the coastline to break the cyclones and the wind force. It will not only mitigate the impact of strong cyclonic winds but also check soil erosion effectively.
- *By improving* the vegetation cover, the water percolation capacity of the soil can be increased.

Major Cyclone Disasters in India

1935	Tropical cyclone killed 30 thousand people.
1942	Cyclone in Odisha and West Bengal killed 40 thousand people.
1943	Rajputana tropical storm killed 5 thousand people.
1971	Odisha eastern coast cyclone killed 9658 people.
1977	Cyclone in Tamil Nadu, Andhra Pradesh and Kerala killed 14,204 people.
1999	Odisha super cyclone disaster killed around 10 thousand people.
2008	Cyclone Nisha of Tamil Nadu resulted 204 fatalities.

15.12 TSUNAMI

A tsunami (pronounced su-nah-me) is a wave train of series of waves generated in a body of water by an impulsive disturbance that vertically displaces the water column. Earthquakes, landslides, volcanic eruptions, explosions, etc., generate tsunamis. These attack coastlines causing devastating property damage and loss of life. Tsunami is a Japanese word with the English translation, "*harbour wave.*" Oceanographers call them as "seismic sea waves." Tsunami also called *tsunami wave train* and is a series of water waves.

Tsunamis can be generated, when the sea-floor abruptly deforms and vertically displaces the overlying water. Tectonic earthquakes are a particular kind of earthquake that are associated with the earth's crystal deformation, when these earthquakes occur beneath the sea, the water above the deformed area is displaced from its equilibrium position, waves are formed as the displaced water mass, which acts under the influence of gravity, attempts to regain its equilibrium. When large areas of the sea-floor elevate or subside, a tsunami can be created. Large vertical movements of the earth's crust can occur at plate boundaries.

Once created, a tsunami resembles the ripples when a pebble is dropped into a pond. However, in contrast to such ripples, tsunami move across the ocean at amazing speeds between 500 to 950 km/hr. A tsunami in the deep open ocean can pass virtually undetected because the height of the wavecrest is usually less than 1 metre, as the distance between the wavecrests is very large ranging from 100 to 700 km. However, upon entering shallower coastal waters, these destructive killer waves get slowed down, and the water begins to pile up the heights that occasionally exceed 30 metres or so, although the largest tsunami ever measured is as high as 65 meters. Therefore, as the crest of a tsunami approaches the shore, it appears as a rapid rise in sea-level with a turbulent and chaotic surface such a great rise in water level on seashores will eventually destroy whatever exist on or near the shores.

15.12.1 Tsunami due to Landslides, Volcanic Eruptions and Explosions

A tsunami can also be generated by any disturbance that displaces a large water mass from its equilibrium position. Submarine landslides, which often accompany large earthquakes can also disturb the overlying water column as sediment and rock slump down slope and are re-distributed across the sea-floor. Similarly, a violent submarine volcanic eruption can create an impulsive force that uplifts the water column and generates a tsunami.

In recent times, one of the biggest tsunami occurred in the Indian ocean on 26 Dec., 2004, badly damaging the Andman and Nicobar group of islands and affecting the coastal towns and villages of Andhra Pradesh, Tamil Nadu and Kerala, besides causing the devastation and catastrophe in Indonesia, Thailand, Malaysia, Sri Lanka and Maldives, where around 2,30,000 people were killed and properties worth millions and millions destroyed. This tsunami occurred as a result of the severs **Sumatra-Andman Earthquake** of magnitude 8.9. This earthquake lasted for a long time as 10 minutes, unlike most large earthquakes that last only for a few seconds.

15.12.2 Mitigation Measures for Tsunami

1. **Steps to be taken before a Tsunami**
   - Identify if your locality is in tsunami hazard prone region.
   - Find out the height of your street above sea-level and distance from the coast.
   - Prepare evacuation routes plan, and regularly practice the evacuation routes.
   - Make an adequate emergency communication plan.
   - Do not erect the buildings near the sea-coast and erect elevated coastal homes.
   - For emergency warning or other informations always use a local radio or television, and always follow necessary instructions issued by local administration.
2. **Steps to be taken during Tsunami:**
   - Evacuate to a safe elevated area.
   - Keep disaster supplies kit on hand.
   - During evacuation take livestock.
   - For updated emergency information always use a local radio or television.
3. **Steps to be taken after a Tsunami**
   - Always use a local radio or television for updated emergency information.
   - Get first-aid for injuries, if required before helping other injured or trapped people.
   - Provide special assistance to infants, elderly and physically infirm people.
   - Stay out of a building, if water remains around it.
   - Take extreme precautions at the time of re-entering the homes or buildings.
   - Wear long pants, a long sleeved shirt and strong and firm shoes.
   - Always use battery-powered lanterns at the time of inspecting homes and buildings. Never use candles.
   - Minutely check foundations for cracks or other damages.
   - Checks for gas leaks, electrical system damage and damage to sewage and water-lines.
   - Watch out for snakes and wild animals that may have come into homes or buildings with the water.
   - Open all the windows, ventilators and doors to help dry the building or home.

15.12.3 Tsunami Warning System (TWS)

Tsunami warning system is a system to detect tsunamis and issue warnings to prevent loss of life and property. It consists of two equally important components, a network of sensors to detect tsunamis and a communication infrastructure to issue timely alarms to permit evacuation of coastal areas.

If there had been an adequate system for detecting deep-sea events, like tsunamis, the under-sea earthquake and resulting tsunami that struck Thailand, India, Sri Lanka and the African coast in December 2004, the appalling loss of life could have been avoided. It is true that the thousands killed in Indonesia were too close to the quake's epicenter to be saved by a tsunami warning system, but then such a system would have given people in Sri Lanka and India, the hardest hit countries, time to flee to higher ground.

The Indian Deep Ocean Assessment and Reporting System (DOARS) have active inputs from Sumatra, Myanmar and Thailand and uses existing software technology to build a virtual reality system for relaying timely alerts on tsunamis.

A Case of Asian Tsunami

The tsunami that hit South-east and South Asia on 26th December 2004, was the biggest ever in the history. This tsunami was triggered by a massive undersear earthquake measuring nearly 9.0 magnitude on the Richter scale that occurred in Sumatra (Indonesia). This resulted the series of tsunamis which moved with a speed of 900 km per hour and hit the Indian Ocean coasts with waves of upto 30 m high.

This tsunami killed more than 2,30,000 people in 14 countries. Indonesia was the hardest hit (1,70,000 dead), followed by Sri Lanka, India and Thailand. The powerful waves destroyed houses and other infrastructure in the coastal areas.

The tsunami also had a severe environmental impact including the following :

- Coastal erosion.
- Contamination of groundwater and soil with salt water.
- Damage of coral reefs and mangroves.
- Adverse impact on habitats of coastal species.

15.13 LANDSLIDES

Landslide or landslip is a geological phenomenon, which includes wide range of ground movement, such as rock falls, deep failure of slopes. Landslides are rock, earth or debris flows on slopes due to gravity. They can occur on any terrain given the right conditions of soil, moisture and the steepness of slope.

Thus, landslides or rockslides are the rapid sliding of large masses of rock, earth or debris that move by force of their own weight down mountain slopes or river banks. Landslides are the natural phenomena that would occur with or without human activity. However, human use and interest has somewhere increased the chances of landslides, such as on hillside developments, because of over-steeping of slopes; and somewhere decreased the chances of landslides due to protective and stabilization measures adopted by man at fragile slopes.

The landslides can also be accelerated by earthquakes, vibrations, etc., or due to saturation of the unconsolidated sediments with water. The road or house building activities near the standing slopes may also cause landslides:

Causes of Landslides

Generally, a landslide is triggered off due to the following reasons:

- Weight of the overlying material and the presence of lubricating material, such as water or creeping of soil down a slope
- Earthquakes, floods and volcanoes
- Landslides occur frequently during the rains
- The buildings that are erected without suitable foundation for a given soil and in sloppy regions
- Steepness of the slope.

Damages Caused by Landslides

- Block river flow, roads and communication lines
- Immense loss of agricultural production and land area
- Causes serious casualties
- Causes flooding which worsens situation.

Elements of Risk

Following are the most common elements at risk:

- Settlements built on the steep slopes
- Settlements built at the toe
- Settlements built at the mouth of the streams
- Buildings that are constructed without suitable foundation for a given soil and in sloppy areas
- Roads, communication lines, electricity lines, etc.

A classic example of a major landslide occurred in India in September 1893, which blocked the *Birahi Ganga River* near Gohana in Garhwal (Uttarakhand). The river valley got filled up by rocks and debris to a depth of about 240 metres, extending for a length of about 3.5 km, thus, forming a kind of dam. The discharge from the drainage area of about 230 sq. km was held back against this obstruction for nearly an year, forming a huge upstream lake. This lake having capacity of 46.6 million cubic metre was breached after about a month, and due to this breach, huge amount of water was released to downstream. The maximum water rises caused in the downstream by this extraordinary flood were observed; starting from a rise of about 80 metres in the vicinity, the water-spread reached a distance of about 20 km in about 25 minutes, where the rise was about 50 metres; and at a distance of about 240 km, the rise was 3.4 metres in about 12 hours.

Mitigation Strategies

Following are the main mitigation strategies :

(i) Proper hazards mapping

(ii) Afforestation of denuded upper slopes with appropriate tree species.

(iii) Erecting retaining walls to stop land from slipping.

(iv) Providing stone lining on fragile slopes of unconsolidated material.

(v) Providing support to the slope with stone-filled wire crates, called *gabions*.

(vi) Providing proper drainage of water through the rocks *e.g.*, infiltration of rain water and spring flows.

(vii) Engineered structures with strong foundations and proper installations of pipes, cables etc. to withstand forces caused by landslides.

(viii) At the time of constructing irrigation canals, roads, etc. due care is to be taken to avoid blockage of natural drainage.

15.13.1 Avalanche

Avalanche is a mass of snow that slides down a mountain slope. Most avalanches result from weather conditions that cause snow on a mountain slope to become unstable. Such disturbances, like heavy wind, earth tremors, explosions, etc. can send the snow sliding down the mountain.

QUESTIONS

1. Define disaster, crisis and disaster management.
2. Mention major objectives of 'disaster management'.
3. Discuss organisational set-up for handling disasters at National, State and District lavels.
4. Write a short note on 'National Disaster Management Authority'. Mention its main functions.
5. Write short notes on the following:
   (a) Pre-disaster preparedness (b) Pre-disaster actions
   (c) Mapping (d) Disaster zoning
6. Explain 'disaster mitigation' in details.
7. Explain why do earthquakes occur? Explain its mitigation strategies.
8. What should be the action plan for safeguarding from earthquakes?
9. Define the term 'flood'. Explain the different causes of floods.
10. What is flood mitigation? Explain the different mitigation measures for floods.
11. Write short notes on the following:
    (a) Flood control measures (b) Non-structural measures for floods
    (c) Flood plain zoning (d) Pre-flood measures
    (e) Measures during flood (f) Post-flood measures
    (g) Landslides
12. What is a cyclone? How is it formed?
13. Explain the impacts of cyclones. Explain the different measures for its mitigation.
14. What is a tsunami? How it is generated?
15. Explain the different mitigation measures for tsunami.
16. Discuss various elements of disaster management.

17. What is "disaster preparedness"? Discuss its elements in detail.
18. Write short notes on
    (a) Disaster prevention.
    (b) Response for disaster or emergency response plan.
    (c) Search, rescue, and relief.
    (d) Training and mock-drills for disaster management.
19. Discuss various steps which should be taken in combating the food disasters.
20. What are the landslides? What are its causes?
21. How will you classify different types of disasters?
22. What do you mean by intensity and magnitude of an earthquake?

PART V : ENVIRONMENT AND DEVELOPMENT

16
Chapter

SOCIAL ISSUES

16.1 INTRODUCTION

A man while living in society cannot detach himself from social issues of life. Since his activities right from birth to death are deeply related to various social issues, man is said to be a social animal. Social issues cover family, associations, institutions and communities. While being in society man has to undertake a number of soio-economic activities, such as agriculture, industry, transport, infrastructure development, such as construction of rail, road, buildings, airports etc. for surviving in the society.

These socio-economic activities contribute much to environmental degradation. Socio-economic developments desired by everyone for having a higher and better standard of living. The land area per person is reducing day-by-day due to growing population. With limited availability of land for cultivation, becomes necessary to use high breed seeds, increasing application of fertilizers, irrigation, and pesticides.

As it has been discussed in previous chapters, these socio-economic activities adversely effect the environmental conditions. Therefore, there is a need of the day to carry out the development without polluting the environment. Development and environment, both are important. One should not place these two things separate.

16.2 UNSUSTAINABLE TO SUSTAINABLE DEVELOPMENT

Unsustainable development (growth) is one in which social issues of life do not progress in a healthy and ethical way. This style of living leads to mismanagement of resources, such as large scale deforestation, indiscriminate construction and expansion of settlement activities, discharging toxic wastes, polluting water, air, soil etc.

The fast expanding greed unfortunately leads to the unsustainable exploitation of natural resources. Excessive and unscientific use of chemical pesticides, fertilizers, minerals also leads to environmental degradation.

With exploding population and rapid depletion and degradation of the natural resource base is significant. The task of meeting the needs of the present generation without eroding the ecological assets of the future generation is receiving top priority by environmental planners.

Until few decades ago, the world looked at economic status alone as a measure of human development. Most countries of North America and Europe, which had become industrialised at an earlier stage are economically more advanced. They not only exploited their own natural resources rapidly, but also used the natural resources of developing countries to grow large economies. Now, even the developed world has begun to realise that their lives were being seriously affected by the environmental consequences of development, based on economic growth alone.

16.3 WATER CONSERVATION AND QUALITY

Rainfall is the only source of fresh water. The water that cannot be used and stored above ground and below ground, runs to the sea. Total rainfall is nearly constant for centuries. However, the population is growing and therefore, per capita availability is going down year after year. The methods of using water in agriculture are generally unscientific. Excessive use of water in agriculture results in waste of water as well as fertilizers. Surplus water percolates down deeper into the soil leaching with its chemicals and nutrients contained in the soil, leading to chemical contamination of ground water. It may be mentioned that out of total water consumption, 83% of the share is of irrigation.

As regards quality of water, it should be aesthetically acceptable (odour, colour, turbidity) and should be safe for human health (bacteriological, viral and other radiation, and microbiological purity) and lastly it should be pleasant in taste. Potable drinking water should have following specifications:

- Total dissolved solids (TDS) 2000 mg/l
- pH value between 6.5 and 8.5
- Chlorides 1000 mg/l maximum
- Total hardness 600 mg/l maximum
- Sulphates 400 mg/l maximum
- Fluorides between 1 and 1.5 mg/1

Over the years, rising population, growing industrialization, and expanding agriculture have pushed up the demand of water. Efforts have been made to collect water by building dams and reservoirs and digging wells. Some countries have also tried to re-cycle and desalinate (removal of salts) water. Water conservation has become the need of the day. The idea of ground-water recharging by harvesting rainwater is gaining importance in many cities.

In the forests, water seeps gently into the ground as vegetation breaks the fall. This groundwater in turn feeds wells, lakes and rivers. Protecting forests means protecting water catchments.

In view of the above, it is necessary to conserve water as well as to protect our water resources from pollution. For this, it is essential that we do not discharge any interacted waste water into our precious water resources, and maintain some minimum flow in all rivers and streams.

For conserving water, we have to eliminate wastage, wastage of water affects all of us in many ways. Water is required to be equitable and fairly distributed so that household use, agriculture and industry all get a share of the water. The overuse and misuse of water due to various activities, and the resulting pollution has led to a serious shortage of potable drinking water.

Water can be same upto large extent by adopting drip irrigation, implementing water distribution system that reduces water losses, stopping the leakage in taps and pipes, lining the irrigation canal to stop seepage etc.

16.4 RAINWATER HARVESTING

Rainwater Harvesting is the collection and storage of rain from roofs or a surface catchment for future productive use. Traditionally, rainwater harvesting has been practised in arid and semi-arid areas, and has provided water for drinking, domestic, livestock, small irrigation and a way to replenish groundwater levels.

Some of **benefits of rainwater harvesting** are as under:

- Increased water availability.
- Checks the declining water table.
- Environmental friendly.
- Improves the quality of groundwater through the dilution of fluoride, nitrate, and salinity.
- Prevent soil erosion and flooding, especially in urban areas.

Methods of Rainwater Harvesting

Following are the common methods adopted for rain water harvesting to recharge the underground reservoirs:

1. On channel storage of storm water in drains and streams by using the grated regulator structures.
2. Development, deepening, and restoration of old village ponds.
3. Creating new water bodies by using old abondoned waste courses.
4. Constructing check dams for storing storm water.
5. Construction of soil conservation bunds to create percolation tanks.
6. Construction of recharge trenches (recharge shafts) in the existing water bodies to accelerate infiltration.
7. Roof-top rainwater harvesting.

Traditional Water Harvesting in Rajasthan (A Case Study)

Traditional water harvesting systems prevailing in Mewar region of Rajasthan share the available water for cultivation. Some of the traditional harvesting systems are discussed hereunder :

1. **Medhbandi** : In Medhbandi a stone embankment is built on a hill slope to help create a level field for cultivation. This controls soil erosion and conserves moisture.
2. **Bandha** : These are the stone check-dams across streams or gullies which are constructed to capture run-off on a stretch of fertile land that is submerged in water during the monsoon. This land after trapping silt not only becomes more fertile but also retains substantial quantities of water in the soil.
3. **Talab** : This region is well-known for its built reservoirs, known as *talabs.* A small reservoir of less than five *bighas* is called *talai,* a medium-sized lake is called *bandh* or *talab,* while a larger lake is called a *sagar* or *samand.*

Roof-Top Rainwater Harvesting

Traditionally, rainwater harvesting has been practised in arid and semi-arid area, and has provided water for drinking, domestic, livestock, small irrigation and for replenishing ground water levels. Now-a-days roof top rainwater harvesting is becoming popular for use for all the above purposes. Here, in the following paragraphs, we are discussing 'rooftop rainwater harvesting' systems.

In this system, all roof and terrace water passes down into a covered tank, where it can be stored for use after the monsoon. In this system, it is difficult to construct large storage tanks.

Another way of using rooftop rainwater harvesting is to collect the rainwater so that it percolates into the ground to recharge wells instead of flowing over the ground into the rivers. This system helps the water-table to rise and the surrounding wells retain water throughout the year. The rainwater before collection should be passed through simple layers of crushed stone/brick bats and sand or charcoal filters for the removal of suspended particles and micro-organisms from the rain run-off being collected. The roofs and courtyards should also be kept as clean as possible at the time of rains.

16.5 WATERSHED MANAGEMENT

As we know that rivers originate in **streams** that flow down mountain and hill slopes. A group of small streams flow down hillsides to meet larger streams in the valley, which form the **tributaries** of major rivers. **The management of a single unit of land with its water drainage system is called watershed management.**

Thus, a watershed is an area bounded by the divide line of waterflow. A watershed, catchment, is all the land drained by a stream of river. A watershed can also be defined as an area from which run-off resulting from precipitation flows past a single point into a large stream, river, lake or pond.

Therefore, watershed management is a technique that has several components including soil and water management, and developing a vegetation cover. The natural drainage pattern of a watershed unit, if managed properly, can bring about the local prosperity by providing year-round supply of water, thereby meeting the water demand in the area.

Watershed management implies an integration of technologies within the natural boundary of a drainage area for optimum development of land, water and plant resources, to meet the people's basic needs in a sustained manner.

Water conservation and soil conservation measures have strong links between them. Reduction of surface run-off can be achieved by constructing suitable structures or by changing land management. This reduction of surface run-off increases infiltration and help in water conservation.

Following are some of the common **watershed conservation practices:**

1. **Construction of Check Dams,** a temporary structure constructed with locally available materials. The main function of the check dam is to impede the soil and water removed from the watershed.
2. **Percolation Pond.** It stores water for livestock and recharges the groundwater. It is constructed by excavating a depression, forming a small reservoir.
3. **Irrigation Tank.** These are the storage structures made of irrigating crops.

Other conservation methods, like contour bunds, contour barriers made by vegetation and stones, contour trenches and contour stonewalls will not only prevent soil erosion but also obstruct the flow of run-off. This obstructed water increases the soil moisture and recharge the groundwater in the area.

16.6 URBAN PROBLEMS RELATED TO ENERGY

In the modern world the per capita energy requirement of man is considered as a measure of quality of life of an individual and that of the society as a whole. The more the energy is consumed, the better the standard of living. That is why, the developed and industrialized countries are the greatest consumers of energy.

As the income increases, the consumer style of living changes and he switches over to better sources of energy. People with higher income envoys the services of many electrical gadgets that increase the per capita consumption energy. Urban centers in hot climates need energy for cooling, and in cold climate for heating. The early systems of fans changed through coolers to air-conditioning, which consumes very high quantities of energy.

New buildings now use large areas covered by glass, which in hot climate adds several degrees to the temperature inside. This requires even more energy to run large central air-conditioning units.

Use of refrigerators, microwave ovens, washing machines, televisions, and alike increase per capita consumption of energy. In addition, enormous lighting and newer methods of transportation and communication also enhances the consumptiuon of energy. In the urban areas, motor vehicles also play an important role in increasing the consumption of energy. Per capita consumption of electricity in India has increased from 176 kWh/year in 1980-81 to about 1000 kWh/year in 2011-12.

In our country, the main sources of energy are fossil fuels, hydel and nuclear power. The coal will remain as main source of energy for several years to come. Through discovery of more oil and gas is in progress, but we do not know what production level can be reached and for how long can it be sustained.

In India, the present petroleum may not last long. Alternative sources of energy must, therefore, be found out, so far the known sources of non-conventional energy are nuclear fusion, solar energy, wind power. As on 31.7.2011, out of total installed capacity of 1,80,358 MW, the share of renewal energy was only 20,162 MW. Thus, renewal energy, such as wind, solar, and others are only about 11%.

As India's economy and population grow, it will need additional energy. This alongwith increased vehicular population will further contribute to existing air pollution problems and urbanization raises the health risks from that pollution. A rapidly increasing population will also lead to an increase in electricity demand, taxing the already straining power generation, transmission and distribution infrastructure and putting pressure on the power sector to add new capacity.

The solution to this problem is to shift from current preference of energy derived from finite, fossil-based fuels to one that is cleaner, low-carbon renewable form of energy. This will help in balancing the increased demand for energy with the need to protect its environment from further damage.

The key issues which should define the shape and future of energy in cities are:

(i) sustainability on long term, the quality and quantity of available alternative/renewable forms of energy and effect on the global environment as a whole
(ii) efficiency in use of energy for human activity
(iii) equitable distribution for all human kind.

16.7 CONSUMERISM AND WASTE PRODUCTS

Increased consumption and waste of used product *i.e.* disposal of industrial, domostic and agricultural waste is a global problem. In the present society, there is tendency of bind consumption, which means consumption without considering and disposal of the residue and wastes. The consumerism culture has converted the society into "throw away society."

20% of the world population living in the industrialised countries consume 80% of the world's resources and produce 80% of wastes. Present day society has a pattern of economic development which encourage people to go on consuming more than they actually need consumerism, causes the wasteful use of energy, material and natural resources for above and beyond that needed for everyday comfortable living. The examples of such wastages are cans; bottles; cartons; paper cups, glasses, plates; napkins etc. Such items are disposed of after one time use.

Therefore, there is need to buy those products which are more productive and efficient, discarding the less efficient old ones which often leads to an almost total waste of energy and material invested in products, such as appliances, cars and other gadgets.

We should pledge to encourage green technologies, reuse, recycling and other measures already discussed in previous chapters.

16.8 ENVIRONMENTAL ETHICS

Ethos, is Greek word from which *'ethic* is derived which means the character of man as described by his actions. Ethics is the science or the theory that guides moral behaviour. It, therefore, forms the basis for distinguishing 'right conduct from wrong conduct.' Environmental ethics deal with issues related to the right of individuals that are fundamental to life and well-being. These concern not only needs of each person today, but also those who will come after us. It also deals with the rights of other living creatures that inhabit the earth.

There is a widespread corruption, greed and dishonesty, prevailing in every field of our so called "civilised world" of today, and more and more people leaving the 'ethics' in our society.

16.8.1 Need for Equitable Utilisation

Nobody can justify that few individuals use resources many times more lavishly than some other individuals, who have barely enough to survive. As discussed above, environmental degradation has been caused primarily because of greed, selfishness and ignorance. The materialistic and **human centre thinking** has largely contributed to the damage caused by man to the environment.

The erosion of ethics, moral and spiritual values over the years is largely responsible for the mess. Such a mess and environmental destruction, if continues unabated, will ultimately destroy the man from the earth, thereby making him as an extinct species.

There is a demand for justice that there should be a more equitable sharing of resources than what exists at present. In this period of economic development, the disparity between the haves and have-nots is widening. The unequal distribution of wealth and access to land and its resources is a serious environmental concern. This results in loss of sustainability in resource management in the rural and forest sectors. Due to this imbalance tribals, especially women and other marginalised sectors of our society, are being left out of economic development.

16.8.2 Need for Gender Equity

Various governments at international level must guarantee their citizens equal protection from environmental risks, regardless of the geographical regions in which they reside. The consumption level of resources have reached in many countries/regions a level that the earth's capacity to bear has been more or less now has adverse impacts across national boundries. All our efforts must, therefore, be made to reduce this desire through education and spread of human values.

The developed countries are responsible for excessive emissions of green house gases, and must take immediate action to stabilize and reduce such emissions.

In India, especially in the rural, tribal, and hilly areas, women work longer hours than men the women collect fuelwork for their home and for sale to nearby urban areas, they collect fodder for their cattle, they travel several kilometers to collect a reasonable clean water and then they cook meals in a smoky unhealthy atmosphere on fuelwood, crop or animal waste. All this makes very hard work everyday throughout the year. But the women has no voice in decision-making on matters related to common resources and unfortunately it is the men who play a decisive role.

16.8.3 Preservation of Resources for Future Generations

As discussed earlier, if we use and misuse resources and energy from fossil fuels, our future generation will be left with nothing for them. The ethical issue must be considered when we use resources unsustainable. Just as our ancestors have left resources for us, it is our duty to leave them behind for our future generations.

This aspect has been discussed in detail under the heading "sustainable development" at the end of this chapter.

16.8.4 The Rights of Animals

The Earth's resources are shared by billions of plants and animals species alongwith human beings. These plants and animals, that share the earth with us, also have a right to live and share these resources and living space. The wild and domestic animals do not have only right to life, but also have the right to a dignified existence. Cruelty of animal is like a cruelty to another human being.

The humans being a single species use and severely exploit the earth's resources. We have forgotten that it is not justifiable to exploit nature and other species beyond a point. We are responsible for recent rapid reduction in the numbers of so many species. Thus, bringing the wild plants and animals close to the brink of extinction is most unfair.

16.8.5 Environmental Education and Awareness about Ethics

There is an important need for creating an ethos that will support a sustainable life-style in society. Every young individual at school and college level is, therefore, being exposed to a course environment. The **valuing nature** and **appreciating the beauty of nature** are the two main aspects that are closely connected with ethical issues.

Environmental education and awareness is essential for bringing back an ethic for nature's conservation. The role of individuals: (*i*) in the conservation of natural resources, and (*ii*) in the prevention of pollution, have already been discussed in the previous chapters, should be practised by all of us.

16.8.6 Traditional Value System and Conservation Ethics

In ancient Indian tradition, the nature was venerated and protected, and people have always valued mountains, rivers, forests, trees and animals. Mountains, forests and rivers have been associated with the names of various 'gods' and 'goddesses' both in the Hindu religion as well as tribal cultures. The *peepal* tree is generated and is not to be cut down.

Similar is the status of Banyan tree in some regions. The *Tulsi* plant is grown in every home and worshipped.

Kailash parvat, Ganga, Yamuna, Narmada, Godavari and most of the other mountains and rivers are also worshipped.

Many plants, shrubs and herbs are used in Indian medicines, which were once available in the wild in plenty. These are now rapidly vanishing.

Many animal species are venerated as being the *vahan* (vehicles) of various gods. The sun god is also worshipped.

Thus, in the past traditional societies several examples are found, which were all a part of ethical values that protected nature. As modern science, which is based on exploitation of nature, spread into India, many of these traditions began to lose their effectiveness as measures in conserving nature.

16.8.7 Environmental Ethics

1. Humans should love and honour the earth, since it has blessed with life and governs their survival.
2. Humans should not hold themselves above other living things.
3. Humans should be grateful to the plants and animals, which provide food for human nourishment and development.
4. Man should not waste resources on destructive weapons and actions.
5. Man should not run after gains at the cost of nature, rather should strive to re-store its damaged majesty.
6. People should not snatch away from their future generations their right to live on a clean and safe planet, by causing environmental pollution.
7. People should limit their consumption. We must leave sufficient food, fuels, metals and other resources for the use of our future generations.

16.9 SUSTAINABLE DEVELOPMENT

"Sustainable Development" is the socio-economic and technical development with a view to secure resources and their distribution within the countries and the world with equity, justice and ecological balance for the existing generation as well as to ensure that the interests of the future generations in these areas are not jeopardised. The various parameters of the sustainable development are protection of environment, development of alternate resources, replenishment of consumed non-renewable resources, equitable distribution of resources within generations and between generations and also within countries and between countries.

The concept of sustainable development is to meet the needs of the present without compromising the ability of future generations to meet their needs. Sustainability also meets, conservation of natural resources and management of biosphere for human needs and use for a definite period of time.

Sustainable development can be defined as that development, which is free from environmental degradation, poverty, and depletion of natural resource base. Sustainable development changes the society from the destructive, exploitative philosophy to the one that adopts long-term protection of the environment and its inhabitants.

Converting forests into cropped and urban land, over-fertilizing the land resources, over-harvesting forests, or avoiding pollution controls and similar harmful practices to meet the needs of growing population and changed lifestyles, are the short term solutions that cannot continue any more.

Sustainability is also for wise use of nature and its resources to provide the greatest sustainable benefit to present generation while maintaining the potential to meet the needs and aspirations of future generations. Economic growth always brings risk of environmental damage, as it puts increased pressure on environmental resources. But policy-makers guided by the concept of sustainable development will necessarily work to assure that growing economies remain firmly attached to their ecological roots and that these roots are protected and nurtured so that they may support growth over the long term.

Environmental protection is, thus, inherent in the concept of sustainable development, as is a focus on the sources of environmental problems rather than the symptoms.

Till about three decades back, it was believed that all the resources are abundant and we must exploit those as much as we can, and little thought was given to the amounts of minerals used or forests cleared or the capacity of our environment to absorb pollutants. It was the problem of environmental pollution, coupled with the shortage of some of the resources which awakened the human society about uncontrolled development.

Sustainable development is not possible without sustainable utilisation of resources. Forest trees though supply valuable timber but there is a limit to this, beyond which the resources will cease to be resource, and not only forests will be gone but also there will be many other undesirable changes in the landscape, soil, climate, flora and fauna, biodiversity etc. Same thing is true with other resources.

Sustainable development can be characterised as "the right to development, which must be fulfilled so as to equitable meet development and environmental needs of present and future generations." Sustainable development, therefore, means economic development and improvement in standards of living, which do not impair the future ability of the environment to provide sustenance and life support for a population.

Important parameters, which can help the world, attain sustainable development and at the sametime protect the environment are:

- Population stabilisation.
- New technologies.
- Efficient use of natural resources.
- Waste reduction and pollution prevention.
- Education and awareness.
- Integrated environmental system management.

The goal of sustainability can be achieved by focussing on areas of efficient utilisation of energy, sustainable use of renewable energy, technologies and reduction of all forms of waste.

Following are the some of the **interpretations of sustainability**:

- Living in harmony with nature.
- Living off the income of the nature, without eroding the capital.
- Realisation that the biosphere is both, for us and for our future generations.
- It is an inter-generational concept seeking equity overtime, and minimisation of disparities between generations.
- Survival of human beings as well as of all other life-forms.
- Maintenance of bio-physical productivity.
- Preservation of environmental quality and ecosystem.
- Inter and intra-generation equity.
- Human survival can only be ensured by preserving the basic resource base.
- Legislative support.
- National and international co-operation.
- Education, training, information and research on various areas related to above mentioned fields for continuous improvement.

Sustainable Construction Activities

Sustainability is the key to any construction activity. Various measures need to be taken to ensure environmental monitoring and to minimise and mitigate environmental degradation in a practical way. The environment during construction activities includes choice of construction methods and technology, appropriate level of Environmental Impact Assessment (EIA), monitoring Environmental Parameters and their Mitigation at right time, launching community development programmes for local population, giving full emphasis on environmental specifications in contractual documents, and detailed monitoring to fulfill environmental obligations by contractors.

Construction activities, generally do contribute to the reduction of natural resources, unwanted side-effect on environment, such as air and water pollution, solid waste, deforestation, erosion/siltation/drainage of soil, dust and noise nuisance, toxic wastes, health hazards/ sanitation and diseases, loss of land and other negative consequences.

The ***environmental effects*** of construction activities can be broadly classified as under. Considering these effects and taking care of them, construction activities can be made sustainable by adopting proper construction management techniques. Implementation of the environment management plan calls for a strong contract and strong supervision. Emphasis need to be given on mitigation measures to reduce the negative impacts and maximum efforts should be made towards enhancing the positive impact of the project.

Types of Environmental Effects

1. **Ecological Effects:**

   (a) Aquatic (fisheries) effects

   (b) Terrestrial effects (Deforestation). It seriously effects on biological diversity.

2. **Physical Effects:**
   (a) Land Degradation. It includes erosion, siltation, drainage, soil etc.
   (b) Water pollution, hydrology, water-table.
   (c) Atmosphere including air, dust, noise.
3. **Social Effects:**
   (a) Health including sanitation and diseases.
   (b) Soico-economic effects including employment, loss of land, agriculture, irrigation, transport, resettlement, local economy etc.
   (c) Aesthetic effects including landscape, refractory.

16.10 SUSTAINABLE ENERGY AND MATERIALS

16.10.1 Sustainable Use (Conservation) of Energy

As we know, the use of energy is increasing day-by-day and according to an estimate the fossil fuel, such as oil, gas and coal reserves will be exhausted shortly. Such energy sources are non-renewable sources. We have already discussed "renewable" and "non-renewable" sources of energy in the chapter on "Energy Resources."

There is urgent need for adopting methods for conservation of energy resources and also to develop and use more and more renewable energy sources. Almost 90 per cent of the world's current energy supply is based on fossil or mineral sources, oil, coal, gas and uranium. Only 10 per cent of the energy is from renewable sources, especially hydropower, solar, wind and biomass.

For sustainable use of energy, the energy crisis can be solved to some extent by proper use of renewable sources of energy. Some of the energy conservation options are:

1. Construction and operating green buildings (sustainable buildings). Salient features of green buildings are:
   - Minimise energy consumption and greenhouse emissions.
   - Protecting the natural environment.
   - Protect and restore local air, water, soils, flora and fauna.
   - These take advantage of natural light, water, sun heat, existing landscape etc.
   - Use a roof top solar array for generating power.
   - Use energy efficient appliances.
   - Install high efficiency lighting system with advanced lighting controls.
   - Use a properly sized and energy efficient heat/cooling system in conjunction with a thermally efficient building shell.
   - Optimize following inter-related aspects:
     - Maximise the availability of natural light.
     - Maximise the use of natural ventilation, instead of blocking it and providing air-conditioning.

16.10.2 Sustainable Use (Conservation) of Materials

As already discussed in the chapter on "Mineral Resources," material resources are exhaustible and non-renewable; and therefore, it is very essential to conserve these resources. Environment is affected adversely due to extraction, processing and disposal of minerals. Large stretches of land are destroyed by mining and are called '*mine spoils*'.

Mining has created some of the largest environmental disaster zones. Each step in mining and processing operations produces large quantities of waste materials.

Minerals can be conserved by bringing efficiency in mining technology. Re-cycling of these minerals can help reducing the waste.

Mineral resources can be conserved by adopting following measures:

1. Economy in use of mineral resources.
2. Making finished products of more durable and long lasting quality.
3. Re-use and re-cycling of metals, as this helps in reducing the waste.
4. Scarce minerals can be saved by substituting them with the minerals which are cheaper and are found in abundance.
5. Search for new deposits.
6. Protection and economy of existing mineral deposits.

QUESTIONS

1. Discuss the necessity of conserving water and its resources, indicating the methods for their conservation.
2. Explain the term rainwater harvesting and its importance in today's context of falling groundwater levels.
3. Enumerate various techniques that are used for rain water harvesting. Describe roof top rainwater harvesting.
4. Discuss social issues related to environment.
5. Write short notes on:
   (a) Unsustainable to sustainable development.
   (b) Benefits of rainwater harvesting.
   (c) Watershed conservation practices.
6. Explain the following in brief:
   (a) Urban problems related to energy
   (b) Traditional value system and conservation ethics.
7. Write a detailed note on urban problems related to energy.
8. What is a watershed management? Explain.
9. Write down a detailed note on consumerism and waste products with reference to environmental degradation.

10. What is environmental ethics? Explain the different types of environmental ethics.
11. Define the term ethics, and what theories are propagated to explain 'ethics'?
12. Describe in detail the environmental consequences of unethical behaviour of human population.
13. Define and explain 'sustainable development' and its importance.
14. "Animals and Plants have rights." Do you agree? Explain.
15. Explain the role of science and technology in sustainable development.
16. What is the meaning of consumerism and how it can be controlled?
17. Explain the following in brief:
    (a) Energy problem of urbanized population.
    (b) Water conservation and water harvesting.
    (c) Preserving resources for future generation.
18. Write down a detailed note on "need for equitable utilisation for resource consumption."
19. Write an essay on "the ethical basis of environmental education and awareness.
20. Write down important 'environmental ethics.'
21. "Unsustainable development is one in which social issues of life do not progress in healthy and ethical way." Discuss in detail.
22. Write a detailed note on sustainable development.
23. Explain the following in brief:
    (a) Need for gender equity.
    (b) Important parameters for sustainable development.
    (c) Interpretations of sustainability.
    (d) Different types of environmental effects.
    (e) Sustainable construction activities.
24. Explain the following in detail:
    (a) Sustainable use of energy.
    (b) Sustainable use of materials.

17

Chapter

ENVIRONMENTAL CHALLENGES

17.1 INTRODUCTION

In the previous chapters, we have discussed that human activities have been causing a great amount of air, water, soil and other forms of pollutions on our earth. Besides depleting and over-utilising and even destructing our available natural resources. Human beings are, therefore, proved to be most powerful disturbers of the environment.

Since the atmosphere is common to all regions and countries of the world, any local major adverse impacts on the atmosphere are likely to expand and become regional or even global to some extent. Thus, the impacts of air pollution can be seen at different scales, such as at local level or at regional level or even at global level.

1. **Local Environmental Problems.** Some local industries or power plant emitting smokes in an industrial area or a town are found to be causing local environmental problems. Such local problems can be corrected by adopting better methods of combustion and smoke collectors etc.
2. **Regional Environmental Problems.** When various air polluting industries or units are throwing huge amount of pollutants unto the atmosphere, they may be carried over to the neighbouring states or countries, making the problem to assume bigger size, thereby making it a regional problem. This situation arises, when a large number of units continue their activities unchecked and uncontrolled, thereby effecting millions of people living in that particular region. Acid rain is an example of such regional pollution problem. Such problems can only be solved with mutual co-operation between the involved states or countries.
3. **Global Environment Challenges.** We do find air pollution problems that have even assumed gigantic proportions, literally engulfing the entire atmosphere in their grip, adversely affecting the entire world, offering global environmental challenges. Such problems can be solved only by the mutual co-operation between all the countries of the world. Such global challenges are: *Global warming* and Ozone depletion.

It has now been realized that environmental problems are not limited to local, regional or national level, but there are several global environmental problems.

Human activities and technological advancements made to enjoy luxurious life have caused severe imbalances and disturbances in the natural environment. Major individual activities, which may cause environmental unbalances may be grouped into: Agriculture, Industrialisation, Urbanisation, Mining, Transportation etc.

17.2 CLIMATE CHANGE

Climate can be defined as an average pattern of weather over the long term in a particular area, climate pattern and seasonal variation are influenced by many factors, like ocean temperature, earth temperature, ocean currents, wind patterns, earth revolution etc. The climate is growing warmer currently mainly as a result of human activities (and behaviour) rather than due to natural changes in the atmosphere.

Following are the key signs of climate change:

1. Increase in global concentration of carbon dioxide
2. Increase in global surface temperature
3. Decline of the arctic sea ice
4. Decrease in land ice
5. Rise of sea-level

Impacts of Climate Change

Climate change can affect crop yield as well as the types of crops that can be grown in certain areas, by impacting agricultural inputs such as water for irrigation, amount of solar radiation that affect plant growth, change in temperature, as well as the prevalence of pests.

Climate change is expected to accelerate current stresses on water resources. The spreading of water scarcity is contributing to food insecurity and heightened competitions for water both within and between countries.

Sea level rises both due to thermal expansion as well as melting of ice sheets. Warming seas also increase intense tropical storms.

Climate change also poses a host of threats to the survival of mankind. Each year many people die from causes attributable to air pollution, diarrhoea resulting from lack of access to clean water supply, sanitation and poor hygiene, many people also die from malnutrition and natural disasters.

Heat waves in Europe and shifting of climate monsoon and other changes are the results of climate change or global warming. However, there is one group of scientists who do not agree to the highly debated climate change.

The Earth Summit (Rio Summit)

The first **Earth Summit** of United Nations Conference of Environment and Development (UNCED) was organised in June, 1992 in *Rio De Janerio*, a city of Brazil for the protection of earth and its environment, maintenance of ecological balance and to enrich biodiversity.

The primary objectives of the conference were to arrive at commonly acceptable agreements and their implementation to tackle the problems of **global warming, depletion of ozone layer and ozone hole, deforestation, biodiversity, weather** and **climatic change, acid rain, sustainable development** etc.

This summit was attended by some 130 Heads of States (countries) and several thousand delegates from over 160 countries. Although several environmental issues were discussed in this summit under a greatly expanded agenda, but finally an agenda called **Agenda-21,** consisting of a comprehensive programme of action for global action covering all areas of sustainable development was drawn at the summit. The treaty called *"Convention on Biological Diversity (CBD)* drafted earlier in Nairobi (Kenya) meeting of February,1992 was also opened up for signatures, as a legally binding agreement. This treaty is aimed at conservation of world's ecosystems and their biodiversity.

This ***Convention on Biological Diversity (CBD),*** thus, provides for conservation of biological diversity, sustainable use of its components, and fair and equitable sharing of the benefits arising out of the utilization of genetic resources. **India is a signatory to this Convention (Treaty).**

Observed Climate Changes in India

Some of the observed changes in climate parameters in India are mentioned hereunder. However, no firm documented changes has yet been established.

1. At the national level increase of 0.4°C has been observed in surface air temperatures over the past century.
2. Through the observed monsoon rainfall at the all-India level does not show any significant trend, regional monsoon variations have been noticed.
3. The records of coastal tide gauges in the North India Ocean for more than 40 years, reports that sea level rise was between 1 to 2 mm/year
4. The available monitoring data on Himalayan glaciers indicates that while recession of some glaciers has occurred in some Himalayan regions in recent years, the trend is not consistent across the entire mountain chain.

India's National Action Plan on Climate Change

There are eight National Missions which form the core of the National Action Plan, representing the multipronged, long-term and integrated strategies for achieving key goals with regards to climate change.

(1) **National Solar Mission** : It is a major initiative to promote ecologically sustainable growth while addressing India's energy security challange. Its main objective is to establish India as a global leader in solar energy.

(2) **National Mission for Enhanced Energy Efficiency** : This mission seeks to strengthen the market for energy efficiency by creating conducive regulatory and policy regime.

(3) **National Mission on Sustainable Habitat** : This mission seeks to promote sustainability of habitat through improvements in energy efficiency in buildings, urban

planning, improved management of solid and liquid waste, improvement towards efficient public transport and conservation through appropriate changes in legal and regulatory framework.

(4) **National Water Mission** : The objective of this mission is to ensure integrated water resource management for conservation of water, minimization of wastage and equitable distribution both across and within states.

It also aims for developing framework for optimum water use and ensuring that a considerable share of water needs of urban area is met through recycling of waste water.

(5) **National Mission for Sustaining the Himalayan Ecosystem** : The objective of this mission is to develop a sustainable national capacity continuously assess the health status of the Himalayan ecosystem. Its objectives are :

(a) Building capacities on climate change,

(b) Develop a coherent database on all knowledge systems,

(c) Detect global environmental changes and project future trends on potential impacts,

(d) Study traditional knowledge systems for community participation in adaptation and mitigation,

(e) Create awareness amongst stakeholders,

(f) Develop regional co-operation to generate a strong knowledge and database for policy interventions.

(6) **National Mission for Green India** : The objectives of the mission are increased forest/ tree cover, improved ecosystem services.

(7) **National Mission for Sustainable Agriculture** : The objective of this mission is to devise strategic plans at the agro-climate zone level so that action plans are drawn in the areas of research and development (R&D), technology and practices, infrastructure and capacity building. It aims to enhance productivity, to facilitate access to information and institutional support, and to promote laboratory support.

(8) **National Mission on Strategic Knowledge for Climate Change** : The objectives of this mission are :

(a) Formation of knowledge networks relating to climate science and facilitating data sharing, development of national capacity for modelling the regional impact of climate change on different ecological zones within the country.

17.3 GLOBAL WARMING

Global warming refers to gradual rise in atmospheric and ground surface air temperature and consequent change in global radiation, leading to climatic change at different levels (local, regional and global). If unchecked, it could alter temperatures, rainfall and sea-levels of the earth.

A rise of 1.5°C in the atmospheric temperature has been reported for the last 75 years. Green House Gases (carbon dioxide, methane, nitrous oxide etc.) have been identified as major factor of global warming. Green House Gases are emitted from industrial establishments, automobiles, domestic appliances etc. and these gases are concentrated in the lower part of the atmosphere.

Global warming is the result of various factors, such as ozone depletion, emission of green house gases, deforestation etc. The international communities are scared of catastrophic adverse effects of future climatic change on different spheres of man and nature *e.g.* deglaciation and sea-level changes, submergence of island nations and major coastal lowlands, global radiation balance, ecological productivity, plant and animal community, human health etc.

Impacts of Global Warming

1. Inundation of coastal areas.
2. Changes in rainfall patterns.
3. Increase in chances of heat waves, floods, hurricanes etc.
4. Melting of ice caps and glaciers.
5. Endanger to wildlife population due to habitat loss.
6. Spread of vector-borne diseases.

17.3.1 Kyoto Protocol

Kyoto Protocol is an International Treaty to control global warming. An international conference was held during December 1 to 10, 1997, in Kyoto (Japan) to discuss global warming and to initiate measures for reducing the green house gases in the atmosphere. Since differences had persisted between the scientists about the reality of the global warming till then, many developing countries were not agreeable to putting checks on their developing activities.

However, the delegates of 84 countries committed to the agreement, under which the industrialised countries committed to a legally binding overall reduction of 5.2% in green house. But, India and China rejected the idea of controlling green house emissions by preventing their growths and industrial development.

Although this agreement, called the Kyoto Protocol, could not come into force, since it could not be ratified by the minimum required 55 countries.

Though the Kyoto Protocol is stated to be weak, atleast a start has been made in the direction to globally recognise global warming and the adverse impacts of excess emissions of carbon dioxide and other green house gases, and to initiate measures to control emissions of such gases.

17.4 GREEN HOUSE EFFECT

Under normal conditions, the temperature at the surface of the earth is maintained by the energy balance of the sunrays that strike the planet earth and heat that is an increase in concentration of green house gases (mainly CO_2), the thick layer of these gases prevent the heat from being re-radiated out. This thick layer of these gases, thus, functions like the glass panels of green house (or the glass windows of motor car), allowing the sunlight to filter through but preventing the heat from being re-radiated in outer space. This is known as green house effect. Thus, most heat is absorbed by the layer of these green house gases and water vapours in the atmosphere, which results in further heating of the earth's atmosphere.

Green house gases in the lower atmosphere, act like the glass of a green house, trapping some of the heat, as it radiates back from the earth into the space. Such gases and their heating effects survived life on the earth, as otherwise, temperature on the earth could be below freezing. Excessive quantity of these gases may cause excessive heat, again to make life impossible.

Carbon dioxide is considered to be the major green house gas. It is responsible for about 60% of the total green house effect, caused by all the green house gases. Other green house gases are : CH_4, NO_X etc.

The continuous excessive burning of fossil fuels (coal, oil and natural gas) and large scale deforestation of forests (which absorb CO_2) by humans have increased the carbon dioxide gas in the atmosphere.

The environmental scientists are worried over the future prospects of global warming, and are making cautioning sounds to stop or to reduce activities, which may lead to emissions of increased green house gases into our biosphere. Therefore, there is an urgent necessity of reducing our consumptions of fossil fuels.

17.5 OZONE LAYER DEPLETION

Ozone is a form of oxygen with three atoms instead of the normal two, and forms a fragile shield scattered in the stratosphere absorbing the sun's ultraviolet radiation, away from the earth's surface. If ozone layer becomes thin, all terrestrial life will be annihilated. The thinning of this layer has generated global concern. The increase in ozone layer depletion will invite the lethal ultraviolet rays from the sun which will increase cancer, eye damage, injure plants and marine life and reduce our immunity to disease.

We must know that atmosphere, from meterological point of view, is divided into five thermal zones—(*a*) the troposphere—upto 16 km from sea-level, (*b*) stratosphere—16 to 40 km, (*c*) mesosphere—40 to 70 km, (*d*) isonosphere—70 to 400 km, and (*e*) exosphere—above 400 km.

Ozone layer is mostly concentrated in stratosphere and is considered as a protective shield and the earth's umbrella because it prevents ultraviolet solar radiation from reaching the earth's surface. Ozone layer (zone of highest concentration of ozone *i.e.* 12 km to 35 km) protects us from one harmful ultraviolet radiations from the sun. Depletion of this ozone layer by human activities may have serious implications and this has become a subject of much concern over the last few years.

In nature, ozone is formed from oxygen by ultraviolet radiations of the sun. These two gases remain in equilibrium condition. But recently oxygen—ozone balance has been upset by the excessive presence of chemical pollutants in the atmosphere. Ozone depletion will cause dreadful diseases, like skin cancer and skin allergies. It may also have hazardous effect on plants and other animals. Continuing ozone depletion and climatic change is a serious threat to the very survival of the human race on the earth. Considering the speed and extent to which ozone is getting depleted, it is important that all nations in the world should move quickly towards arresting further depletion.

Man-made chemicals especially **Chlorofluorocarbons** (CFC) adversely affect the ozon-layer also known as ozonosphere. Freon gas used in refrigerations is also a chlorofluorocarbons, these chemicals are also used in air conditioners and plastic foams.

It has been estimated that for every 1 per cent increase in ultraviolet radiation, there is a 3 per cent rise in skin cancer and one lac cases of catract. Ultraviolet radiation also compromises the immune system and could aggravate the spread of disease. It also damages crops and marine ecosystems enough to reduce the world food supply.

As a part of **Global Efforts,** the first global conference on the depletion of ozone layer was held in Vienna (Austria) in 1985, the year, the British scientists discovered a hole, in ozone layer as large as that of the United States in South Pole.

Tropospheric Ozone

The ozone gas, as explained above, in the stratosphere protects us from the harmful ultraviolet radiations from the Sun. On the other hand, ozone is also formed in the lower atmosphere, known as troposphere, through chemical reactions. The ozone gas in this layer is causing deadly lung diseases. Thus, ozone is a **destroyer** as well as **protector.**

As we know that temperature decreases with increasing altitude in the troposphere (upto 16 km from the earth surface), while it increases with increasing altitude in the stratosphere (above 16 km upto about 40 km). This rise in temperature in stratosphere is caused by the ozone layer. The ozone layer has two important and inter-related effects:

(a) Firstly, it absorbs ultraviolet light and thus, protects all life on the earth from harmful effects of radiation, as explained in previous paragraphs.

(b) Secondly, by absorbing the ultraviolet radiation the ozone layer heats the stratosphere, causing temperature inversion. The temperature inversion limits the vertical mixing of pollutants, thereby causing the dispersal of pollution over larger areas and near the earth's surface. Some of the pollutants, therefore, when enter the stratosphere, remain there for years and react with the ozone and deplete it.

The production of ozone in troposphere is linked to atmospheric chemistry. In the troposphere, critical gases are: nitrogen oxide (NO, NO_2, NO_3), free hydrogen/oxygen radicals (OH, HO_2), hydrocarbons (methane and non-methane) and carbon monoxide (CO). Ozone formation near the earth's surface creates pollution problems. Ozone and other oxidants are formed by light dependent reactions between NO_2 and hydrocarbons. Ozone is also formed by NO_2 under ultraviolet radiation effect. These pollutants cause photochemical smog. Photochemical smog is highly oxidising polluted atmosphere and is primarily the result of industrial gases, domestic fires and the motor vehicle exhaust. This pollution causes eye irritation and reduced visibility. Since photochemical smog is formed only during night or cloudy days, the word smog is coined by combining smoke and fog.

Photochemical smog adversely affects plants, human health and materials. It interferes with respiratory system and causes **asthma** and other lung problems. The smoke and the particulate matters, like fog, dust, soot etc. in the smog reduce the visibility, damage crops and livestock, and cause corrosion of metals, stones, building materials, paint surfaces textiles, paper, leather etc.

Ozone layer depletion increases ultraviolet levels, which may increase skin cancer, cataract formation in eyes, adversely effects the crops, and other health risks to human beings and other organisms.

Thus, while higher levels of ozone, in the stratosphere protects us, it is harmful when it comes in direct contact with us and plants at the earth's surface adversely effects the crops, and other health risks to human beings and other organisms.

17.6 PROTECTION OF OZONE LAYER

As discussed above, ozone is formed by the action of sunlight on oxygen and forms a layer at 15–50 kms above the surface of the earth. Ozone in this upper atmosphere absorbs the ultraviolet radiation, preventing it from reaching the earth's surface. The ozone is considered a pollutant at ground level and have serious implications for humans. It adversely effects on human health, crop production, on climate, and is responsible for several adverse effects of global warming.

Various surveys have indicated that, the present damage in ozone layer is due to emissions occurred during last about one century. The study also found that chlorine atoms are responsible for the ozone depletion. Research also indicated that human made chemicals were significantly damaging atmospheric ozone. The United States, have banned chloroflurocarbons (CFCs) in spray cans, CFCs provided coolant for air-conditioning and refrigerators, foams for insulation and mattresses.

First of all in early 1980s, scientists have detected a thinning of ozone layer (known as ozone hole) in the atmosphere above Antarctica. Similar phenomenon is now being detected in other places as well, including Australia.

This finding in an unprecedented global co-operation, culminating into the Montreal Protocol 1987. The global community then started making serious efforts for 'Protection of ozone layer'.

17.6.1 Montreal Protocol (For Protection of Ozone Layer)

'Montreal Protocol' is an international treaty for controlling global ozone depletion. In 1987, the international negotiators met in Montreal, to adopt some legally commitment, wherein the Montreal Protocol was adopted on 16.09.1987. This was followed by **Helsinki Meet** in May 1989, **London Conference** in June 1990, and **Copenhagen Meeting** held in November 1992, to bring certain amendments in the original protocol.

India signed the Montreal Protocol on 17.09.1992. USA agreed to contribute 25% of the amount set out for creating a special fund worth 240 million dollars, to help developing nations switch to CFC-free technology based on the use of HCFCs (Hydro-chlorofluoro-carbons). This HCFCs have an ozone depleting potential of 0.12 as against 1.0 for CFCs. Thus, the HCFCs cause only 12% damage than that caused by the CFCs.

At present, 183 nations have become parties to the Montreal Protocol. With the signing of this Montreal Protocol, **a treaty for the protection of the ozone layer**, the use of CFCs was to be banned after which the ozone layer is expected to slowly recover.

The major achievement of the protocol was to set targets and dates to reduce the production of ozone depleting substance (ODS). Consequent to this effort the consumption level of ODS is reduced to one-third in about 7 years.

17.7 ACID RAIN

Acid rain is commonly defined as "a condition in which natural precipitation becomes acidic after reacting with pollutants in the air." Acid rain is formed when industrial emissions of gases including sulphur dioxide, nitrogen oxides and hydrogen chloride combine with water droplets in the atmosphere. This chemical reaction forms weak sulphuric acid and nitric acid that cause acid rain.

The ill-effects of acid rain can be found on vegetation, soil, marine resources, monuments as well as on man. The ill-effects of acid rain have no national limits, as nature knows no geographical boundaries.

The term acid rain refers the fall of acids mixed with rain and snow. The main cause of acid rain are the gases emitted by the industries, petroleum refineries and motor vehicles which pollute the atmosphere. In other words, acid rain simply means fall-out of acids caused by sulphur dioxide and nitrogen oxide with rainfall and thus, increase in the amount of acidity of rainwater.

The main sources of acid rains are oxides of sulphur and nitrogen which are produced mainly by combustion of fossil fuels, power plants, smelters, automobile exhausts, domestic fuels etc. These oxides are swept up into the atmosphere and can travel thousands of kilometres. The longer they stay in the atmosphere, the more likely they are to be oxidised into aids. Sulphuric acids and nitric acids are the two main acids, which then dissolve in the water in the atmosphere and fall to the ground as acid rain or may remain in atmosphere in clouds and fogs.

Acid rains have assumed a global ecological problem, because oxides travel a long distance and cover much larger areas far away from the source of pollutants.

The purity of water is generally measured by pH value. The water with pH value of 7.0 is called neutral water, whereas the pH value below it makes the water acidic and above the neutral point makes the water alkaline. The water becomes more injurious when the pH falls below 4.0.

Chemical Acid Rains—A Secondary Pollution

Acid rain is formed when industrial emissions of gases including sulphur dioxide (SO_2), nitrogen oxides (NO_x) or hydrogen chloride (Hcl) combine with water droplets in the atomoshere. The resultant rainfall tends to become more and more acidic with increasing concentration of these pollutants. This extra acidity in rain water is caused due to the formation of the *secondary pollutants,* like sulphuric acid (H_2SO_4) and nitric acid (HNO_3) due to the reaction of water vapour with SO_2 and NO_x gases respectively.

The greater is the concentration of these pollutants, the greater would be the acidity, and hence, lesser the pH value of the resultant rain. When the pH value of the rain water falls to 5.6 or below, the rain is specifically termed as *acidic.*

(1) $SO_2 + H_2O \rightarrow H_2SO_3 + O \rightarrow H_2SO_4$

(2) $2NO + O_2 \rightarrow 2NO_2 + H_2O + O \rightarrow 2HNO_3$

Acid rain can convert forests into wastelands by damaging trees, plants and crops; acidify fresh waters of lakes, reservoirs, wells, and underground sources. It can also induce respiratory illnesses in humans, poison wildlife, and inflicts irreversible damages on buildings to deface the old monuments.

Effects of Acid Rains

Acid rains affect human community adversely, and create complex problems and their impacts are far reaching. Some of the adverse effects of acid rains are explained hereunder:

(i) Increased soil acidity significantly lowers the productivity of soil, as it destroys mineral elements and other nutrients of the soil.

(ii) Forests of Canada, the USA, Germany and many countries of middle Europe have been damagd due to acid rains.

(iii) Land to acid rains flora and fauna are also affected adversely.

(iv) Acid rains have been identified as main factors of the **'deaths of lakes'** which mean destruction and death of all acquatic lives including plants and animals in lakes, ponds and rivers. Because of acid rains most of the lakes in Scandinavian countries (Norway, Sweden) have lost their biological communities and are now being termed as **'biologically dead lakes'**.

(v) Affects human health.

(vi) Acid rains also corrodes buildings, monuments, statues, bridges etc.

Acid rains are carried away by winds to elsewhere, where precipitation takes place. Thus, oxides are produced at one place, and these effect at other place by turning into acids, may be at thousands kilometers away. Therefore, acid rain is not a local problem of a particular country, rather it is an international problem, and should be tackled at global level. The two such victims are Canada and Sweden. Canada gets acid rains from petrochemical units in North America. Heavy winds pick up acid rain from industries in Britain and France to Sweden.

In India also, industrial areas with the pH value of rainwater below or close the critical value have been recorded in Delhi, Nagpur, Hyderabad, Pune, Mumbai and Kolkata. This is due to sulphur dioxide from coal based power plants and petroleum refinery. The situation is likely to worsen in future due to increased installation of thermal power plants and consequent increase in coal consumption. Therefore, there is urgent need for proper monitoring to provide timely warnings about acidification of our environment.

17.8 EL NINO

EL Nino is defined as the 'disruption of the ocean-atmosphere system in the tropical pacific, having important consequences for weather around the globe.' This involves the unusual warm conditions that appear in the coastal areas near Peru and Ecuador, South America. Actually, EL Nino is a term used to describe the warm southward current of the

Pacific Ocean that appears every December of the year. However, it has now become more severe and intense over the past few years, and once occurring, it affects the climate for one year around the globe. It is now happening every 3 to 7 years.

Disturbance of the El Nino to the weather system is identified by the several of the normal weather conditions in the Eastern and Western Pacific caused by sufficiently warm and persistent ocean currents. During this event, the winds that usually blow from east to west collapse and get reversed to blow from west to east, thus, changing the climate pattern of the Western Pacific and the Eastern Pacific. The Western Pacific which is usually moist, warm and high in rainfall is changing to an arid and low rainfall conditions, thus, bringing droughts to south-eastern Asia, India and southern Africa. While the Eastern Pacific, usually arid, cold and low rainfall is reversed to moist, warm and high in rainfall, leading to heavy rainfall in the areas.

17.9 WASTELAND AND ITS RECLAMATION

Wastelands are highly degraded lands which remain barren and unfit for any useful cultivation. Wastelands include those lands, whose biological producivity (plants and vegetations) and usefulness has decreased substantially. Generally, land which produces less than 20% of its optimum biological productivity, are considered as wasteland.

Forms of Wasteland

A degraded land may be in any one of the following forms:

1. Degraded forests
2. Village common land/grazing grounds
3. Land on roadsides, canal sides, and railway line side
4. Marginal agricultural land
5. Saline-alkaline lands
6. Water-logged lands
7. Ravine lands
8. Abandoned mining area
9. Sanddune, alpine pastures, cold deserts.

Classification of Wastelands

Wastelands can be classified as under:

1. Marginally deteriorated lands. These are easily reclaimable.
2. Partially deteriorated lands. These are reclaimable with some difficulty.
3. Severely deteriorated lands. These are reclaimable with extreme lands.

Causes of Formation of Wasteland

1. Natural causes, such as poor rainfall or wind aided excessive soil erosion.
2. Continued intensive and unsustainable agricultural activity on fragile, marginal and poor soil thereby destroying the fertile soil converting it into a complete wasteland.

3. Mining activity, massive construction activity and dumping of industrial waste.
4. Over-grazing due to over-population of cattles.
5. Over-population of wild herbivores animals due to reduced numbers of wild carnivores.
6. Unplanned, careless and intensive irrigation practices, that may turn the soil into saline lands.

Reclamation Measures

1. Identification of specific strategies for reclaiming the wasteland with the help of government officials and NGOs, and environmental scientists.
2. Testing of soil and selection of appropriate crops, for fodder and trees, identify suitable plant species according to the availability of resources for their aftercare after planting.
3. To undertake agro-forestry or farm forestry. Planting of trees that provide fodder, fuel, shade, fruit and other products.
4. To undertake suitable measures to store rainwater and land can then be used for horticulture or fruit orchards.
5. To undertake precautions and measures to control soil erosion.
6. Local people should be provided with new technological advances in irrigation and other expertise for improving productivity without creating unsustainable pattern of development.
7. Guidelines be provided to control water-logging.
8. To involve local people and to create awareness about the importance of rainwater harvesting, agro-forestry, protection of vegetation, and conservation of water.

QUESTIONS

1. Enumerate and explain the various types of human activities that have brought large scale disturbances to various ecosystem and the environment in general.
2. Differentiate between local, regional and global environmental challenges.
3. Write detailed notes on the following:
   (a) Acid rain
   (b) Ozone depletion
   (c) Effects of acid rains
4. Write short notes on the following:
   (a) Global Warming
   (b) Kyoto Protocol
   (c) Impacts of global warming
5. What is a wasteland? Describe measures that you would suggest for reclamation of wastelands. Also discuss causes of formation of wastelands.
6. Discuss the following in brief:
   (a) Earth Summit (b) Green house gases
   (c) Ozone layer depletion (d) El Nino

7. Write short notes on:
   (a) Effects of Acid Rains
   (b) Forms of Wastelands
   (c) Agenda-21
   (d) Convention on Biological Diversity
8. What is a climate change? What are its impacts?
9. What do you understand by 'Green house effect'?
10. What do you mean by protection of Ozon layer? What efforts are being made in this respect?
11. Write a detailed note on "India's National action plan on Climate Change."

18

Chapter

Human Population and the Environment

18.1 POPULATION GROWTH AND POPULATION EXPLOSION

Population is the number of individual organisms of the same species in a given area. Population dynamics refers to trends of population growth with time. Population growth results from the interaction of three variables: births, deaths and migration. For the world, growth occurs only when there are more births than deaths. Whereas, for individual countries, migration is also a factor for population growth.

The human population of the world has increased very fast over the past about 200 years, and most particularly during last about four decades. The total global population was about only one billion in the year 1800, which rose to about 1.6 billion in the year 1900, and about 6 billion by the end of year 2000. If the rate of increase of world population is not checked, it will be about 8 billion by 2025, and 11 billion by the year 2045. At present world population is about 7 billion. This unprecedented growth of human population at an alarming rate is usually referred to as **population explosion.**

The population grew rapidly during Industrial Revolution, not because the birth rate increased, but because the death rate began to fall. Death rate fell as new farming and transportation technologies enhanced the food supply and reduced the impact of famine. All these factors put together contributed towards the improvement in public health and living standards, with the result death rate has fallen at a faster rate. Developments in the field of medicine has further reduced the death rate, with the result, the world's population is growing by 1 billion people every dozen years.

Distribution of Population

The distribution of population is uneven, which is clear from the following facts:

1. About 90% of the population live in the Northern Hemisphere.
2. The continent of Asia including Russia has nearly two-thirds of the world's population.

The land area is only 30% of the earth's surface, and of that area 80% is not conducive to human settelement. This includes deserts, the polar regions, tropical rainforest, tundras, etc. Thus, only about 6% of the earth's area is suitable for human settlement. Most of the population lives in coastal areas, river basins and cities. Atleast 40% of the world's population lives within 100 km of the coast. Many of the megacities of the world are on the coast.

Population Growth Measurement

Population growth can be measured as under:

1. **Crude birth rate** : Number of live births per 1000 people in a population in a particular year.
2. **Crude death rate** : Number of deaths per 1000 people in a population in a particular year.
3. **Animal Population growth rate** = $\frac{\text{Crude birth rate} - \text{Crude death rate}}{1000} \times 100$

Growth rate (in 2018) in some of the countries	
World	1.15
India	0.99
China	0.37
U.S.	0.59
Italy	(–) 0.15%

Population Growth Pattern in India		
Year	Population (in million)	Annual growth%
1941	319	1.33
1951	361	1.25
1961	439	1.96
1971	548	2.20
1981	683	2.22
1991	844	2.11
2001	1027	1.93
2011	1027	1.93
2019	1370 m (approx.)	0.99

Most Populous Countries in the World (2020)

Rank	Country	Population (millions)
1.	China	1439
2.	India	1380
3.	U.S	331

4.	Indonesia	273
5.	Brazil	212
6.	Pakistan	221
7.	Bangladesh	164
8.	Nigeria	206
9.	Russia	145
10.	Japan	126

Major Cities in the World (2020) UN Estimates

Rank	City	Population (millions)	
		Metropoliton Area	City Proper
1.	Tokyo (Japan)	37.2	13.5
2.	Delhi (India)	29.0	16.7
3.	San Paulo (Brazil)	21.7	12.2
4.	Mumbai (India)	24.4	12.5
5.	Mexico city (Mexico)	20.9	8.9
6.	New York (USA)	19.3	8.4
7.	Shanghai (China)	26.3	24.1
8.	Kolkata (India)	14.8	4.5
9.	Dhaka (Bangladesh)	14.5	8.9
10.	Los Angeles (US)	13.3	4.0

Major Cities in India (2020) Municipal

Rank	City	Population (millions)
1.	Delhi	16.7
2.	Mumbai	12.5
3.	Kolkata	4.5
4.	Chennai	6.7
5.	Bangalore	8.4
6.	Hyderabad	7.0
7.	Ahmedabad	5.6
8.	Pune	3.1
9.	Surat	4.4
10.	Kanpur	2.8
11.	Jaipur	2.7
12.	Lucknow	2.5
13.	Nagpur	2.2
14.	Patna	2.1
15.	Indore	1.8

18.2 POPULATION GROWTH AND ENVIRONMENT

In early days, humans were afraid of dense forests, wildlife, rivers, oceans, lightning and thunder. They worshipped different aspects of the nature, mountains, rivers, winds, sun, forests, oceans etc., and they had to adopt themselves to the environment. With the development of tools, metals and learning about the use of fire, they started feeling their impact on environment. Then, came the agricultural revolution during 17th and 18th centuries, and industrial revolution during later half of 18th and first half of 19th centuries.

Increase in population and economic growth, sharp rise in agricultural production, rapid urbanisation and industralisation are increasingly affecting the natural environment. Though environment is keeping man alive but he is steadily destroying it. Man being the most intelligent animal of all the animals and technologically most advanced, tries to make use of the environmental resources and to make the environment influences favourable for him. In doing so man affects and modifies the environment and thus, environmental conditions are changed by human activities.

Over-exploitation of exhaustible resources, use of those resources whose generation is a very slow process (like soil), continuous exploitation of those natural resources whose reserves have reached critical limit and are on the verge of exhaustion etc. are a few examples of rapacious use of natural resources by greedy man. Such human actions make the environment sick and degrade the environmental quality to such an extent that it endangers the very existence of human beings.

The environmental crisis being faced by the earth, can be attributed to the manner in which the industrial and technological development has taken place in the world. Man's impact on the environment has resulted in pollution of environment, which affects air, water, land and also the organisms in the biosphere. When harmful substances contaminate the environment in large quantities, the ecosystem is unable to absorb them and they accumulate in the system, resulting in degradation of environment.

Many environmental problems are the result of over-population, as it creates scarcity of resources, natural hazards, leads to degradation of environment, excessive exploitation of forests, and high environmental health risk.

The increasing population need more land to live, thereby reducing forests; will have more cattle to graze needing conversion of forest land into pasture land, will need more goods, which will increase industrialisation and commercialisation, will need more fuels and energy to meet our modern lifestyles. All such activities are likely to cause severe adverse environmental impacts. In the near future, it will be impossible to meet the demands for food from the existing agro-system. We shall, therefore, need more land and water for agriculture for increasing the production of food and fibre.

As discussed in the previous chapters, over-population is the main cause of various environmental problems, such as:

(i) Depletion of resources including forest, soil, mineral, energy, clean water

(ii) Hunger/malnourishment

(iii) Poor basic sanitation

(iv) Pollution

(v) Global climate changes, like global warming, green-house effect, depletion of ozone, acid rains

(vi) Loss of biological resources and biodiversity

(vii) Desertification and land degradation.

Environmental problems being faced can be divided into following categories:

1. **Type I. Environmental Problems:** These are problems arising out of lack of development, such as deforestation due to fuel wood extraction, unhygienic slums, soil erosion, lack of sanitation and polluted water supply etc. If the pace of development is slow, the type-I problems will intensify.
2. **Type-II. Environmental Problems:** These are those problems which are caused by lack of environment safeguards in the development projects, such as air and water pollution, hazardous waste disposal, water-logging in irrigation projects, pesticide pollution, fertilizer run-off into streams etc. If sufficient investments are not made in development projects to compensate for environmental degradation, these problems will increase.
3. **Type-III. Environmental Problems:** These are global environmental problems, such as global warming, ozone depletion, marine pollution, loss of genetic diversity, automobile pollution etc. These problems are created by the current pattern of development, lifestyle and energy use mix etc.

18.3 FAMILY WELFARE PROGRAMME

Considering the problems related to over-population and alarming situation due to population explosion, there is a need for effective 'Family Welfare Programme (FWP) so as to develop action-plans and strategies to increase public understanding of how rapid population growth can be controlled. Collective vision about health care, family planning and women's education at the community level build a basis for action.

Family planning allows couples to decide on their family size and to provide suitable time gap between their children, by preventing conception of females. Modern science has provided several birth control techniques including the use of 'oral contraceptive pills' and intra-uterine devices like 'copper Ts' for women, and the use of condoms for men. India has been using permanent sterilization for males as well as females.

India's family planning programmes started in 1951. In 1970's, the forced family planning campaign by the Government resulted in a turmoil all over the country. While taking other measures as a family welfare programme, in 1978, the Government raised the legal minimum age of marriage from 18 to 21 years for men; and from 15 to 18 years for women. In 1981, the family planning programme has been re-named as the **family welfare programme.**

18.4 WOMAN AND CHILD WELFARE

1. Women Welfare

A large number of women, particularly in developing countries, work extremely hard to nurture their families and suffer a lot. In most of our rural areas, a housewife usually gets up very early and works very hard through out the day, as she feeds and milk her cow or livestock, cooks food for the family, prepares and sends the children to school, goes to forest to collect fuel wood, bring drinking water, washes clothes and utencils etc. Inspite of doing such a hard work, women usually suffer gender discrimination and bed treatment at home, at work place, in matrimony, in inheritance and in public life. The many cases of abduction, dowry deaths, rapes, domestic violence and mental torture of women is a testimony of bad treatment being given to her.

Since the human rights of women are more often violated than respected. To check and control the atrocities and injustice being committed towards the young girls and women in our male dominated society, the national and international community has initiated some measures and enacted certain legal laws, giving legal rights to women and setup some grievance redressal mechanisms.

The voluntary women groups and NGOs are also now coming into existence to take up the cause of suffering poor women and to take effective legal and other measures for the protection of women's rights and dignity. The UN and other international agencies are now also working in various countries for the general welfare of women, and for the protection of their human rights.

The Human Rights Commission at National as well as State levels are working to help giving some justice to the suffering of women. A full-fledged Union Ministry for Women and Child Development is working for the welfare and upliftment of women, including health care, family planning, education and creation of awareness in relation to the general welfare of women and children.

Many innovative programmes have been launched for generating employment, improving income and creating awareness among women. The ultimate goal is to make women economically independent and self-salient. Improvement in health-care of females during childhood, adolescence, pregnancy, middle and old age should take precedence.

Now-a-days, there is an increased participation of women in economical, social, and political life. More attention is being paid towards the women empowerment, and to the role of women both, in the society and family life.

2. Child Welfare

It has been estimated that each year about eleven million children worldwide die from the effects of disease and adequate nutrition. Most of these deaths are in the developing countries. 0% childhood deaths in developing countries can be attributed to 6 main causes, namely: pneumonia, respiratory diseases, diarrhoea, measles, malaria and malnutrition.

These diseases and mortality is the prevalent problem among the infants and children upto 5 years, the **child labour** afflicts the younger children between the age of 5 and 14 years. These children work for wages in developing countries.

Most of the children of poorer families could not get education and normal childhood. Some are even confined and beaten, or reduced to virtual slavery. Some of the children are even abducted and forced to work, or even beg.

The children, though, carry out their work with speed and precision, yet they are paid a very less amount. This system of cheaper child labour creates problem of unemployment for adults, as adults demand their rightful minimum prescribed wages. Some of the worst cases of child labour in India are found in the silk industry, matches and fire-cracker units, and carpet and *sari* weaving industries. Tamil Nadu, Karnataka and U.P. has large number of bonded labourers.

Efforts have been made at the international level to provide some special rights to children, called **Childhood Rights** so as to reduce their exploitation and to help them in getting free nutritious diet and education.

Some of the major child welfare schemes in India are:

1. Integrated child development services. Its objectives include: to improve nutritional and health status of children; and to reduce incidence of mortality, and malnutrition.
2. Balwadi nutrition programme for providing early education to children of 3-5 years age.
3. Child labour eradication scheme for shifting child labour from hazardous jobs to schools.
4. Mid-day meal scheme for school children.
5. Legislative measures. The Child Labour (Prohibition and Regulation) Act, 1986 prohibit the employment of children under the age 14 in 13 occupations and 57 processes, including weaving and since 1999, sericulture and *zari-making*.

18.5 HIV/AIDS

AIDS (Acquired Immune Deficiency Syndrome) today has reached such an alarming proportion that it should concern each one of us.

It is a condition in which the body's immune system is weakended and is less able to fight certain infections and diseases. It is caused by infection with HIV (Human Immunodeficiency Virus). Aids, in itself, not kills its victims, but the people with HIV, only some will develop AIDS as a result of their infection.

This virus is passed from one person to another through blood to blood and sexual contact. Infected pregnant women can pass HIV to their babies during pregnancy or delivery, as well as through breastfeeding HIV spreads rapidly when many people use the same needle to inject drugs. In this way, just one infected person spreads HIV to many others.

HIV is not transmitted by day-to-day contact in any social setting. It is not passed by shaking hands, hugging etc. One cannot become infected from a toilet seat, dishes, drinking glasses, food. HIV is not an air-borne or food borne virus, and it does not live long outside the body.

Magnitude of the Problem: According to ILO estimates, forty two million people around the world today have AIDs. Out of these, 65% are in economically productive age group of 15 to 49 years. India ranks second in the world in terms of number of infectecd cases which are estimated around five million. With better treatment and early detection, infected people are living longer. Hence, the need to preserve their work capacity to maximum possible is very important for the individual as well as for the enterprise. ILO estimates that by the year 2020, AIDS will have consumed 10% to 30% of today's total workforce.

Industry Response to HIV/AIDS: The response of the industry to the problem of HIV/AIDS has to be multi-pronged. Ideally, various components of the response should be:

1. **Establish a clear Workplace HIV/AIDS Policy:** This ensures a consistent and standardised response by the organization. A comprehensive HIV/AIDS Policy should address the issues, such as confidentiality, discrimination issues, work disruptions and most important prolonged productivity of the employee-patient.
2. **Train Managers and Supervisors:** Managers and supervisors should be thoroughly trained in the HIV/AIDS Policy guidelines. Training must include clear guidance on the medical facts. The minimal risk of transmission in the workplace must be highlighted.
3. **Employee Education:** Workplace education programmes are inexpensive and cost-effective ways to minimize risks, protect workers' lives and save money on healthcare costs. HIV/AIDS Education Programmes can be integrated into the ongoing Health Education Programmes. A thorough Workplace AIDS Education Programme should communicate the following things:
   - *The medical facts:* How HIV is not transmitted, emphasizing on the facts that casual contact cannot spread it and that it is virtually impossible to contact in the workplace.
   - Guidelines on company policies and benefits.
   - *The legal facts:* Importance of confidentiality.
   - *The psychosocial issues:* How to respond to a co-worker with AIDS, how to live and work with HIV/AIDS, information about where to get oneself tested etc.
   - Take home information for employees' families about their protection from HIV/AIDS.

18.5.1 Frequently Asked Questions (FAQs) about HIV/AIDS

1. **What is HIV?** HIV (Human Immuno Deficiency Virus) is the virus that causes AIDS.
2. **How can I tell if I'm infected with HIV? What are the symptoms?** Blood test is the only way to know, if you are infected. The warning signs of HIV infection include: rapid weight loss, persistent dry cough, recurring fever or profuse night sweats, profound fatigue and unexplained weakness, swollen lymph glands, diarrhoea

lasting more than a week, chest infection, blotches on or under the skin or ulcers in the mouth.

3. **How long after a possible exposure should I wait to get tested for HIV?** Blood test will be positive after two to twelve weeks (the average is 25 days) of exposure.
4. **Where can I get tested for HIV infection?** The nearest Pathology Laboratory can at least collect the blood sample. All District Hospitals and Government Medical Colleges normally have approved/recognised testing facilities.
5. **What if I test positive for HIV?** If you test positive for HIV, early medical treatment and a healthy life style can help you survive well longer. Prompt medical care may delay the onset of AIDS. There are now many medications available to treat HIV infection and help you maintain your health. TB and HIV co-exist in the same individual quite often and both can worsen each other rapidly in such cases. Smoking cigarettes, drinking too much alcohol, or using illegal drugs (such as cocaine) can weaken your immune system. You must reduce or stop using these substance. Not having (abstaining from) sex is the most effective way to avoid transmitting HIV to others. If you choose to have sex, use a condom to help protect your partner from HIV.
6. **How is HIV passes from one person to another?** The pre-dominant mode of HIV transmission is through heterosexual contact, the second most common mode being drug injecting. Earlier, blood transfusion and blood product transfusion were also major causes, but blood safety measures are now in place to minimise such transmission.
7. **Which body fluids transmit HIV? What about saliva, tears and sweat?** Body fluids containing high concentrations of HIV are: Blood, semen, vaginal fluid, breast milk and other body fluids containing blood. Contact with saliva, tears, or sweat does not result in transmission of HIV.
8. **How well does HIV survive outside the body?** HIV does not survive well outside the body, so environmental transmission is not possible. Drying of HIV-infected human blood or other body fluids reduce the risk of environmental transmission to essentially zero.
9. **Are health care workers at risk of getting HIV on the job?** For them, main risk is accidental injuries from needles and other sharp instruments that may be contaminated with the virus; however, this risk is small (less than 1%).
10. **How HIV is not spread?** You don't get HIV from the air, food, water, insects, animals, dishes, knives, forks, spoons, toilet seats etc. You don't get HIV from faeces, nasal fluid, saliva, sweat, tears, urine, or vomit, unless these have blood mixed in them. You can help people with HIV eat, dress, even bathe, without becoming infected yourself, as long as you follow some common sense precautions. HIV does not spread by everyday social contact with an infected person. Nor is it possible to get HIV by shaking hands, touching, hugging, sharing utensils, toilet seats or bedding. Coughing and sneezing do not spread HIV. One does not get HIV from mosquito bites.

18.5.2 Common Misperceptions about HIV Transmission

Households of HIV/AIDS Patients: Transmission of HIV between family members in a household is very rare. To prevent this, exposure to the blood of infected persons must be avoided. For example, gloves should be worn during contact with blood or other body fluids. Cuts or sores on exposed skin of both the caregiver and patient should be covered with bandages. Hands and other parts of the body should be washed immediately after contact with blood or other body fluids. Surfaces soiled with blood should be disinfected appropriately (using dosim hypochloride etc). Sharing of razors and toothbrushes should be avoided. Needles and other sharp instruments should be used only, when medically necessary.

Workplaces: There is no known risk of HIV transmission to co-workers or consumers form contact in industries. Instruments intended to penetrate the skin (tattooing and acupuncture needles, ear piercing device) should be used once and disposed off or thoroughly cleaned and sterilized. Razors should be used for only one client and then disposed off.

18.6 ENVIRONMENT AND HEALTH

In previous chapters, we have already discussed the impacts of environmental degradation on human health. Environment and human health are very closely related, since any adverse impact caused on the environment well surely cause adverse impacts on the health of human beings.

A healthy environment is the one which contains minimum pollution and provides good health to its inhabitants; whereas an unhealthy environment is the one which causes adverse impact on human health. Besides, causing various types of diseases in the human body, a polluted environment may result in serious climatic changes, which may cause adverse effects on human health. The health of humans depend on sufficient amounts good quality food cleaner, safe drinking water and adequate shelter. These requirements can be fulfilled only through clean environment.

The deterioration in the environment has increased the incidence of many diseases leading to many kinds of stresses on human health, temperature and humidity, the state of vegetation, affect the vectors of disease. The environment also affects our output of work, our mental alertness, change in attitude and responses.

18.6.1 Environmental Factors and Human DIseases

Diseases Caused by Air Pollution

The air pollution have suspended particulate matters in the air and their presence is of considerable importance to public health. Some of the major air-borne diseases are:

1. **Common Cold.** Respiratory syncytial virus causes it. The symptoms include irritation of nose, headache, sneezing, leaky nose, sore throat and slight fever.
2. **Cold Influenza.** It is a serious respiratory diseases, causing destruction of cells of upper I respiratory tract, bronchi and trachea. Common indicators of cold influenza include fever, sore throat, sneezing and body pain. Further complications by presence of bacteria in air may lead to pneumonia.

3. **Chicken Pox.** It is highly infectious disease, which is indicated by low fever, rash, vesicles and pustules over the skin that is eventually shaded as dark brown scales.
4. **Tuberculosis (T.B.).** The germs of T.B. spread mainly through cough spray, dried contaminated dust, sputum, contaminated articles and from the physical contact with a patient suffering from lungs TB.
5. **Silicosis.** This disease is caused when silica particles in the air get into human body.
6. **Bronchitis and Asthma.** Continuous exposure to air pollution may lead to bronchitis and asthma. This is caused mainly when organic dust enters into human body causing allergic reaction.

Diseases Caused by Water Pollution

Use of contaminated water leads to several water-borne diseases or gastroenteritis diseases. Some of the major water-borne diseases are:

1. Cholera
2. Typhoid
3. Dysentry
4. Ineffective hepatitis
5. Intestinal disorders.

Food-borne Diseases

Contaminated food leads to various diseases. This includes the following:

1. **Bolutism.** This leads to double vision and cause difficulty in speaking. Involuntary muscles become paralyzed and paralysis spreads to the respiratory system.
2. **Viral food poisoning.** Several viruses are transmitted through food and cause infection in human body. It affects the liver causing jaundice.
3. **Fungal food poisoning.** Such food leads to indigestion and illness called **myco-toxicosis.**

Deficiency Diseases

Primary health-care is the key to attainment of health for all, based on essential elements of primary health-care. Education concerning health problems and the methods of preventing and controlling them, proper nutrition, safe water and basic sanitation, maternal and child health-care immunization against the major infectious diseases. This may lead to following common deficiency related diseases:

1. Avitaminosis.
2. Anaemia.
3. Rickets and osteomalacia.
4. Teeth and gum diseases.

Disease Caused by Improper Disposal of Solid Wastes

Improper disposal of society's wastes (organic solid wastes) may create unhygienic and insanitory conditions, which under warm and moist conditions may cause breeding of (*a*) rodents like rats, (*b*) insects like flies, and mosquitoes which help in spreading deadly diseases.

The **diseases caused by insects.** Among the disease carrying insects the main enemies of man are mosquitoes, flies, fleas and lice. *Diseases transmitted by mosquitoes* include malaria, filaria, dengue, yellow fever etc.

Diseases transmitted by flies include cholera, typhoid, dysentery etc.

Diseases transmitted by fleas is plague. The most common fleas in India are those which live on the bodies of rats, as their external parasites. Wild rodents, such as squirrels, gerbils, marmots and wild guinea pigs are also the reservoirs of plague, and could infect the rat population with the plague bacteria, thereby threatening human population.

Diseases caused by lice. Lice are small wingless insects, leading an obligatory parasitic life upon its host. It transmits certain dangerous diseases, such as the typhus, and the relapsing fever, besides the usual irritation caused by sucking of blood by the lice.

18.7 HUMAN RIGHTS

Keeping in mind its main aim of **human dignity,** the United Nations set up the **U.N. Commission of Human Rights** in 1946. This commission wrote an important document, known as the **Universal Declaration on Human Rights (NNDHR),** which was approved by all the members of U.N. General Assembly on 10th December, 1948. This day is, therefore, celebrated as the **World Human Rights Day.** This declaration expressed the hope that people would learn to respect the rights and dignity of others.

While discussing environmental issues, which are closely linked to human rights, in previous chapters, we deliberated on the equitable distribution of environmental resources, utilisation of resources, conflicts between people and wildlife etc.

Human Rights Legislation generally contains:

1. *Security rights* against curve.
2. *Liberty rights.* It protects freedom in areas, such as belief and religion, association, assembling and movement.
3. *Political rights* to participate in politics by expressing, protecting, voting and serving in public office.
4. *Due process right* that protect against abuses of the legal system, such as imprison without trial, secret trials and excessive punishments.
5. *Equity rights* that guarantee equal citizenship, equality before the law and non-discrimination.
6. *Welfare rights* (also known as economic or social rights) that require provision of education and protection against severe poverty and starvation.
7. *Group rights.* It provides protection for groups against ethnic genocide and for the ownership by countries of their national territories and resources.

Four important rights: (*i*) right to life, (*ii*) right to be free from slavery, (*iii*) right to be free from torture, and (*iv*) right to be free from retroactive application of penal laws cannot be pushed aside even during national emergency.

The United Nations has set up a number of bodies to monitor and study human rights under the leadership of the UN High Commissioner for Human Rights (UNHCHR).

Status of Human Rights in India

In addition to human rights granted by United Nations, the Constitution of India has also granted to its citizens several **Fundamental Rights** which can be legally enforced, such as the:

(i) Right to Equality,

(ii) Right to Freedom,

(iii) Right against Exploitation,

(iv) Right to Freedom of Religion,

(v) Cultural and Educational Rights, and

(vi) Right to Constitutional Remedies.

In order to keep a watch on frequent violations of the human rights, and to enforce the United Nation's Universal Declaration of Human Rights, the Government of India, under the provision of **Human Rights Act** has established a **National Human Rights Commission (NHRC)** at the national level, while *State Human Rights Commissions (SHRCs)* have been established at the state levels.

18.8 VALUE EDUCATION

To understand the meaning of 'value education', we must first understand the meaning of education. The education is required to make people read, write, think, understand, and then act appropriately like a knowledgeable person on various social, economic, cultural, and other issues of our life. It is one of the most important tools required for bringing about socio-economic and cultural progress of a community or a country.

The **value education** must be oriented to impart moral, social, cultural, and above all human values. An educated person should have compassion, patience, forgiveness, self-control, honesty, integrity, sincerity, truthfulness, consciousness and courage to face bad situations and circumstances. A well-educated person, in any case must be above all types of greeds, should be God fearing, and should extend help to the poor, weak and downtrodden. The value education must, therefore, be aimed at building a person's overall character.

Value education should also give full knowledge nature and natural laws, and should apply his knowledge for the total protection of his surrounding environment (biotic as well as abiotic world) in all his activities of daily life.

Objectives of Value Education

The report of the Joint Study on Moral Education in Asian Countries (JSMEAC) has listed the following objectives of value education:

1. Full development of Child's personality in its physical, mental, emotional and spiritual aspects.

2. Inculcation of good manners, and responsible and co-operative citizenship.
3. Developing respect for the dignity of the individual and the society.
4. Inculcation of a spirit of patriotism and natural integration.
5. Developing a democratic way of thinking and living.
6. Developing tolerance towards and understanding of different religions and faiths.
7. Developing a sense of human brotherhood at the social, national, and international level.
8. Helping the children to have faith in some supernatural power and order, that is supposed to control this universe and human life (this applies to the communities of believers).
9. Enabling children to make moral decisions on the basis of sound moral principles.

The report further emphasizes **certain specified values at different school levels** as under:

(a) At the **elementary level,** good manners, orderliness, punctuality, discipline, obedience, piety, cleanliness, cooperation, truthfulness, honesty, kindness etc.

(b) At the **secondary level,** patriotism, dignity of labour, justice, tolerance, human brotherhood, dignity of the individual, democratic spirit, understanding of other religions, international understanding, strength of character, ability of moral judgement etc.

Value Based Environmental Education

We have already discussed about 'environmental ethics' in the chapter on "Social issues." In continuation to this, we shall discuss here about environmental education. The environmental education to be imparted to the children should also be made value-based by incorporating following aspects:

1. The basic **human value.**
2. **Social values** like love, compassion, tolerance, and justice.
3. Promote our **cultural and religious customs and values.** For example, treat rivers, mountains, forests and their products as sacred etc. may be highlighted.
4. Emphasise on the **ethical values** of being earth-centre rather than human-centric.
5. **Global values.** The entire natural environment is common to the entire world. All natural occurring over the earth are inter-dependent and inter-linked with special bonds of harmony.
6. **Spiritual values.** This includes principles of self-restraint, self-discipline, contentment, reduction of needs, freedom from greed etc.

If all the above values are incorporated in our environmental education system at school level, it will help in attaining our goals of sustainable development and environmental conservation, as it will bring a total change in our thinking, our attitudes and lifestyles.

QUESTIONS

1. Is India a over-populated country? Comment.
2. Explain population explosion and how it can be checked. What are the reasons for the population explosion?
3. What can be the possible consequences of over-population in a country like India?
4. Suggest some of the measures to control over-population in India. Mention the different methods of birth control.
5. Explain the term HIV and AIDS. Give causes of AIDS and preventive measures.
6. Discuss as to how the global population has increased during past four centuries?
7. What is family welfare programme? Explain in detail.
8. Discuss family planning measures being used to check population growth.
9. Discuss the environmental impacts of increasing population.
10. How does environmental degradation affect human health? Describe briefly the health effects of polluted air, water and soil.
11. Explain the term 'human rights'. What is the status of human rights in India?
12. How do human rights help in protecting the environment, and what are the environmental rights available to Indians?
13. Describe the necessity of imparting value education.
14. What is value education? How value based environmental education can be imported?
15. Write a detailed note on HIV/AIDS. Explain the social and environmental impacts of this disease.
16. Explain the following in brief, with reference to the related environmental issues:
    (a) Women welfare
    (b) Child welfare
    (c) Child labour
17. Write an essay on "population growth and environment."
18. Write short notes on:
    (a) Disease caused by improper disposal of solid wastes.
    (b) Different human rights as per legislation.
    (c) Status of human rights' in India.
19. Explain the following in brief:
    (a) Objectives of value education.
    (b) Different specified values at different school levels.
    (c) Value-based environmental education.

19

Chapter

RESETTLEMENT AND REHABILITATION (R&R)

19.1 INTRODUCTION

Development projects generally require the displacement of people from areas, where they live and work to other resettlement locations. Such relocation causes economic, social and cultural disruption to the affected individuals. Adequate policy and its implementation is necessary to minimise or reverse the negative effects of compulsory relocation (also known as involuntary resettlement) on individuals and on the national economy.

The projects involve acquisition of public (government and forestland) and private land from title holders located in urban or rural areas. The acquisition of land and consequent displacement have potential impacts on the social, economic, cultural and environmental attributes of the affected population with specific impacts on their productive assets, sources of income, habitat, community structure, social relations, cultural identity, traditional authority and also their potential for mutual help.

A Social Impact Assessment (SIA) and Rehabilitation Action Plan (RAP) is prepared for the affected and displaced population. Land requirement for the project should be kept minimum and limited for accommodating different project components and other allied works. Land required should be sufficient to create the necessary facilities and infrastructure and other activities under the project.

Some Terms

1. **Displaced Person.** Displaced person are those, who have been ordinarily residing in the impact area or practising agriculture, carrying on any trades, occupation, or working for livelihood in the area likely to be affected from the development projects.

2. **Displaced Family.** Displaced family includes husband, wife, children and other persons dependent on the head of the family, *viz.*, widowed mother. Certain people feel that every major son/daughter should be treated as a separate family.

3. Resettlement. Resettlement may be defined as final movement of displaced persons to a new location site after getting full compensation of their land and properties. The stages of resettlement are:

(i) Compensation for land, house, trees and other properties coming under impact.

(ii) Allotment of agricultural land and house plots at new site.

(iii) Free transport for bringing household material to new site.

(iv) Payment of ex-gratia, rehabilitation grant, subsistence allowance, development assistance.

(v) Civil amenities provided at new site.

(vi) Land rights given for allotted land.

4. Rehabilitation. Rehabilitation is a re-storative attempt to recover the feature of the natural state. It includes various measures and strategies to develop resettlement site ecologically sustainable. The Resettlement and Rehabilitation (R & R) provisions should be decided/designed on the basis of detailed socio-economic and cultural milieu of the affected/ displaced population.

19.1.1 Avoiding or Minimising Resettlement

In recognition of the hardship and human suffering caused by involuntary resettlement objective should be that whenever feasible, involuntary resettlement must be avoided or minimised, and alternative development solutions must be explored. If it is unavoidable, a careful and systematic advance planning is necessary. Because involuntary resettlement dismantles a previous production system and way of life, all resettlement programmes must be development programmes as well. It must be ensured that the productive base and income-earning ability of these re-settled people is improved—so that they share the benefits of the new development and are compensated for transitional hardship—or atleast helped to attain the standards, they would have achieved without relocation. On this aspect, socialists suggest following:

(a) People subjected to relocation are prone to develop the syndrome of settler dependency, if paternalistic help policies are applied. Because such policies discourage self-mobilisation and undermine the settler's commitment to self-support and development, government assistance should be an interim measure.

(b) After relocation of the affected families a socio-economic study should be carried out (*i*) to ensure that the earning capacity of the people have improved after relocation—and atleast not deteriorated, and (*ii*) to see that their relations with host communities are not bad.

19.1.2 Development-Oriented Approach to Resettlement

In the past resettlement was being dealt as a mere hasty physical removal of people out of the path of the flooding reservoir water or of the coming highway. Ideally, it may have approach as an opportunity for the construction of systems of production and human settlements that would represent a development in the standard of life of those affected, as well as in the regional economy of which they are a part.

This difference in perspective leads to different approaches to conceptualizing, designing, planning, financing and implementing resettlement. When certain development projects make the resettlement of people unavoidable, it must be ensured, after a transition period, the displaced people regain atleast their previous standards of living and that so far as possible, they be economically and socially integrated into the host communities.

In pursuing such integration, the major objective is to ensure that settlers are provided opportunities to become established and economically self-sustaining in the shortest possible period, at living standards that atleast match and if possible, improve upon those before resettlement. For this purpose, 'With-and-without-the-project' type of analysis is applied of assessing general returns aiming to re-establish resettlement at levels comparable or better than those that would have been reached without the project-induced resettlement.

Whether rural or urban; the future-looking approach calls for re-constructing settlements that are residentially superior to, and not just replicas of, those existing previously. It should be recognised that new settlements are expanding socio-cultural systems, whose collective needs will increase overtime.

Development-oriented resettlement means that social infrastructure, school and health services, access to employment opportunities, and if possible, the housing-plot allotment and dwellings should be planned to meet the needs of re-settlers' growing families, taking into account atleast first and second generation in the settlement. The same applied to infrastructural systems, such as feeder roads, water supply, or services system like marketing networks.

In order to accomplish these objectives, the planning of resettlement should be integral part of the preparation of the main project that caused the resettlement *i.e.* of the planning, appraising, financing, and supervising of this main project.

In view of the above, the resettlement of those displaced must be undertaken based on a well-prepared 'settlement plan' aimed at using the opportunities created by relocation towards resettlement and development. To be successful, resettlement planning and implementing require core scrutiny of the essential needs of the settler. The planners need to think about the economic and cultural characteristics of the population to be moved.

The planning for resettlement should begin as early as possible because delays in the resettlement have led in the past to serious cost overruns, not only in the resettlement operation itself but also in the overall project completion.

19.2 SOCIAL IMPACT ASSESSMENT (SIA)

The loss of private assets resulting in loss of income and displacement has made social impact assessment an important input into the project design, while initiating and implementing developmental intervention an understanding of the issues related to social, economic and cultural factors of the affected people is critical in the formulation of an appropriate rehabilitation plan. A detailed SIA is carried out incorporating social analysis and participatory processes into project design and implementation to make it responsive to social development concerns.

SIA is also helpful in enhancing the project benefit to poor and vulnerable people, while minimizing or mitigating concerns, risks and adverse impacts.

The project impact on the human environment for the purpose of proper designing of rehabilitation policy and programmes should be studied keeping in view of the following:

(i) Impact on villages/cities/towns, land, families.

(ii) Magnitude or impact.

(iii) Categories of affected populations *e.g.*, land owners, joint holders, encroachers, landless labourers, artisans, etc.

(iv) Ecological relationships of the population.

(v) Socio-economic and cultural environment of the affected villages/cities/towns, individual ownership/Common Property Resources (CPRs), economic structures, health, food habits, education, services and relationship with host population.

The **main objective** of the study is to ensure that project addresses the adverse impact on the livelihood of the people and that nobody is left worse off after implementing the Rehabilitation Action Plan (RAP) and those affected have access to project benefits, both, during project constriction as well as operation. To be specific, the objectives of the SIA study are:

1. To carry out a socio-economic, cultural and political/institutional analysis to identify the project stakeholders and social issues associated with the project.
2. To assess the extent of land acquisition/appropriation and other losses and undertake the census of potential project affected people.
3. To develop a RAP in consultation with the affected people and project authorities.
4. To identify likely occurrence of HIV/AIDS resulting from the influx of outside labourers and others develop a strategy to reduce their incidence.
5. To develop a consultation framework for participatory planning and implementation of proposed mitigation plan.

The main **focus areas of the SIA study** are:

1. *Stakeholders Analysis*

(i) Identify stakeholders—project decision-makers, beneficiaries, affected community, project influencer (*e.g.*, NGOs and civil society), project implementers, project facilitators (other agencies) and identifying their stake in the projects;

(ii) Identifying key formal and informal institutions operating at village and sub-regional levels and assessing their role in community decision-making processes; and

(iii) Assessing local capacities in terms of participation in planning, implementation, supervision and monitoring.

2. *Review Legal Framework.* Review of relevant national and state legislation and regulations pertinent to the land acquisition, resettlement involvement of project stakeholders. Attention should be paid:

(i) to the laws and regulations governing the social safeguards and project implementation, and

(ii) on the access to and exclusion of groups to the services and opportunities provided by the project.

3. *Vulnerable Population.* The living pattern of vulnerable population (including tribes, scheduled castes, woman, landless etc.) in the project area and assess whether they are involved in community decision-making process.

19.3 APPROACH AND METHODOLOGY FOR SIA STUDY

Approach and methodology mainly consists of quantitative and qualitative tools and techniques. The study is conducted in three phases, as discussed here under:

Phase-1: Pre-Survey Activities

1. **Collection and Review of Project Literature.** This phase intends to familiarize with the concerned and important stakeholders to identify and collect the available literature and to scope the activities. This involved two pronged approach, *viz.*:

(a) Discussions with project implementing authorities and other concerned;

(b) Collection of available Project Affected Persons (PAPs) database and other relevant project literature. Consultations should also be held with concerned village revenue officials to update the ownership of land and its utilisation pattern by referring to Records of Right (ROR) or *Jamabandi* Registers. Main literature consulted includes: Environmental assessment report, Draft R & R policy; policy on land acquisition and others; land acquisition plans, if any; project details etc.

2. **Rapid Preliminary Field Visits.** This is a familiarizing field activity and is conducted as a part of a ground truthing exercise. This is primarily observatory exercise and entails briefing by the concerned. This provide the basis for field research preparation and help in testing the questionnaires and checklists.

Thus, both the above mentioned review of project literature and rapid preliminary field survey help in finalizing the study instruments and inception report detailing the final methodology and workplan.

Phase-2: Survey Activities

1. **Census and Socio-Economic Survey.** The census survey of all the project-affected persons is conducted in this phase. The survey, Inter-alia, should assess the impacts of the project, the socio-economic condition, and living standards of affected persons. Information about following are collected during the survey:

1. Socio-economic conditions of affected persons.
2. Family structure and number of family members.

3. Education levels.
4. Occupation and income levels.
5. Asset levels.
6. Loss of assets due to the project.
7. Accessibility to the community resources.
8. Perceptions on the resettlement and rehabilitation measures.
9. Perceived income restoration measures.
10. Grievances of affected persons.
11. Awareness and knowledge levels on HIV/AIDS.
12. Willingness to participate in the project.

2. Qualitative Survey. Qualitative survey may not always reveal the true facts, particularly when assessed for poor and vulnerable sections. The qualitative survey includes focus group discussions and in depth interviews with various sections of people, such as women, knowledgeable persons, community leaders basically to elicit their expectations and suggestions, which will support and provide additional information collected through quantitative survey.

3. Assessment of Livelihood Losses. The study identify people losing their livelihood directly or indirectly. Loss of common resources should also be assessed and develop appropriate entitlements to mitigate the impact. For this, an analysis is undertaken for the resources available and the resources lost. Consultations are also necessary with people losing properties/common resources, knowledgeable persons and village heads for indentifying the rehabilitation strategies for those losses by way of training requirements for income generation and other remedial and restoration measures.

4. Review of Legal Provisions. Relevant national and state legislations and regulations are reviewed. The study is also conducted to ensure that sufficient implementation arrangements and its capacities are available for delivering the R & R services and timely implementation of Rehabilitation Action Plan (RAP).

Phase-3: Post Survey Activities

1. Analysis of Data. The information collected through structured interviews and questionnaires is systematically coded, validated, analysed and tabulated. Wherever required, the observations are supported from the information collected through desk research.

2. Preparation of Study Report. SIA study report is the outcome of all the above activities and also the results of the SIA and RAP report, which provide the socio-economic risks involved and strategy to minimize the risks of the programme with special consideration on the vulnerable, and develop participative monitoring mechanism.

19.4 LAND ACQUISITION AND ITS IMPACTS

A developmental project involves acquisition of public and private land from title holders, located in the project area. The acquisition of land and consequent displacement have potential impacts on the social, economic, cultural and environmental attributes of the affected population with specific impacts on their productive assets, sources of income, habitat, community structure, social relations, cultural identity, traditional authority and also their potential for mutual help.

Acquisition of land results into following main impacts:

(i) Loss of land (private and Government land).

(ii) Uses of private land (Agriculture, residential, barren, others).

(iii) Uses of government land (Barren, grazing, river, reserve forest, private occupation and others).

(iv) Loss of structures (Commercial, residential, mixed and others with details about pucca, semi-pucca, kutcha).

(v) **Loss of Trees.** Acquisition of land for the project also resulting in loss of trees both, on private and government land. The number of different species of trees lost which includes fruit bearing, fodder based and timber based should be determined.

(vi) Land holdings before and after proposed land acquisition with special mention of marginal and landless farmers.

(vii) The number of affected and displaced households.

(viii) Project affected family and population as per R&R policy/package.

Loss of Vanpanchayat Land. Vanpanchayats are carved out of civil (protected forests under the jurisdiction of the Revenue Department). This land mostly is allotted to a certain users (called self-help groups) generally to the landless people to protect and develop the forestland and use forest produce for the benefit of the identified users. However, with time the land is being used by the total village community *i.e.* both land-holders and landless. This land is earmarked and allotted by the forest department and is managed by the Management Committee elected by the General Body (Forest Users Group). The committee is headed by the Sarpanch.

An assessment is made for loss of income for the affected villages from acquisition of Grazing land and Vanpanchayat Land used for grazing, collection of fodder and fuel wood and collection of timber. The usage and collection of fodder and fuel wood is done by all the villagers mostly daily. To mitigate the losses from Vanpanchayat land, the Forest Department in co-ordination with the local Vanpanchayat can undertake fodder and tree plantation in the vacant area, available in respective Vanpanchayat of the affected village.

Measures for minimising the land acquisition impacts are mentioned hereunder:

1. Site selection for infrastructure and other facilities is done in such a manner which involve minimal land acquisition of private land.
2. Efforts should be made to avoid displacement of habitation/settlement centres and the project activities are planned in such a way that they do not disturb the main habitation.
3. Efforts should be made to have maximum barren land.
4. The project infrastructure locations are planned in such a way that the existing approach roads are used and laying of new approach roads is kept bare minimum to avoid private land acquisition.

19.5 IMPACT ON VULNERABLE GROUPS

Development induced displacement has lot of consequential impact and bring changes in the lives of Project Affected-Persons (PAPs). These changes are economic, social and cultural in nature. If changes are adverse, then certain vulnerable sections of the society, like women, disabled, destitute, orphans, widows, unmarried ladies, abandoned women, tribal, old age persons etc., are worst hit. Therefore, it is imperative to study the socio-cultural and economic aspects of such under-privileged section of the society, among the affected community particularly in the light of their vulnerability to changed situation. These are not provided or cannot be immediately provided with alternative livelihood and who are not otherwise covered as part of the family and are generally not aware of their rights and privileges or even of their entitlements.

It has generally been noted that such group would experience most adverse socio-economic impacts, such as loss of common property resources in terms of fodder, fuel wood and loss of access to some of these resources. Similarly, the other backward and vulnerable sections of society, such as SCs and STs are also affected and need special focus in the project.

Women are required to be integrated in the project as full-fledged participant taking part in all the stages of the project starting from planning through implementation and even in the post-project stages. Participation from women could be sought through allowing them taking part in the consultation process. Women are consulted and invited to participate in group-based activities to gain access and control over the resource as a part of the RAP implementation. This can be ensured by initiating Self-Help Group formation in each of the village. It is also advisable that for doing the social impact study, each field team should include atleast one women investigator/facilitator.

19.6 STAKEHOLDERS' PARTICIPATION AND CONSULTATIONS

Participation is a process, through which stakeholders influence and share control over development initiatives and the decisions and the resources, which affects them, are identified. The effectiveness of R&R programme is directly related to the degree of continuing involvement of those affected by the project. Comprehensive planning is required to assure that local government, NGOs, host population and project staff interacts regularly, frequently and purposefully throughout all stages of the project.

A *Stakeholder* in the project is any individual, group, agency or organisation affected by a project arid/or with concern or interest in a development project and its outcome, or in common resources impacted by a development project. A stakeholder should be treated as a 'Partner in Development' and not as an opponent of the project.

The effects of a project on stakeholders can be adverse or beneficial, direct or indirect, sooner or later. The identity of the stakeholders, the timing of their involvement and the activities they may be encouraged and enabled to undertake, depend largely on the type, size and location of a project as well as on the insightfulness, creativity, sensitivity and dedication to public, involvement of the developer and the members of the public involvement Facilitation Team.

Generally speaking, those most directly affected by a project are known as the key stakeholders—they are at the greatest risk, they feel the impact most intensely, they benefit the most from opportunities; hence, they should be the first to be involved. Since local people will be stakeholders over most of the life of a project, their involvement and participation from the beginning is crucial to project success. Stakeholders involvement begins with the identification of affected communities and group, followed by the formation of representative committees, followed by identifying the core issues arising and to be followed. Stakeholders involvement in planning, the mitigation of project impacts, and in monitoring is especially important.

Participation of Re-settlers

Since resettlement planning implies critical decisions regarding the future of the displaced groups, it is incumbent upon to seek the participation of the re-settlers. The affected population should be consulted directly or through their formal and informal leaders, representatives, or non-government organisations with respect to social and economical aspects of various alternatives being considered for resettlement. This will improve the understanding of their needs, resources and preferences, and help reduced the reluctance to move. It also accelerates the subsequent transition to and integration within the new settlements.

Host Population

Besides re-settlers themselves, the other major actor in resettlement process is the host population living in the receiving areas. Generally, when host population is omitted, then during implementation lot of problems arise. In such a case, risk is that population density in receiving areas will increase suddenly to levels above the carrying capacity of the land and natural resources available to both, hosts and incomers on a sustainable basis. Serious conflict may arise as increased demands are placed on land, water, services etc.

The resettlement agency should also anticipate that felling of jealousy is likely to be aroused among hosts, if superior services and housing are provided to the settlers. If possible, education, water, health and other services should be made available for both groups and a suitable social climate generated for their integration. Equitable treatment should be accorded as far as possible to both, hosts and re-settlers.

Through consultation people are properly informed and consulted about the project, their situation and preferences, are allowed to make meaningful choices. This serves to reduce the insecurity and opposition to the project which otherwise are likely to occur. The project will therefore, ensure that the affected population and other stakeholders are informed, consulted, and allowed to participate actively in the development process. This should be done throughout the project during preparation, implementation and monitoring of project result and impacts.

In order to get representation of all the segments of affected population, a number of group consultations may be held in every village. Separate group meetings may be held for women and vulnerable sections. The consultation programme should include the following:

1. Heads of households likely to be impacted.

2. Household members *i.e.* affected land-owners and their family members.
3. *Clusters of PAPs*. The participants in these consultations also include indirectly affected population other than directly affected population.
4. *Villagers*. These are consulted by way of public meeting and small group meetings.
5. Village panchayats.
6. Local NGOs or voluntary organisations.
7. Government agencies and departments.
8. Agencies/departments may also be consulted to identify and list the income generating opportunities, training opportunities etc.

It is also a good practice to open a public information centre(s) to provide actual information about policies and social issues to the people in a regular manner.

19.7 SOCIO-ECONOMIC ISSUES

While planning for R&R package, following major socio-economic issues need consideration:

The majority of the displaced are rural and poor because new projects are generally brought to the most underdeveloped, poorest areas, where infrastructure is largely lacking and where land and political costs are lowest. The remote locations of many dam sites are often inhabited by indigenous people, ethnic communities, or pastoral groups, which explains as to why issues pertaining to tribal and cultural differences are so prominent in resettlements.

In the urban areas, the demographic and occupational composition of affected people is different where people to be relocated are employed in industry and services. Small business people, like shopkeepers, artisans, food-stall owners and vendors are larger proportion than in rural areas. Since the affected people in urban areas are socially and culturally different from those in rural areas, and depend on different income sources; they have different demands, and therefore, need difference resettlement package. Urban people include tenants and other vendors or workers coming from nearby villages for earning their bread and butter.

Women may experience the adverse consequences of resettlement more strongly than men. This is because compensation payments are usually paid only to the heads of households, converting the collective assets of the family to cash in male hands and leaving women and children at higher risk of deprivation.

Projects causing displacement have many times more beneficiaries than victims (adversely affected population). Large number of people gain significant advantages from projects that entail re-settling people.

19.7.1 Land Related Issues

Following are major land related issues which are of most importance for effective resettlement in rural areas:

1. **Eligibility.** Land acquisition laws provide compensation only to formal owners of land. However, in most project areas, a large number of people without formal title to land

but dependent on it, are affected. These can include tribal groups whose ancestral lands were never titled, while in urban areas, they often include established squatter neighbourhoods and informal urban settlers. Therefore, the customary land rights should be recognised as valid and eligible for land compensation/other benefits.

2. Land Valuation. Displaced persons should be compensated for their costs at full replacement prior to the actual move, rather than on 'depreciated or book value' which is often significantly less than the replacement value of the lost asset. Therefore, while deciding R & R Policy and package, it should be considered whether it is possible to pay enough compensation to enable a person to buy a comparable property, or whether cash compensation should be supplemented by provision of replacement assets (*i.e.* an apartment or shop).

3. Land Quality. In many projects, the land proposed for resettlement is surplus government land, little suitable for farming. Therefore, it should be ensured that the resettlement areas should be able to support viable farming systems with irrigated land.

4. Institutional Ability to Locate Land. There are many agencies and procedures involved in acquiring land. The assistance of projects and institutions must be made available to re-settlers to enable their relocation on land, rather than payment of simple cash. In India, land resources which is surplus and can be cultivated is scarce, and per capita land ratios are decreasing. Generally, projects acquire land as per land acquisition laws and market mechanisms are rarely used for replacement land, nor are market prices paid by the project agencies to obtain it for re-settlers. In view of this, following approaches for acquiring adequate land should also be tried:

(a) Organisation carrying out the project should find one or more large resettlement sites, to which rural communities or urban neighbourhoods move more or less in their entirety, or as natural social subgroups.

(b) Land purchase committees are convened to help individuals to negotiate fair prices for land and to help re-settlers assess its technical suitability. This method was tried in Gujarat in Sardar Sarovar project and have had considerable success, where land purchase committees included local NGO and government representation.

(c) Allow re-settlers or their representatives to find land themselves. This limits the burden on the agency, and allows the complex personal strategies and knowledge of individuals to come into play. In Upper Indravati Project (in Odisha state) resettlement unit put compensation and ex-gratia funds into blocked account for re-settlers, to be released at the time of the land transaction negotiated by the resettlers.

5. Job Creation or Self-Employment. In urban resettlement and also in non land-based package in rural areas, creation of jobs or self-employment is to be considered and substantial numbers of jobs are created by an expanding economy. Jobs can be promised in new local industries of projects to re-settlers. Small scale credit programmes should be implemented. More promising are the strategies that make immediate use of project created new resources (*e.g.*, reservoir fisheries, tourism, commercial opportunities etc.) for the benefit of the re-settlers.

19.8 PREPARATION AND APPRAISAL OF RESETTLEMENT PLAN

During project identification, it is important to identify whether resettlement will be required and the size of the population that will be affected, and to start preparing for the their relocation. While taking investment decision, its intensity should be known, so that the decision-making authorities can weigh this aspect during decision-making process, and not dealt with as an after-thought, when the key decisions had already been taken.

The possibility that the magnitude and complexity of involuntary resettlement might be better handled under a special full scale resettlement project, rather than as a subsidiary component in the main project, may be considered at this stage. Preparation of the resettlement should begin at the sametime as preparation of other project components.

The resettlement plan should incorporate three distinct sets of activities concerning:

(i) The preparation of the affected groups for the shifting,

(ii) The transportation of the displaced to the new site; and

(iii) The integration of the displaced into the new community.

During project preparation, as already discussed, it is important that affected population be informed, consulted and involved, directly or through their formal and informal leaders, representatives, and through NGOs with respect to the various alternatives being considered for resettlement.

The project which cause significant resettlement need appraisal after the resettlement plan and its time-table is prepared. The basic objective of the appraisal of the proposed arrangements for resettlement should be to ascertain whether they are adequate and feasible and whether, when implemented, they will relocate the affected people without undue hardship, provide them with support services, and enable them fully rebuild their livelihood. This incorporates the elements usually reviewed under Social Impact Assessment (SIA) or Environmental Impact Assessment (EIA).

The **appraisal of the plan** for implementation includes:

(i) Properties to be destroyed by the primary project has been inventoried and valued;

(ii) Land in the receiving sites has been identified and is available and accessible to the settlers, and

(iii) The number of people choosing from among the various alternative proposals is adequately known and these alternatives are technically and economically sound.

The appraisal should also be assessed the overall policy in respect of resettlement; as well as the legal and organisational framework within which the resettlement will be executed. At this stage, institutional mechanism for implementing the resettlement work should also be decided.

R&R Costss

R&R costs are required to be reflected in appraisal reports and project completion reports. The following grouping could be considered to ensure uniformity in project evaluation:

1. Compensation Costs. Mainly the costs of land, houses and other assets acquired, and compensation payment for lost incomes. This also includes the costs of censuses, surveys and studies carried out in the project area.

2. Resettlement Costs. These include costs of land acquision in the resettlement area, new housing and local infrastructure, necessary studies, transportation of people and their belongings, transitional income payments, investments to compensate host population affected by the resettlement, and efforts to protect the environment in the resettlement area.

3. Rehabilitation Costs. These include the costs of activities to support income re-storation and improved living conditions of the affected population beyond compensation and resettlement components. It also includes additional local development activities, new services, agricultural extension, training, employment creation and loans.

4. Administrative Costs. These include costs of offices, technical assistance, community participation activities, communication etc.

5. Costs of Lost Public Assets. Replacement costs over and above the infrastructure built for the local resettlement components. This will include regional/national roads, bridges, and any other infrastructure that has to be re-built because of the project.

Profile of Project Affected Population

Resettlement plan is prepared based on the socio-economic survey which include the profile of Project Affected Population. This profile should cover the following aspects in the Social Impact Assessment report:

1. Demographic Profile (Gender based details of the population).

(i) Rural/Urban population

(ii) Population density

(iii) SC/ST and others

(iv) Literacy

(v) Employment and occupation

(vi) Economic status (land holding, house holding)

2. Socio-economic Profile

(i) *Work Participation.* Cultivators, agricultural labour, household industry/artisan, marginal workers.

(ii) *Land-use pattern.* Forest land, cultivable land, barren land, uncultivable/waste land. Grazing land, horticultural land, others.

(iii) *Land holding pattern.* Marginal farmers owning less than 0.5 Ha, 0.5 to 1 Ha, 1 to 2 Ha, 2 to 4 Ha, 4 to 10 Ha and more than 10 Ha.

(iv) *Asset ownership.* Type of house, electricity connection, cooking facility, drinking water facility. Domestic durable ownership, like LPG stove, furniture, Radio/TV, Telephone/ cell phone, bicycle/scooter/motor bike, car, tractor, refrigerator, electric fan.

(v) *Livestock ownership.* Cows, bullocks, buffalos, goat, sheep, poultry.

(vi) *Income level.*

3. Socio-Demographic Profile

(i) *Details of villages to be affected.*

Total no. of village, total no. of families (SC/ST, others);

Total population (SC/ST, others).

(ii) *Village wise, land details.*

Name of village, total land, land coming under project area, main occupation of villagers (agriculture, service, labourers, business).

(iii) *Age profile (Sex wise).* 0 to 5 yrs; 5-14,14-18, 18-25, 25-35, 35-60, above 60 years.

(iv) *Social groups.* Religion-wise, Caste-wise, Nuclear family or individual.

(v) *Literacy.* Illiterate, literate without formal education, primary school, middle school, high school, senior secondary/intermediate, graduate, post-graduate and other specialisation.

(vi) *Details of families to be displaced.* Land oustees only, homestead oustees only, land and homestead oustees.

(vii) *Infrastructure affected.* Education, industrial development; drinking water, communication, roads, electricity, sanitation.

(viii) *Expenditure pattern.* Expenditure on food, clothing, health, education, social functions agriculture, animal husbandry, etc.

(ix) *Cultural sites.* Places of worship, Archaeological sites/monuments, Anthropological sites.

(x) *Uses of Vanpanchat land.* Type of uses with frequency.

4. Health Profile

(i) Existing health.

(ii) Disease surveillance.

(iii) Awareness on HIV.

19.9 MITIGATION MEASURES

Based on the Social Impact Assessment (SIA) study, various mitigation measures must be involved in the following steps:

Strategies for Developing the Package

Development package for resettlement must offer sufficient opportunities and resources for their economic and social re-establishment as self-sustaining producers or wage earners. The two basic strategies may be pursued for economically and socially re-establishing those dislocated from rural settlements are:

(a) Land-based strategies, and

(b) Non-land-based strategies

Land based strategies must provide economic opportunities to re-establish the displaced population as agricultural producers, rural artisans etc. Such economic opportunities arise from building into the agricultural re-development package, based on project-funded activities such as land reclamation, irrigation schemes, agricultural intensification, tree crops development, fisheries, commercial or social forestry, vocational training, off-farm employment, and other kinds of lasting income-generating activities.

The resettlement plan for rural populations should start by establishing the basic indicator: the amount of land necessary on a productive basis. Even when land is available, however, alternative non-land-based strategies may be needed for some of those displaced. Such alternatives become imperative in situations of extreme land-scarcity. Opportunities in that case need to be opened up for those displaced to re-establish themselves in other sectors of the local or regional economy.

In **urban and semi-urban settlement**, those displaced usually depend on non-land-based sources of livelihood (*e.g.*, the service sector, industrial employment, self-employment etc.) but sometimes they may possess also some farming lands. In this situation approach should take into account, in addition to their housing their access to employment opportunities and when warranted to some land for farming or gardening. Depending upon local (rural, urban or semi-urban) circumstances, a combination of land-based and non-land-based strategy is required.

19.9.1 R&R Policy of the Project

The objectives of R&R policy should be to ensure that the population displaced by a project receives benefits from it. As we have already discussed, involuntary resettlement is an integral part of project design and therefore, should be dealt from the earliest stages of project preparation, taking into following policy consideration:

1. Involuntary resettlement should be avoided or minimised where feasible, exploring all viable alternative project designs.

2. When displacement is unavoidable, resettlement plans should be developed. All involuntary resettlement should be conceived and executed as development programmes. Displaced persons should be:

(a) Compensated for their losses at full replacement cost prior to actual move;

(b) Assisted with the move and supported during the transition period in the resettlement site; and

(c) Assisted in their efforts to improve their former living standards, income earning capacity, and production levels, or atleast to re-store them. Particular attention should be paid to the needs of the poorest groups to be re-settled.

3. Community participation in planning and implementing resettlement should be encouraged. Social and cultural institutions of re-settlers and their hosts (Host communities receiving resettlers) should be supported and used to the greatest extent possible.

4. Re-settlers should be integrated socially and economically into host communities so that adverse impacts on host communities are minimised. The best way of achieving this integration is for resettlement to be planned in areas benefitting from project and through consultation with the future hosts.

5. Land, housing, infrastructure, and other compensation should be provided to the adversely affected population, indigenous groups, ethnic minorities and pastoralists who may have customary rights to the land. The absence of legal title to land by such groups should not be a bar to compensation.

19.9.2 R&R Programme

R&R programme should include following activities:

1. *Physical Rehabilitation (Resettlement)*
   (a) Compensation payment.
   (b) Agriculture land allotment.
   (c) House plot.
2. *Socio-economic and Cultural Rehabilitation*
   - Providing resettlement grants, like payments/allotment of land to other major dependent members of the family, subsistence allowance for a certain period say one year, ex-gratia payments, insurance cover, transport expenses, payment for seed and fertilizer.
   - Preference in employment on the project.
   - People should be helped in visiting new site in advance for selection of the resettlement site.
   - Host and displaced person relationship should be developed.
   - People should be involved in the rehabilitation process.
   - Religious deities are to be shifted to the R & R sites and installation should be made in the similar environment, where possible.
   - Access of the tribals or forest dependent people to forests, wildlife and water vital to their sustenance and reproduction.
   - Special programmes for self-employment/income generation.
   - Civic amenities, like primary school, panchayat ghar, dispensary, drinking water, religious place, approach road, electrification, drainage, state transport stand may be provided depending upon population and existing facilities in host villages or in nearby areas.
3. *Ecological Rehabilitation*
   - Different plantation programmes, like agro-forestry, community forestry, home gardens, agricultural boundary side plantation should be given due emphasis at rehabilitation site.
   - Providing careful handling and sustainable use of natural resources and Common Property Resources (CPRs).
   - Running programme for Biogas/Smokeless *chulhas*.

19.10 REHABILITATION ACTION PLAN (RAP)

A Rehabilitation Action Plan (RAP), based on Social Impact Assessment (SIA) is prepared, which apart from covering mitigation plan for adverse impacts, also details out legal policies and regulations, institutional arrangement, monitoring plan, timeframe for implementation and budget for implementation.

These plans normally specify what social measures and institutional arrangements are to be incorporated into the project (resettlement project), how they will be carried out and monitored, and how social changes, that may occur as the project is implemented will be incorporated (*e.g.*, through dispute resolution mechanisms and adjustments and fine tuning in the implementation).

Elements Considered in Planning Resettlement

1. **Baseline Information and Studies Required for Planning**

A. *Information on the Departure (submergence) Area:*

- Census of population to be displaced.
- Description of production system(s) existing in the area and of the main social and ethnic groups.
- Inventory of public or common areas, infrastructure, and other productive or social resources, shared or divided.

B. *Information and Design for Arrival Areas (relocation sites):*

- Identification of new settlement sites.
- Survey of host population at/around new settlement areas and their resource use partners.
- Feasibility studies. It includes adequate water sources, soil capability, grazing lands, fuel-wood sources, site preparation requirements etc.

2. **Definition of Policy and Legal Framework**

- Defining resettlement policy and development policy indicating entitlement to replacement land and other productive resources, house re-construction, cash compensation, overall development objectives.
- Local definition of the rights of displaced population, resettlement grants or awards, institutional assistance for land replacement, technical and financial assistance for bringing new resources into production, rights of landless and those in non-agricultural sectors, legal grievance procedure.

3. **Organisation Capability for Resettlement**

- Designation of department/agency responsible for resettlement and for implementation of development plan.
- Expected assistance/co-operation required from departments, ministries etc.
- Strategy for securing the participation of the displaced and host communities.

- Training of staff engaged in implementation of R&R, training of representatives of re-settlers and host population.

4. **Resettlement Plan for the New Sites**
   - Development objectives for the displaced groups, alternative development strategies among which families may choose.
   - Development packages proposed for new site (agriculture intensifications, irrigation, land reclamation/preparation etc; livestock improvement plans, opportunities for women, fuel wood production, main farm models anticipated; expected benefits and likelihood of re-storing/improving farmer income/living standards, financial analysis of each package.).
   - Urban development, if necessary, permanent employment opportunities outside agriculture, development of small industries, services, investment required for creation of new jobs.
   - Compensation for lost property and hardship to resettlers, recommended uses of compensation for re-establishment of economic potential.
   - Habitat provisions; re-construction of houses, public and common installations, potable water system, drainage and water disposal system.
   - Support for the social organisation of re-settlers.
5. **Transfer Arrangements**
   - Information dissemination among the displaced and host people regarding policies, package, legal provisions, grievance procedure, schedule of displacement, transport means etc.
   - Maintenance arrangement during transfer period.
   - Mobilisation schedule.

19.11 LEGAL FRAMEWORK

Adequate provisions for protection of environment and forest are made in the Constitution of India. Article 47 provides for protection and improvement of health. Article 48A is directed towards protection and improvement of environment and protection of forest and wildlife. Article 51 (A) says, it is the duty of every citizen to protect and improve natural environment. Following the UN Conference on Human Environment (Stockholm, 1972), a constitutional amendment (42, 1976) inserted relevant provisions for environment protection in Constitution in Part IV– Directive Principles and Part IV-A– Fundamental Duties.

19.11.1 Environment Related Acts

In order to ensure sustainable development, the Government of India has enacted various Acts and Legislations. Prominent among these is the Environment (Protection) Act, 1986 through which the Government has acquired wide powers for protecting the environment. Some other Acts related to Water and Environment are: Water (Prevention and Control) of Pollution Act, 1974 (amended in 1988), Water (Prevention and Control) of Pollution (Cess) Act,

1977 (amended in 1991), Forest Conservation Act, 1980 and Environmental Impact Assessment Notification of 1994 (amended in 1997) and the Ministry of Environment & Forest's Notification of Jan. 1997, constituting the Central Ground Water Authority.

The Water (Prevention and Control of Pollution) Act, 1974 seeks to maintain or re-store "wholesomeness of Water" and Central and State Pollution Control Boards have been established under this Act. According to the Water Cess Act, 1997, boths, Central and State Governments have to provide funds to the Boards for implementing this act. The Forest Conservation Act, 1980 provides for compensatory afforestation to make up for the diversion of forest land to non-forest use. The Environment (Protection), 1986 was enacted in 1986 for the protection and improvement of human environment. The EIA Notification of 1994 has made the environmental clearance mandatory for all new projects and expansion/modernisation of existing projects covering 29 disciplines. Its amendment in 1997 has made it mandatory to hold environmental public hearing before according to environmental clearance. It was later superceded by EIA Notification, 2006 (already discussed).

Here, in this section we shall discuss the legal requirements for the land acquisition process and the Resettlement and Rehabilitation (R & R) policy arid the entitlements for affected families. The R & R policy of the project should be based on the National R & R policy (NRRP), 2007 and the experience gained over the years. It should fulfill the objectives of R & R policy as laid down earlier. This R & R policy of the project should recognise the need to support restoration of livelihoods of adversely affected peoples and lays down norms for rehabilitating the affected people and broadly outline the approach and institutional framework to achieve its objectives.

19.11.2 Legal Process of Land Acquisition

Land Acquisition Act, (LA Act), 1894 is followed for acquisition of private properties under the project. Acquisition under this law is a comprehensive process and involves issuance of various notification informing affected persons as well as general public regarding impending acquisition of private land for public purpose. The process is presented below:

Table : Land Acquisition Process for the Project

STAGE-I
Section-4 (1) • Publication of a preliminary notification by the Government that land in a particular locality is needed or may be needed for a public purpose or for a company.
STAGE-II
Section-6 (1) • Declaration of intended acquisition by Government. **Section-7 & 8** • Collector to take order from the Government for land acquisition and land to be marked out, measured and planned.

STAGE-III

Section-9

- Public notice and individual notices to persons interested to file their claims for compensation.

Section-11 & 12

- Enquiry into claims and declaration of final award by the Collector/Commissioner.

Section-16

- Taking possession of the land by the Collector.

Section-17

- Possession before award in case of emergency.
- Under this section, the Government has been given special powers to acquire land without making the award. This power can be used only in case of emergengy. The Collector can acquire the land 15 days after the publication of the notice under Section 9, but not before.

Although the process under the LA Act, is exhaustive some of the limitations of the Act are listed below:

1. The Act only deals with compensation and not rehabilitation of people, whose lands have been acquired.
2. The Act only considered payment of compensation to those individuals, who have legal ownerhip rights.
3. The Act does not specify any compensation to landless labourers, forestland cultivators, forest produce collectors, as they do not have any land record in the record of rights.

19.11.3 World Bank's Resettlement Policy

The basic elements of the Bank's resettlement policy are:

- Involuntary displacement should be avoided or minimized wherever feasible, because of its disruptive and impoverishing effects.
- Where displacement is unavoidable, the objective of Bank policy is to assist displaced persons in their efforts to improve, or at least re-store, former living standards and earning capacity. The means to achieve this objective consist of the preparation and execution by the borrower of resettlement plans as development programs. These resettlement plans are integral parts of project designs.
- Displaced persons should be: (*i*) compensated for their losses at replacement cost, (*ii*) given opportunities to share in project benefits, and (*iii*) assisted in the transfer and in the transition period at the relocation site.
- Moving people in groups can cushion disruptions. Minimizing the distance between departure and relocation sites can facilitate the re-settlers' adaptation to the new

socio-cultural and natural environments. The trade-offs between distance and economic opportunities must be balanced carefully.

- Resettlers' and hosts' participation in planning and implementation should be promoted. The existing social and cultural institutions of re-settlers and their hosts should be relied upon in conducting the transfer of re-establishment process.
- New communities of re-settlers should be designed as viable settlement systems equipped with infrastructure and services, able to be integrated in the regional socio-economic context.
- Host communities that receive resettlers should be assisted to overcome possible adverse social and environmental effects from increased population density.
- Indigenous people, ethnic minorities, pastoralists, and other groups that may have informal customary rights to the land or other resources taken for the project, must be provided with adequate land, infrastructure, and other compensation. The absence of legal title to land should not be grounds for denying such groups compensation and rehabilitation.

19.11.4 National Rehabilitation and Resettlement Policy (NRRP)-2007

Compulsory acquisition of land for public purpose, including infrastructure projects displaces people, forcing them to give up their home, assets and means of livelihood. In addition, displacement has other traumatic psychological and socio-cultural consequences. The Government of India recognizes the need to minimize large scale displacement to the extent possible and, where displacement is inevitable, the need to handle with utmost care and forethought issues relating to Resettlement and Rehabilitation (R & R) of Project Affected Families (PAFs).

Government of India has enacted a comprehensive rehabilitation and resettlement policy to safeguard the rights of involuntarily displaced population due to any development projects. The policy establishes the framework for extending rehabilitation assistance to PAFs, over and above the compensation for assets provided under LA Act and provided a broad canvas for an effective consultation between PAFs and the project authorities. The policy essentially addresses the need to support the rehabilitation efforts of the resource poor sections, namely, small and marginal farmers, SCs/STs and women who have been displaced. It imparts greater flexibility for interaction and negotiation so that the resultant package gains all-round acceptability in the shape of a workable instrument.

Objectives of the Policy

(i) To minimize displacement and to identify non-displacing or least-displacing alternatives;

(ii) To plan the R & R of PAFs, including special needs of tribals and vulnerable sections;

(iii) To provide better standard of living to PAFs; and

(iv) To facilitate harmonious relationship between the requiring body and PAFs through mutual co-operation.

Salient Features of National Policy on R & R

- The assistance proposed in the policy is minimum. States where R&R packages are higher than proposed in the policy are free to adopt their own package.
- Option is given to the affected families to take a lump-sum amount determined by the appropriate Government authority in lieu of the R&R benefits listed under the policy.
- The policy applies to projects displacing 500 families or more in plain areas and 250 families or more in hilly area, Desert & Drought Prone (DDP) Blocks and areas mentioned is Scheduled V and VI of the Constitution.
- The benefits under the policy would be applicable to all project affected families–both belonging to Below Poverty Line (BPL) and others. Poverty line changes from time to time.
- Allotment of agricultural land or cultivable wasteland to the extent of actual land loss subject to a maximum of one hectare of irrigated land or two hectares of un-irrigated land/cultivable wasteland subject to availability of Government land in the district.
- Allotment of house-sites free of cost to those Project Affected Families (PAF), whose houses have been acquired.
- One time grant of ₹ 25,0007- for house construction to families, Below Poverty Line (BPL) is recommended.

19.12 INSTITUTIONAL ARRANGEMENTS

For implementation of RAP timely establishment and involvement of appropriate institutions would significantly facilitate achievement of the objectives of R & R programme. For efficient implementation, it is essential to have skilled and trained staff for which there is need to have capacity building and training staff. The main R & R institutions include the following:

1. Land Acquisition Group

For land acquisition purposes, a land acquisition group should be established in the project. The group should be formed before initiating notification under section 4 of LA Act and should continue till the land acquisition process is completed and land mutated/leased in the name of the project. This group undertakes all the necessary survey and identifies the project locations and prepares the land plan schedules. This group interact with the State Government for all matters regarding land acquisition. This group is also responsible to ensure vacation/physical possession of the entire acquired land.

2. Administrator for R & R

In case of major projects, administrators for R & R are appointed by the State Government. The R & R works carried out under his directions and guidance. He is assisted by the officers and employees as the appropriate Government may provide. The roles and responsibilities of the Administrator include the following:

(i) Approval of award passed by land Acquisition Officers.

(ii) Oversee the disbursement of compensation for land acquired.

(iii) Hold consultations with PAPs and ensure the interest of adversely affected PAFs of vulnerable sections.

(iv) Verification and approval of PAFs eligible for various entitlements.

(v) Review the implementation of RAP and community development works.

3. Project Social Group

This group works under the Project Head and co-ordinates with the corporate office on matters related to R & R policy, budget and progress monitoring. The broad functions of the group include the following:

(i) Co-ordinate the implementation with the assistance of NGO.

(ii) Co-ordination and close interaction with the State Government authorities during the preparation and implementation of the RAP.

(iii) Conduct constant dialogue and regular meetings with the concerned State Authorities during the implementation.

(iv) Co-ordinate RAP implementation with the Village Development Advisory Committees (VDACs).

(v) Monthly review and monitoring the progress and incorporate corrective measures, if required.

4. Public Information Centre (PIC)

To maintain transparency and keep PAFs informed, PICs are set up under the project. These are very useful for the local community in general and project affected people in particular, as they act as first line of interaction with the project authorities for any information related to the project. They also act as an interface for other government agencies and non-government agencies for any information related to the project. They are also helpful in addressing the representations, complaints, grievances and suggestions of the public quickly.

5. Village Development Advisory Committee (VDAC)

A VDAC is constituted after initiating the land acquisition process through notification under section 4 of LA Act. The committee helps to ensure that the affected persons are meaning fully consulted and provided opportunities to participate in the planning, and implementation process of the R&R programme.

Some of the important roles and responsibilities of this committee include:

(i) Represent the village in consultation and participation process.

(ii) Identify the development activities to be taken up for the village in terms of improving the accessibility and availability of basic infrastructural facilities for the village such as roads, drinking water, health, education, sanitation etc.

(iii) Co-ordinate for identification of income generation schemes for the PAFs.

(iv) Co-ordinate and participate in the verification and upgradation of the RAP.

(v) Monitor the disbursement and utilisation of the R&R assistances.

(vi) Co-ordinate and assist in identification and relocation of common property resources.

(vii) Represent the grievances of the PAFs to the appropriate body for suitable remedial measures.

This committee should be supported by the NGO(s) and also the Social Workers engaged under the Project Social Group by way of giving them guidance in preparation of their development plans and others.

6. NGO(s)

For the efficient and smooth implementation of the RAP, a local NGO should be engaged, who will implement the RAP in co-ordination with the project social group and VDAC. The NGO should help in implementing various components of the RAP, particularly the consultation process, verification of the PAPs, disbursement of compensation and assistance, counsel for the productive use of compensation and assistance amounts, monitor the use of compensation and assistance amounts.

19.13 TRAINING AND CAPACITY BUILDING

Though revenue department does the land acquisition, all the preparation of land acquisition plans are undertaken by the land acquisition group of the project and it requires a constant follow up on various related issues. In addition, other R&R works, like the identification and verification of PAFs, issuing of identity cards, development of resettlement sites, if any, the proponent staff is responsible. Therefore, to enhance capabilities, the project staff should be sent on exposure visits to other projects as well as training workshops in R & R.

Some of the courses of R&R are:

(i) Post-graduate diploma in Participatory Management of Displacement, Resettlement and Rehabilitation.

(ii) Capacity Building for R&R.

(iii) R&R in Development Projects.

Some of the training programmes which can be conducted are:

(a) *Overview of social issues.* Social issues, methodology followed for SIA, Entitlement matrix and detailed R & R policy.

(b) *Land Acquisition.* Legal and operational issues and LA Act-1894.

(c) *Social Impact Assessment* (SIA).

(d) *Public consultation in R&R.*

(e) *Rehabilitation.* Issues in economic rehabilitation.

(f) *Resettlement issues.*

(g) *Monitoring and Evaluation of R&R.*

19.14 GRIEVANCE REDRESSAL MECHANISM

Various provisions under LA Act enable the affected persons to represent their cases to Land Acquisition Officer or even refer to court for redress and seek higher compensation. However, as this process is time consuming and does not cover the issues related to rehabilitation assistance, which result in delay of implementation of projects. Therefore, a grievance redressal system should be adopted in the R&R policy of the project.

This system should be simple. Aggrieved person when not satisfied should first approach the local VDAC constituted under each village under the project. If aggrieved person is not satisfied by action taken by the VDAC, he can represent his case to the Project Social Group. In case, if he is still not satisfied, he can represent his case to the head of the project. The concerned NGO engaged for RAP implementation should assist the aggrieved person in representing his case.

However, it is suggested that a separate Grievance Redressal Cell may be set-up which should be headed by a person of repute from the district *viz.*, a retired District Judge or a retired Collector or a Principal of a college etc.

QUESTIONS

1. What is Resettlement and Rehabilitation? Explain.
2. What do you understand by "Involuntary Resettlement"?
3. Displacement of persons due to construction of projects results in a conflict between national long term interests with the interests of immediately adversely affected people.
4. "Resettlement should have development oriented approach." Discuss this statement.
5. Discuss strategies for developing a R&R package.
6. Explain various objectives of a good R&R policy.
7. With reference to R&R, write short notes on:

   (a) Participation of re-settlers (b) Host population

   (c) Preparation and appraisal of 'Resettlement Project'.
8. What is rehabilitation action-plan? Describe as to how these should be drawn?
9. Discuss the various elements of R&R costs.
10. Describe elements included in good R&R programme.
11. What measures would you take for R&R, considering socio-economic and cultural aspects?
12. What activities a good R&R programme should include?
13. Discuss land related issues which are important for effective resettlement of displaced persons.
14. Explain basic elements of World Bank's Resettlement Policy.
15. Explain the importance of National Policy on R&R. What are its objectives? Explain its salient features.

16. Discuss Social Impact Assessment for project affected families.
17. Discuss methodology for SIA study.
18. Explain impacts of land acquisition.
19. Impact of displacement on vulnerable groups.
20. Discuss about the importance of stakeholders participation and consultation for framing an effective R&R programme.
21. Explain various socio-economic issues which need considerations while planning R&R package.
22. Explain as to how a "Rehabilitation Action Plan" is prepared.
23. What institutional arrangements are required for proper handling the R&R issues of an infrastructural project.
24. Write short notes on:
    (a) Training and capacity building for R&R.
    (b) Grievance redressal mechanism.
    (c) R&R policy of the project.

PART VI : ENVIRONMENTAL PROTECTION

20

Chapter

Environmental Protection

20.1 INTRODUCTION

Increasing population, increasing demands of each person, decreasing raw material resources and increase in waste are serious threat to survival of human race. How the water, energy and several other resources are destroyed, maintained, dissipated or conserved, depleted or restored has significant implications for human health. It is the legitimate requirement of a technological process to meet the needs of the society without harming ourselves.

It is expected that population of the earth will rise to 8 billion by 2025. If this population is not controlled, the earth would succumb to overload. Green area reserves need to be maintained to safeguard against complete exhaustion of the natural elements (soil, atmosphere and water).

The industrial and other developmental activities are increasing in number and also the pollution. The vegetation area on the earth is being on the decrease due to population growth and associated developmental activities which is a threat to our environment. Therefore, steps need to be taken urgently to restore the quality of environment, for waste management and control, developing technology for pollution abatement.

The **Environmental Protection** is necessary to protect the people from physiological damage, annoyance, irritation, discomfort possible from offensive conditions of polluted water, air and land; provision of environment which adds to comfort, pleasure and better productivity; and to safeguard the balances in the earth's ecosystems and to conserve natural resources. Environment protection measures will pay back in terms of money, economy, productivity, social justice and cleaner environment. To make this dream true, common man, workers, technical persons, scientists, government authorities, legislators, planners, economists all have an important role to play.

20.2 INTERNATIONAL EFFORTS (CONVENTIONS/SUMMITS) ON ENVIRONMENTAL PROTECTION

Important conventions/summits held on environment protection are mentioned hereunder:

1. **Ramasar Convention (1971).** It was signed in the Iranian city of Ramsar and is one of the oldest Inter-governmental accord for preserving the ecological character of wetlands.
2. **World Climate Conference (1979).** The first world climate conference was held in Geneva from 12th to 23rd February, 1979 and a declaration was issued to appeal the nations to take full advantage of the available knowledge of climate, improve the knowledge significantly, to foresee and prevent potential anthropogenic changes in climate that might be adverse to the well-being of humanity.
3. **Toronto Conference on Changing Atmosphere (1988).** This conference was organised in 1988 by the Canadian Government in response to the Brutland Commission's call for action on climate change in order to improve awareness about chemical changes in the atmosphere and global community's ability to mitigate critical anthropogenic atmospheric changes which influence the human and natural resources. The conference made a consensus call for: (a) reduction of CO_2 emissions by 20% of 1988 levels by the year 2005 and by 50% as soon as feasible, (b) adoption of a target of 10% improvement in energy efficiency by 2005, (c) measures to promote inter-governmental co-operation and public awareness.
4. **The Earth Summit (1992).** A United Nations Conference on Environment and Development held at Rio de Janerio in 1992 is called The *Earth Summit,* as it was attended by the top experts and government bosses in the field of environment of 178 countries. At this conference, principle for sustainable development was adopted. A **convention on climate change** was also signed by the industrial nations who committed themselves to cut down emissions of CO_2 and other green house gases. **Convention on biodiversity** was also signed at this conference. The document issued at the Earth Summit, containing guidelines on strategies for pollution control to disposal of hazardous wastes, is called **Agenda 21.**
5. **Kyoto Protocol (The Earth Summit + 5).** It was held in Kyoto (Japan) in Dec., 1997. About 160 countries participated at the United Nations Framework Convention on Climate Change (UNFCCC). In this convention, it was agreed to reduce green house gas emission by 5% from 1990 level over the first commitment period 2008-2012. The protocol allows for achieving commitment of industrialised countries for emission reduction in flexible manner by collaborating through bubbling, joint implementation, emission trading and Clean Development Mechanism (CDM).

   The conference was convened to assess the progress made in five years since the Earth Summit of 1992, hence, it was also called as "The Earth Summit+5." This Summit was a step further in the direction of environmental protection, resources conservation and sustainable development and it carried out a detailed assessment of progress achieved since The Earth Summit, 1992.

6. Convention for the **Prevention of the Ozone layer-Vienna, 1985.** This was held to protect the human health and environment against adverse effects from human activity.
7. **Framework Convention on Climate Change, New York, 1992.** To achieve stabilisation of green house gas concentration in the atmosphere at a level that would prevent dangerous anthropogenic interference with the climate system.
8. Convention on the **Protection and Use of Transborder Water Courses and International Lake, Helsinki, 1992.** To protect and manage in an ecological sound manner, transboundary surface waters and ground waters. To prevent, control and reduce water pollution.
9. **U.N. Convention to Combat Desertification** in countries experiencing serious drought and Desertification, 1994.
10. **World Summit on Sustainable Development, Johannesburg (South Africa), 2002.**
11. **Bali Summit on Climate Change (2007).** The conference was held in Bali, Indonesia. It was attended by 180 countries. The summit resolved "deep cuts in global emissions". It also discussed achievements of earlier protocols.
12. **Paris Agreement (2017).** The Paris Agreement sets out a global framework to avoid dangerous climate change by limiting global warming to well below 2ºC and pursuing efforts to limit it to 1.5ºC. It was signed by 194 countries and ratified by 188 in 2020.

20.3 GOVERNMENT EFFORTS ON ENVIRONMENTAL PROTECTION

Various important policies, acts, rules, notifications, agreements on Environment Legislation and Indian Standards are listed below:

1. **Policies.**
   - Policy statement on Abatement on Pollution (1992).
   - National Conservation Strategy and Policy Statement on Environmental and Development (1992).
   - National Forest Policy (1998).
   - Wildlife Conservation Strategy (2002).
   - National Environment Policy (2004).
2. **Environment Acts.**
   - The Indian Forest Act, 1927.
   - The Water (Prevention and Control of Pollution) Act, 1974.
   - The Forest (Conservation) Act, 1980.

- The Air (Prevention and Control of Pollution) Act, 1981.
- The Environment (Protection) Act, 1986.
- The Motor Vehicle Act, 1988.
- The Wildlife (Protection) Act, 1972.
- The Biological Diversity Act, 2002.
- The National Environment Tribunal Act, 1975.
- The Insecticides Act, 1968.

3. **Environment Rules.**
   - The Environment (Protection) Rules, 1986.
   - The Hazardous Wastes (Management and Handling) Rules, 1989 & 2003.
   - The Forest (Conservation) Rules, 1981.
   - The Bio-Medical Waste (Management and Handling) Rules, 1998.
   - The Municipal Solid Wastes (Management and Handling) Rules, 2000.
   - The Noise Pollution (Regulation and Control) Rules, 2000.
   - The Air (Prevention and Control of Pollution) Rules, 1982.
   - The Water (Prevention and Control of Pollution) Rules, 1975.
   - The Biological Diversity Rules, 2004.
   - The Wild Life (Protection) Rules, 1995.
4. **Environment Notification.**
   - Environment Impact Assessment Notification-2006 of MOEF (Ministry of Environment and Forests), Govt. of India.
   - Fly Ash Notification.
5. **Guidelines.**
   - Guidelines for River Valley Projects, MOEF (1981), Govt. of India.
   - Guidelines for Sustainable Water Resources Development and Management, 1992, Central Water Commission.
   - National Guidance Manual on Environmental Impact Assessment (2003), NEERI, on behalf of MOEF.
6. **Internal Agreements to which India is a Signatory.**
   - The Convention on Wetlands of International Importance (The Ram Sar Convention), 1991.
   - The Convention on International Trade in Endangered Species of flora and fauna (CITES), 1973.
   - The Framework Convention on Climate Change, 1992.
   - The Convention for Conservation of Biological Resources, 1992.
   - The Rio Declaration on Environment and Development and the Agenda 21, 1992.

20.4 ENVIRONMENTAL ORGANISATIONS/INSTITUTIONS

There are large number of Institutions/Organisations that deals with environmental protection and conservation. Here, some of the well-known organisations are mentioned.

20.4.1 International Organisations

1. **Earthscan.** An agency founded by UNEP in 1976 that commissions original articles on environmental matters.
2. **Environmental Protection Agency (EPA).** This is an Independent Federal Agency of the US Government and deals with the protection of environment by Air, Water, Solid Wastes, Radiation, Pesticides, Noise etc.
3. International Union for Conservation of Nature and Natural Resources (IUCN).
4. United Nations Educational, Scientific and Cultural Organisation (UNESCO).
5. United Nations Environment Programme (UNEP).
6. World Commission on Environment and Development (WCED).
7. **The Coalition Against Wildlife Trafficking** (CAWT) aims to focus public and political attention and resources on ending the illegal trade in wildlife products.
8. **The International Tropical Timber Organisation (ITTO)** is an intergovernmental organisation under UN promoting the conservation and sustainable management, use and trade of tropical forest resources.
9. **United Nations Forum on Forests (UNFF).** The Economics and Social Council of the United Nations established this UNFF in the year 2000, with the objective to promote "the management, conservation and sustainable development of all types of forests."
10. **The International Union for Conservation of Nature and Natural Resources** has its mission 'to influence, encourage and assist societies throughout the world to conserve integrity and diversity of nature and to ensure that any use of natural resources is equitable, and ecologically sustainable.' Its priority areas are : Biodiversity, Climate change, Sustainable energy, Human well being, and Green economy.
11. **Global Tiger Forum.** Formed in 1994, with its secretariat at New Delhi, is intergovernmental and international body compaigning to save tigers worldwide.

20.4.2 National Organisations

1. Advisory Board on Energy (ABE).
2. Animal Welfare Board.
3. Botanical Survey of India (BSI).
4. Central Arid Zone Research Institute.
5. Central Pollution Control Board (CPCB).
6. Central Zoo Authority.
7. Centre for Mining Environment, Dhanbad.

8. Centre for Science and Environment (CSE), New Delhi.
9. Compensatory Afforestation Fund Management and Planning Authority (CAMPA).
10. Environmental Protection Council.
11. Forest Survey of India (FSI).
12. Geological Survey of India (GSI).
13. Indian Council of Forestry Research and Education, Dehradun.
14. Indian Forest Academy, Dehradun.
15. Institute of Forest Management, Bhopal.
16. Mining of Environment and Forest (MoEF).
17. National Afforestation and Eco-development Board.
18. National Bamboo Mission.
19. National Clean Energy Fund.
20. National Committee on Environmental Planning and Co-ordination.
21. National Environmental Engineering Research Institute (NEERI).
22. National Ganga River Basin Authority (NGRBA).
23. National Lake Conservation Plan (NLCP).
24. State Pollution Control Boards (SPCB).
25. Tata Energy Research Institute (TERI).
26. The National Biodiversity Authority.
27. Wildlife Crime Control Bureau (WCCB).
28. Wildlife Institute of India (WII), Dehradun.
29. Zoological Survey of India (ZSI).

20.5 PUBLIC AWARENESS

In previous chapters, we have discussed the awareness about the ethics and traditional ethical awareness. Here, in this section we are discussing about the public awareness with regards to environment. Upto some decades ago, for a common man, environment means public health, sanitation and tree plantation. Today, the subject has assumed a great significance and has widened covering a large number of subjects, as covered in the book.

But due to human activities, the earth's natural resources are rapidly depleting and our environment is being increasingly degraded as seen in previous chapters. The prevention of environmental degradation must become a part of our lives. This is possible only through public awareness. Mass media should help in influencing public opinion. Some NGOs are working hard towards environmental protection. The Government and its various departments/ organisations are also working for creating interest in environmental protection and conservation of nature and natural resources.

For the purpose of creating public awareness following environmentally significant days are observed:

Table : Environmental Calender

Date and Month		Event
02 February	—	World Wetlands Day
28 February	—	National Science Day
8 March	—	International Women's Day
21 March	—	World Forestry Day
22 March	—	World Water Day
22 April	—	Earth Day
22 May	—	International Biodiversity Day
05 June	—	World Environment Day
17 June	—	World Day to Combat Desertification and Drought
1–7 July	—	Van Mahotsava Week
11 July	—	World Population Day
16 September	—	Ozone Layer Preservation Day
22 September	—	Car Free Day
October (First Monday)	—	World Habitat Day
1–7 October	—	Wild Life Week
4 October	—	Animal Welfare Day
14 November	—	National Children's Day
21 November	—	World Fisheries Day
01 December	—	World AIDS Day
02 December	—	Bhopal Tragedy Day
11 December	—	International Mountain Day
14 December	—	National Energy Conservation Day
23 December	—	Farmers Day

During public awareness, the public has to be educated about the fact that by degrading our environment we are actually harming our ownself. In this context, education, special attention must be paid to school going children and women.

The main **objectives of environmental awareness programmes** should be:

(i) improving the quality of environment,

(ii) creating an awareness among people on environmental problems and conservation, and

(iii) creating such an atmosphere so that people are fit enough to participate in decision-making.

Various laws on environmental issues are existing in our country, but their implementation will be effective only with public support and participation for which an awareness and clear understanding of key environmental problems is most essential.

For developing an environmental sensitivity can only grow through major public awareness compaigns, for which major tools are:

Electronic media, press, school and college education, and adult education.

20.6 ENVIRONMENTAL EDUCATION AND TRAINING

While discussing 'Value Education' in the chapter on 'Human Population and the Environment' we have studied the education system related to values in respect of environment. Here in this section, we shall discuss about the environmental education system.

Environmental education is a multi-disciplinary approach to the study of problems related to maintaining a livable earth. The main aim of environmental education is to make everyone vigilant and aware about environmental problems.

The lay public in rural, tribal, slim and urban areas, women, students and teachers in schools, colleges and universities as well as planners and decision-makers, and R&D workers need to be educated about environment.

Radio, television, newspapers, books, cinema, posters, street plays etc. are the most effective means of environmental education. Many social and scientific organisations are doing a good work to generate interest and awareness in environmental issues among rural and urban population.

Objectives of Environmental Education

1. Awareness and sensitivity to the total environment and its allied problems.
2. Knowledge and experience of environment and its associated problems.
3. Attitude *i.e.* acquire a set of values and feelings of concern for the environment.
4. Skill for identifying and solving environmental problems.
5. Ability to evaluate environmental measures.
6. Provide opportunity for active participation towards resolution of environmental problems.

Environmental Education in India

1. Formal Environmental Education

Its main aim is to:

(i) improve quality of environment,

(ii) create an awareness on environmental problems and conservation,

(iii) create an atmosphere so that people participate in decision-making and develop the capability to evaluate the development programmes.

Stages of education levels

(a) Primary school stage,

(b) Lower secondary stage,

(c) Higher secondary stage,

(d) College stage,

(e) University education at postgraduate level:

   (i) Environmental engineering

   (ii) Conservation and management

   (iii) Environmental health

   (iv) Social ecology

2. Non-formal Environmental Education

Groups may be constituted as under:

1. Adult education
2. Rural youth
3. Children activities
4. Tribals/forest dwellers
5. NGOs
6. Public representatives
7. Training senior executives/administrators
8. Foundation course for probationers
9. Research and development programmes.

20.7 GREEN BUILDINGS (SUSTAINABLE BUILDINGS)

Constructing and operating buildings require enormous amounts of energy, water and materials and creates large amounts of waste. Where and how they are built, affects the ecosystem around us in several ways, and the buildings themselves create new indoor environment that present new environmental problems and challenges. The green building leads the way to reduce the environmental impact.

A 'green building' is a building which adopts the best practices of environmental technology for optimizing the use of natural resources for providing the clean and cost effective working atmosphere. The "Green building" also called "sustainable building" or "high performance building" is a building having a significantly reduced impact on the earth's resources compared to conventional building. It is a building that is healthier and more comfortable for its occupants consequently enhancing productivity. It is a structure that is built or used in an ecological and resource efficient manner and reduce the overall impact to the environment.

Although new technologies are constantly being developed to complement current practices in creating greener structures, the common objective is that green buildings are designed to reduce the overall impact of the built environment on human and the natural environment by:

- Efficiently using energy, water, and other resources.
- Protecting occupant health and improving employee productivity.
- Reducing waste, pollution and environmental degradation.

General principles adopted for the green building are to:

(i) Minimise the consumption of materials and maximise their re-use and also by using re-cyclable materials.

(ii) Minimise energy consumption and green house gas emissions.

(iii) Protect the natural environment. Utilizing minimum natural resources and preserving them for future generations.

(iv) Minimize the consumption of mains water and maximise recycling of treated waste water.

(v) Adopt environmentally sound and healthy best practices, during construction and occupancy.

(vi) Creating a quality building that is commercial viable.

(vii) Minimise pollution and damage to environment.

(viii) Protect and restore local air, water, soils, flora and fauna.

Most green buildings are high-quality buildings, they last longer, costless to operate and maintain and provide greater occupants satisfaction.

Reducing Environmental Impact

Green building practices aim to reduce the environmental impact of buildings. Buildings account for a large amount of landuse, energy and water consumption, and air and atmosphere alteration. Considering the statistics, reducing the amount of natural resources and the amount of pollution given off is seen as crucial for future sustainability.

Following are the **elements of green building design,** which help in reducing the environmental impact:

1. **Siting and Orientation.** Select a site to take advantage of natural light, water, sun heat, mass transit system, existing landscape etc. Select plants that have low water and pesticide needs, and use compost and mulches.
2. **Energy Efficiency.** Use a rooftop solar array for generating power. Use energy efficient appliances. Install high efficiency lighting systems, with advanced lighting controls. Use a properly sized and energy-efficient heat/cooling system in conjunction with a thermally efficient building shell. Minimize the electric loads from lighting, equipment and appliances. Consider alternative energy sources.

3. **Material Efficiency.** Select sustainable construction materials and products. Use dimensional planning and other materials efficiency strategies so as to reduce the amount of building material needed and cut construction costs. Re-use and recycle construction and demolish materials. Adopt construction waste management techniques.
4. **Water Efficiency.** Design for dual plumbing to use recycled water for toilet flushing or a grey water system that recovers rainwater or non-potable water for site irrigation. Minimize waste water by using ultra low-flush toilets, low-flow shower heads and other water conserving fixtures. Install point-of-use hot water heating systems for more distant locations. Use state-of-the-art irrigation controllers and self-closing nozzles on hoses.
5. **Occupant Health and Safety.** Good overall environment reduces the respiratory disease, allergy, asthma and enhance work performance. Many building materials and cleaning/maintenance product emit toxic gases, therefore, it is advisable to choose construction materials and interior finish products with zero or low emissions to improve indoor air quality. Provide adequate ventilation and a high-efficiency, in-duct filtration system. Install adequate ventilation in bathrooms, allow proper drainage including that from the roof, and design other building system to control humidity.
6. **Ventilation and Air Quality.** Ventilation is necessary to control the concentration of contaminants that may produce objectionable odours or toxic effects. Ventilation with untreated outdoor air is commonly used in non-air-conditioned building to reduce discomfort from over□heating in summer and to prevent excessive relative humidities in winter. It is also used to improve the occupational environments and to discharge heat, moisture and contaminants arising from industrial processes.
7. **Optimize** following inter-related aspects:
   (i) Maximize the availability of natural light.
   (ii) Maximize the use of natural ventilation, instead of blocking it and providing air-conditioning.
   (iii) Maximize the potential productivity of its inhabitants keeping acceptable temperature and airflow to have a reasonable level of comfort.

   By managing the demand for lighting and air-conditioning, the size of the systems required and the extent to which they need to be operated can be reduced.
8. **Environmental Aspects.** Minimize internal as well as .external pollution and other environmental damages. Conserve non-renewable energy and scarce materials. Use renewable energy and materials that are sustainably harvested.

   Green building brings together a vast array of practices and techniques to reduce and ultimately eliminate the impacts of buildings on the environment and human health. It often emphasizes taking advantage of renewable resources, *e.g.*, using sunlight through passive solar, active solar, and photo voltaic techniques and using plants and trees through green roofs, rain gardens, and for reduction of rainwater run-off. Many other techniques, such as using packed gravel or permeable concrete instead of conventional concrete or asphalt to enhance replenishment of groundwater, are used as well.

While the practices, or technologies, employed in green building are constantly evolving and may differ from region to region, there are fundamental principles that persist from which the method is derived; Siting and Structure Design Efficiency, Energy Efficiency, Water Efficiency, Materials Efficiency, Indoor Environmental Quality Enhancement, Operations and Maintenance Optimization, and Waste and Toxic Reduction. The essence of green building is an optimization of one or more of these principles. Also, with the proper synergistic design, individual green building technologies may work together to produce a greater cumulative effect.

On the aesthetic side of green architecture of sustainable design is the philosophy of designing a building that is in harmony with the natural features and resources surrounding the site. There are several key steps in designing sustainable buildings: specify 'green' building materials from local sources, reduce loads, optimize systems, and generate on-site renewable energy.

Energy Conservation Building Code (ECBC)

- ECBC set minimum energy efficiency standards for design and construction
- ECBC encourage energy efficient design or retrofit of buildings so that:
  — It does not constrain the building function, comfort, health, or the productivity of the occupants
  — Has appropriate regard for economic considerations (lifecycle costs *i.e.,* construction + energy costs are minimized)

Bureau of Energy Efficiency (BEE)

Energy conservation Act, 2001 has created BEE as the nodal agency at the center and State Designated Agencies (SDAs) at the state level to implement the Act. The Act provides that:

(i) BEE would take suitable steps to prescribe guidelines for energy conservation building codes.

(ii) Central Government can prescribe conservation building codes, and direct owners/ occupiers to comply with them.

(iii) State Government can modify the code in response to local climate conditions.

BEE was set up in March, 2002 under the provisions of Energy Conservation Act of 2001 to provide a legal framework for the government's energy efficiency initiatives in the country.

20.8 CLEAN DEVELOPMENT MECHANISM (CDM)

As per global agreement under **Kyoto Protocol** adopted during the conference of the parties to the UN Framework Convention on Climate Change (UNFCCC) in Dec, 1997, the

Developed Countries (known as Annexure 1 countries) committed to reduce the Green House Gases (GHG) by 5.2% as compared to the level in 1990. Kyoto Protocol as on 23rd March, 2005 has been ratified by 146 countries including 37 Annexure I countries accounting for 61.6% of 1990 level CO_2 emission. Different countries have different targets for the 5 year period of 2008-2012 (first commitment period). For example, EU countries commit to reduce their emission by 8%, USA by 7% and Japan by 6% form base-year emission in 1990. It is estimated that around 2 to 2.5 billion tonnes of CO_2 emission reduction would be needed to meet Kyoto Protocol Targets by 2012.

The flexible Clean Development Mechanism (CDM) that has been built into the protocol is two-fold, *viz:* (*i*) it is a case of Joint Implementation (JI), where the industrialised "Annexure 1" countries can trade emission with each other; and (*ii*) there is a provision of CDM, whereby the industrialized nations can claim Certified Emission Reductions (CER) from developing countries. The provision of CDM, therefore, encourages the industrialised nations to transfer technology to the developing countries that would reduce emissions.

To explain as to how this is possible under CDM, let us consider an industrialised national 'A', who finds that it is not able to meet its emission reduction target. It then invests in a project in a developing country 'B', in a way that the project would reduce carbon emissions to a greater degree than in the normal course, in country 'B'. This means that a high emission establishment working in country 'B', would be replaced by a low emission establishment with the financial and technological assistance provided by the industrialised country 'A'. Carbons saved in this manner can be claimed by country 'A', as its own emission reduction.

Thus, the CDM gives the developing country an opportunity to attract Foreign Direct Investments (FDIs) for replacement of its old CO_2 emitting plants with low emission-high technology plants. India must, therefore, grasp the opportunities that have become available to it under the CDM incentives.

The Kyoto Protocol has recognised following 6 main Green House Gases (GHG), which are eligible to be claimed for benefits under CDM:

1. Carbon dioxide (CO_2).
2. Methane (CH_4).
3. Nitrous oxide (N_2O).
4. Hydrofluoro carbons (HFCs).
5. Perfluoro carbons (PFCs).
6. Sulphur hexafluoride (SF_6).

Certified Emission Reduction (CER). One tonne of CO_2 reduced through a CDM project when verified by a designate operational entity and issued by the Executive Board (EB) of the UNFCC is known as a CER, which is tradable in the carbon market.

20.9 CARBON CREDITS

A carbon credit is a commercial term used to assign a value to a metric tonne of green house gas emissions. When you purchase a carbon credit from a third party, you are helping to fund a project that will either reduce or sequester a metric tonne of carbon emission from atmosphere in another location. As a result of purchasing a carbon credit from a third party, you are effectively 'offsetting' some or all of your own carbon emissions.

Carbon credits are a key components of national and international emissions trading schemes that have been implemented to mitigate global warming. The credits can be exchanged between businesses or bought and sold in international markets at the prevailing market price. A 'carbon credit', sometimes referred to as an offset, is a permit to emit a specified amount of GHG, usually expressed in metric tonnes of carbon dioxide equivalent (CO_2e).

Carbon dioxide equivalent (CO_2e) is a term used to compare global warming potential of various GHGs. For example, the contribution of methane to global warming is rated as 21, which means that an emission of one metric tonne of methane is equal to emission of 21 tonnes of carbon dioxide.

The Kyoto Protocol allows for carbon credits to be earned for sequestration projects, such as the planting of trees. For example, forestry companies that replant deforested areas could potentially earn offset credits, as forests absorb carbon dioxide, as they grow. A key issue in this area is that the projects must represent an actual reduction of emissions compared to "business as usual."

Carbon Offsetting

Carbon offsets are credits for reductions in greenhouse gas emissions made at another location, such as wind farms which create renewable energy and reduce the need for fossil fuel powered energy. Carbon offsets are quantified and sold in metric tonnes of carbon dioxide equivalent.

Buying one tonne of carbon offsets means there will be one less tonne of carbon dioxide in the atmosphere than there would otherwise have been. The carbon offsetting is the fastest way to achieve the deepest reductions within businesses and also delivers added benefits at the project site.

For a carbon offset to be credible, it must meet essential quality criteria, including proof that it is additional.

QUESTIONS

1. What is environmental protection? Why it is necessary? Explain.
2. Explain the various efforts made at international level to protect the environment.
3. Explain the different efforts made by the Government at national level to protect the environment.
4. Name different important organisations/institutions working for environment protection at national and international levels.
5. What is the necessity for public awareness on environment? What are the steps to be taken to generate public awareness on it?
6. Mention the names of five important days of environmental significance.

7. Describe the importance and objectives of environmental education.
8. Discuss the possible levels of environmental education in the country.
9. Give an account of organisations and agencies engaged in management of environmental issues.
10. What do you understand by Green Buildings? Discuss various elements of green building design which help in reducing the environmental impacts.
11. Write short notes on:
    (a) Energy Conservation Building Code.
    (b) Bureau of Energy Efficiency.
12. (a) What is "Clean Development Mechanism"?
    (b) Explain 'Carbon Credit' system.
13. Write short notes on :
    (a) Coalition against Wildlife Trafficking.
    (b) Carbon Offsetting.
    (c) Carbon Credits.

21

Chapter

ENVIRONMENTAL LEGISLATIONS

21.1 ENVIRONMENTAL LEGAL FRAMEWORK

As discussed earlier, development and environment conservation should go hand in hand. Environmental legislations/laws/regulations of a country for planning and execution of projects determine the standards for location and operation of these projects that must be respected and the government authorisations/approvals/permits that must be obtained before the works of project begin or during the various phases of execution or operation of projects.

Before according to environmental and forest clearances to projects, MOEF requires various environmental management plans, which include Rehabilitation and Re-settlement Plan, Catchment Area Treatment Plan, Compensatory Afforestation Plan and collection of data of flora, fauna, forest, wildlife protection and rehabilitation etc.

Attention was focussed on environmental issues during 1972 UN Conference held at Stockholm, which brought an awakening towards environmental impacts of various development activities. Various environmental impacts of these projects were noticed, which included the degradation of land, effect on flora and fauna and displacement of people.

Provisions in Constitution of India

Adequate provisions for protection of environment and forest are made in the Constitution of India. Article 47 provides for protection and improvement of environment and protection of forest and wildlife. Article 51(A) states that it is the duty of every citizen to protect and improve natural environment. Following the UN conference on Human Environment (Stockholm, 1972), a constitution amendment (H_2 of 1970) inserted relevant provisions for environment protection in constitution in Part IV—Directive Principle and Part IVA—Fundamental Duties.

Enactment of Environmental Acts

In order to ensure sustainable development of water resources, prevention and control of pollution and conservation of forests, wildlife, biodiversity and natural ecosystems, the

Government of India have enacted various Acts and legislations. Ministry of Environment and Forest (MoEF) is the model ministry for formulation and implementation of the policies relating the protection and conservation of the environment.

Major legislations directly dealing with the protection of environment are:

1. The Wildlife (Protection) Act, 1972.
2. The Forest (Conservation) Act, 1980.
3. The Water (Prevention and Control of Pollution) Act, 1974.
5. The National Environment Appellate Authority Act, 1977.
6. The Air (Prevention and Control of Pollution) Act, 1981.
7. The Environment (Protection) Act, 1986.
8. The Public Liability Insurance Act, 1991.
9. The National Environment Tribunal Act, 1995.
10. The Biological Diversity Act, 2002.
11. The Hazardous Wastes (Management and Handling) Amendment Rules, 2003.
12. The Indian Fisheries Act, 1897.
13. The Factories Act, 1948.
14. The Ancient Monuments and Archaeological Sites and Remains Act, 1958.
15. The EIA Notification, September 2006.

21.2 ENVIRONMENT (PROTECTION) ACT, 1986

Concern over the state of environment has grown the world over since the 'sixties. The decline in environmental quality has been evidenced and biodiversity, excessive concentration of harmful chemicals in the ambient atmosphere risks of environmental accidents and threats to life support system.

With the international obligations cast on the member countries from the declaration adopted by the U.N. Conference on Human Environment and subsequent monitoring done by the U.N. Environment Programme (UNEP), the necessity of exercising an effective Central Control on preventing the environmental pollution become more and more important after the conference decisions in 1972.

The India enacted water (Prevention and Control of Pollution) Act, 1974, and the Air (Prevention and Control of Pollution) Act, 1981, but these acts were not considered sufficient for protection of the environmental, as these Acts were related to only water and air pollution. There was a need for an Act to check and prevent other kinds of pollution, such as those caused by the hazardous industrial wastes and chemicals, biomedical wastes, plastic wastes, municipal solid wastes etc. and to control noise pollution etc.

The **objective of this Act are:** The improvement of environment and the prevention of hazards to human beings, other living creatures, plants and property. This Act supplements and reinforces the water, and Air Pollution Control Acts.

The **scope of this Act** is very vast, since the term "environment" under this Act, includes water, air, land and inter-relationships which exist among these; and human beings and other living creatures, plants, micro-organisms and property.

Other important terms as defined in the Act are:

1. **"Environment Pollution"** means any solid, liquid or gaseous substance present in such concentration as may be, or tend to be, injurious to environment.
2. **"Hazardous Substance"** means any substance or preparation which, by reason of its chemical or physico-chemical properties or handling is liable to cause harm to human beings, other living creatures, plants, micro-organisms, property or the environment.

Other Important Provisions of the Act

Section 3(1) of the Act empowers the Central Government to take all such measures, as it deems necessary or expedient for the purpose of protecting and improving the quality of the environment, and preventing, controlling and abetting environmental pollution.

Section 5 of the Act gives authority to the central Government to issue direct written orders including order to close, prohibit or regulate any industry, operation, or process, or to stop or regulate the supply of electricity, water or any other service.

Section 7 of the Act prohibits discharge of environmental pollutants in excess of the prescribed regulatory standards.

Section 8 of the Act prohibits handling hazardous substances except in compliance with the prescribed regulatory procedures and standards.

Section 10 of the Act empowers the central Government to enter and inspect any industrial plant, equipment, object, or to conduct any reach or to seize any equipment, plant, records etc. relating to the commissioning of an offence punishable under this Act or under the Rules made thereunder, or where such seizure is necessary to prevent or mitigate environmental pollution.

Section 11 of the Act authorise the Central Government or its designated officers to take samples of air, water, soil, or other substance from any factory, premises or other place for analysis.

Section 15 of the Act provides for penalties for contraventions of the provisions of the Act and the rules, orders and directions made thereunder.

Section 16 of the Act deals with offences committed under this Act by companies, where in the company's director, manager, secretary, or any other officer, with whose consent or connivance a particular offence is committed, shall also be held responsible.

Section 17 of the Act makes the Head of a Government Department responsible for any offence committed by the department under this Act. However, he will not be responsible, if it is proved that the offence was committed without his knowledge.

Section 20 of the Act authorise central government, in relation to its functions under this Act, from time to time, require any person, officer, State Government or other authorities to furnish to it or any prescribed authority or officer any reports, returns, statistics, accounts and other information.

Section 25 of the Act empowers the central Government to make rules by notification in the Official Gazette, for carrying out the purposes of this Act.

21.3 THE AIR (PREVENTION AND CONTROL OF POLLUTION) ACT, 1981

The Origin of the Act lies in the United Nations Conference on the Human Environment held Stockholm in 1972. In this conference, decisions were taken to take appropriate steps for the preservation of the natural resources of earth which among other things include the preservation of the quality of air and control of air pollution.

The **main objectives of this Act** are:

1. To provide for the prevention, control, and abatement of air pollution.
2. To provide for the establishment of Central and State Boards with a view to implement the Act.
3. To confer the powers to implement the provisions of the Act on the Boards and assign to them functions relating to pollution.

The **scope of this Act** is to preserve the quality of air and control of air pollution where 'air pollution' have been defined as the presence of air pollutant in the atmosphere.

Other Important Terms as defined in the Act:

1. **"Air Pollutant"** means any solid, liquid, or gaseous substance (including noise) present in the atmosphere in such concentration as may be or tend to be injurious to human beings or other living creatures or plants or property or environment.
2. **"Central Board"** means the Central Pollution Control Board, constituted under section 3 of the Water (Prevention and Control of Pollution) Act, 1974.
3. **"Emission"** means any solid or liquid or gaseous substance coming out of any chimney, duct or flue or any other outlet.
4. **"State Board"** means in relation to a State in which the Water (Prevention and Control of Pollution) Act, 1974 is in force and the State Government has constituted for that State.

The **main provisions under this Act** are:

Chapter II (Sections 3 to 15) deals with the provisions related to Central and State Boards for the prevention and control of air pollution.

Chapter III (Sections 16 to 18) deals with the Powers and Functions of the Boards.

Functions of Central Pollution Control Board (CPCB)

The main function of the Central Board is to implement the legislations that has been created to improve the quality of air, and prevent and control air pollution in the country. Other functions mentioned in the Act includes:

1. Advise the Central Government on matters concerning the improvement of air.
2. Co-ordinate the activities of the State Board and resolve disputes.
3. Provide technical assistance and guidance to the State Boards.
4. Plan and organise the training of persons to be engaged.
5. Lay down standards for the quality of air.
6. Collect and disseminate information in respect of matters relating to air pollution.
7. Perform such other functions, as may be prescribed in the Act.

Functions of the State Pollution Control Boards (SPCB)

Its important functions are:

1. To advise the State Government on matters concerning the prevention, control or abatement of air pollution.
2. To inspect any control equipment, industrial plant or manufacturing process and to give directions to such persons, as it may consider necessary.
3. To inspect air pollution control areas.
4. To lay down standards for emission of air pollutants into the atmosphere from industrial plants and automobiles.
5. To perform such other functions, as may be prescribed or entrusted to it by the Central Board or the State Government.

Section 37 to 43 (In Chapter VI) of the Act deals with the penalties and procedures for different offences by Companies and Government Departments.

21.4 WATER (PREVENTION AND CONTROL OF POLLUTION) ACT, 1974

This Act is an Act to provide the prevention and control of water pollution and the maintaining or re-storing wholesomeness of water and for the establishment of Boards for the purpose.

The **main objectives of the water Act** are to provide for prevention, control and abatement of water pollution and maintenance and restoration of the quality of water. The Act is designed to assess pollution levels and punish polluters. The Central Government and State Governments have set up Pollution Control Boards (PCBs) to monitor water pollution.

Some **important terms** is defined in the Act are:

1. **"Pollution"** means such contamination of water or such alternations of the physical, chemical or biological properties of water or such discharge of any sewage or trade effluent or of any other liquid, gaseous or solid substance into water (whether directly or indirectly) as may, or is likely to create a nuisance or render such water harmful or injurious to public health or safety, or to domestic, commercial, industrial, agricultural or other legitimate uses or to the life and health of animals or plants or of aquatic organisms.

2. The word **"trade effluent"** as used above, includes any liquid, gaseous or solid substance which is discharged from any premises used for carrying on any industry, operation or process or treatment and disposal system, other than domestic sewage.
3. **"Stream"** includes:
   (i) river,
   (ii) water course (whether flowing or for the time being dry),
   (iii) inland water (whether natural or artificial),
   (iv) sub-terrain waters,
   (v) sea or tidal waters to such an extent or, as the case may be, to such point as the State Government may, by notification in the Official Gazette, specify in this behalf.

Important Provisions under the Act

Section 3 of the Act, provides for the constitution of a Central Board (CPCB).

Section 4 of the Act, similarly provides for the constitution of State Boards (State Pollution Control Boards, SPCBs) by the respective State Governments.

The **functions of CPCB**, as defined in Section 16 are:

Its main function is to promote cleanliness of streams and well in different areas. In addition to this, the CPCB has to perform the following functions:

1. to advise the Central Government on any matter concerning the prevention and control of water pollution,
2. to co-ordinate the activities of the State Boards and resolve disputes among them,
3. to provide technical assistance and guidance to the State Boards,
4. to plan and organise the training,
5. to organise through mass media, a comprehensive programme,
6. to collect, compile and publish technical and statistical data relating to water pollution,
7. to laydown or modify the standards for a stream or a well.

Functions of State Boards (State Pollution Control Boards, SPCBs) as per Section 17 of the Act are:

1. to plan a comprehensive programme and secure the execution thereof,
2. to advise the state government on any matter concerning the prevention, control or abatement of water pollution,
3. to collaborate with the central board in organising the trainings and organise mass education programmes,
4. to collect and disseminate information relation to water pollution,
5. to inspect sewage or trade effluent, works and plants for the treatment of sewage and trade effluents,
6. to lay down or modify effluent standards and to classify waters of the state,

7. to evolve economical and reliable methods of treatment of sewage and trade effluents,
8. to evolve methods of utilisation of sewage and suitable effluents in agriculture,
9. to evolve efficient methods of disposal of sewage and trade effluents.

Chapter VII (Sections 41 to 50) of the Act deals with the penalties and procedure. Penalties are charged for acts that have caused pollution. This includes failing to furnish information required by the Board, failing to inform the occurrence of any accident or other unforseen act. An individual or companies (organisations) that fails to comply with the directions given in the law, can be convicted or punished with imprisonment for a term upto three months or with a fine of ₹ 10,000 or both and in case failure continues, an additional fine of ₹ 5000 every day. If a person who has already been convicted for any offence is found guilty of the same offence again, he would be punished with imprisonment for a term not less than two years, extending upto seven years with fine.

21.5 THE WILDLIFE (PROTECTION) ACT, 1972

1. **The Wildlife Protection Act, 1972,** amended in 2002 and in 2006, provides for 'the protection of wild animals, birds and plants, and for matters connected there with or ancillary or incidental there to, with a view to ensuring the ecological and environmental security of the chordates and invertebrates, and also includes their young and eggs. It also provides protection of specific endangered species listed in Schedule I, II, III and IV, and the protection of all species in designated protected areas. **Protected areas** categories include national parks, sanctuaries, conservation reserves, community reserves and tiger reserves. The Act restricts entry into a Sanctuary and National Park, if further states that no person shall destroy, exploit or remove any wildlife from a national park/sanctuary or destroy or damage the habitat of any wild animal.

The Act deals with the declaration of National Parks and wildlife sanctuaries and their notification.

The Act provides for appointment of a Director of Wildlife Preservation, and other officers by Government of India, appointment of a Chief Wildlife Warden and other officers by the State Governments, Constitution of National Board for Wildlife by Government of India, Constitution of State Boards for Wildlife and several other officers of the State Government.

The primary duty of the ***National Board*** for wildlife is to promote conservation and development of wildlife and forests, by such measures as thinks fit. The National Board is also supposed to advise the Government of India and State Governments on the ways and means of promoting wildlife conservation and to effectively control poaching. The National Board is also supposed to carry out impacts assessment of various projects and activities on wildlife or its habitat.

The primary duty of **State Board** is the selection and management of the areas to be declared as protected areas; and in formulation of the State policy for the protection and conservation of the wildlife and specified plants.

Section 51 of the Act deals with the **penalties** a person who breaks any of the conditions of any license or permit granted under this Act shall be guilty of an offence against this Act. The offence is punishable imprisonment for a term extending upto three years, or a fine of ₹ 25000 or both. An offence committed in relation to any animal specified in Schedule I or Part II of Schedule II, such as the use of meat of any such animal, or animal articles like a trophy shall be punishable with imprisonment for a term not less than one year or may extend to six years and a fine of ₹ 25000. In case of a second or subsequent offence of the same nature, the term of imprisonment may extend to six years and not less than two years with a penalty of ₹ 10,000.

21.6 FOREST CONSERVATION ACT, 1980

This Act was enacted by the Government of India in the year 1980 and amended in 1988, to provide for the conservation of forests and for matters connected therewith.

Historical Background: Before discussing "The Forest Conservation Act", we must first understand its background. The Indian Forest Act, 1927 consolidated all the previous laws regarding forests. This Act was enacted to consolidate the law relating to forests, the transit of forest produce and the duty leviable on timber and other forest produce.

The Act of 1927 gave the Government and Forest Department the power: (*i*) to create Reserved Forests, and the right to use **Reserved Forests** for Government use alone, (*ii*) to create Protected Forests in which the use of resources by local people was controlled, (*iii*) to create village forests to be controlled by the village community.

This Act remained in force till the 1980, when it was realised that protecting forests for timber production alone was not acceptable. Protecting the services that forest provide and its valuable assets, such as biodiversity was realised. Hence, the Forest Conservation Act of 1980 was enacted.

The Forest Conservation Act of 1980 was primarily enacted to control deforestation. It ensured that forestlands could not be deserved without prior approval of the Central Government. The enactment of this Act is helping in protecting the forests, which are also the natural habitats of wildlife.

In order to enable the enactment of Conservation Act of 1980, the subject of "forest" was transferred from the **State List** of the Indian Constitution to the **Concurrent List.** This transfer was made vide 42nd Constitutional Ammendment Act, 1976.

Penalties:

1. **Penalties for offences in Reserved Forests.** No person is allowed to make clearings or set fire to a Reserved Forest. Cattles not permitted to trespass into the reserved forests. Felling, collecting of timber, quarrying or collecting any forest produce is punishable with imprisonment for a term of six months, or with a fine which may

extend ₹ 500 or both.

2. **Penalties for offences in Protected Forest.** A person who commits any of the offences like felling trees, stripping the bark or leaves of the trees, setting fire to such forests, kindling a fire without taking precautions to prevent its spreading, dragging timber, permitting cattle to damage any tree, shall be punishable with imprisonment for a term, which may extend to six months or with or fine upto ₹ 500 or both.

When a forest offence has been committed pertaining to forest produce, the produce together with all tools used in committing such offence may be seized by a forest officer or police officer.

21.7 NATIONAL GREEN TRIBUNAL (NGT)

Government of India established a 'National Green Tribunal' on 18 October, 2010, headed by a former Judge of the Hon'ble Supreme Court. This is a tribunal exclusively dedicated to environmental issues. This tribunal was established by an Act of Parliament, named as "National Green Tribunal Act of 2010", and have zonal benches across the country to try all matters related to and arising out of environmental issues.

The tribunal was established for effective and expeditious disposal of cases related to environmental protection and conservation of forests and other natural resources including enforcement of any legal right relating to environment and giving relief and compensation for damages to persons and property and for matters connected therewith for incidental thereto.

The tribunal also consist of members who are experts in the field of environmental and related sciences. The tribunal is empowered to issue directions for the compensation and restitution of damages caused from actions of environmental negligence.

21.8 GREEN HIGHWAYS POLICY-2015

Green Highways (Plantation, Transplantation, Beautification and Maintenance) Policy-2015 have been framed for India's about 100,000 kms of National Highways. The Highways carry about 40% of traffic load. The Ministry of Highways has decided to develop all of existing National Highways and 40,000 kms of additional roads in the next few years as Green Highways.

The vision for Green Highways is to develop eco-friendly National Highways with participation of the community, farmers, NGOs, private sector, institutions, government agencies and the Forest Department. The objective behind this is to reduce the impacts of air pollution including dust as trees and shrubs along the highways act as natural sink for air pollution and arrest soil erosion at the embankment slopes. The plants along highway median strips and along the edges reduce the glare, which sometimes are responsible for accidents.

The community involvement in tree plantation directly benefits local people by generating employment.

The Punchayats, NGOs, and other Self Help Groups (SHGs will be involved in the process of planting and maintenace.

The plant species selected will be region specific depending on local conditions such as rainfall, climate and type of soil.

1% of the total project cost of all highways projects will be kept aside for the highway plantation and its maintenance.

In the Green Highway Policy, the provisions about the responsibilities attached have also been clearly defined. The monitoring of the plantation status has been included as an integral part of the policy. The strong monitoring mechanism is in place by using ISRO's satellite system.

QUESTIONS

1. Give salient features of the India's Environment (Protection) Act, 1986.
2. Give the historical background and necessity of the Environment (Protection) Act, 1986.
3. Explain the provisions of the Constitution of India in relation to environment and its related issues.
4. What are the legal rights of the Indian citizens in respect of the clean environment? Explain.
5. Write notes on objective and scope of:
   (a) Environmental Protection Act, 1986.
   (b) Air (Prevention and Control of Pollution) Act, 1981.
6. Briefly explain the salient features and provisions of the Air (Prevention and Control of Pollution) Act, 1981.
7. Define water pollution. Explain the functions of Central Pollution Control Board, and State Pollution Control Board.
8. Write short notes on:
   (a) Penalties as per Water (Prevention and Control of Pollution) Act.
   (b) Wildlife (Protection) Act, 1972
9. Write a detailed note on "Forest Conservation Act, 1980".
10. Explain the background for enactment of Forest Conservation Act, 1980.
11. Write short notes on:
    (a) National Green Tribunal (NGT)
    (b) Green Highways Policy—2015

PART VII : ENVIRONMENTAL MANAGEMENT

22

Chapter

ENVIRONMENTAL IMPACT ASSESSMENT (EIA)

22.1 INTRODUCTION

Generally, Environmental Impact Assessment is the study to predict the effect of a proposed activity on the environment. Environmentalists have defined EIA in various ways. Some of the more important definitions are mentioned below:

1. A study of probable changes in various socio-economic and bio-physical characteristics of the environment, which may result from a proposed or impending action.
2. An activity that aims at establishing quantitative values for selected parameters, which indicate the quality of environment before, during and after the proposed activity.
3. An instrument to identify and assess the potential environmental impacts of a proposed project, evaluate alternative and design appropriate mitigation, management and monitoring process.
4. A process by which a development is assessed in terms of its likely impact upon the **baseline** environment, to enable decision makers to determine the acceptability of a development prior to its approval.
5. A systematic integrated evaluation of both the positive and negative impacts of a project on the natural environment; on beneficial uses of the environment, including man-made structures, amenities and facilities; and on the socio-cultural environment.
6. The United Nations Environmental Programme (UNEP) defines Environmental Impact Assessment (EIA) as 'a tool used to identify the environmental, social and economic impacts of a project prior to decision-making'.

22.2 AIMS OF ENVIRONMENTAL IMPACT ASSESSMENT

The EIA is means of avoiding environmental disturbances, due to a proposed project, that are much more difficult and expensive to correct after their occurrence than before.

Today, there is a world-wide evidence that man cannot ignore the quality of the environment. Thus, environmental issues must be addressed as soon as possible during project planning. In case a project is expected to have detrimental impact on the environment, there should be no hesitation in abandoning such a project at an early stage.

The objectives of an EIA can be summarised as follows:

1. To identify, predict and evaluate the economic, environmental and social impacts of any developmental activity,
2. To provide information on the environmental consequences for decision making and
3. To promote environmentally sound and sustainable development through the identification of appropriate alternatives and mitigation measures.

22.3 HISTORY OF ENVIRONMENTAL IMPACT ASSESSMENT

Considering the negative effects on environment of the industrial and other developments the concept of EIA came into existence. Consequently, many of the more advanced planning systems around the world have considered the issue of any development's impact upon the environment in one form or the other.

First of all in the United Kingdom (UK), the first Town and Country Planning Act (1947) enabled the local planning authority to take environmental factors into consideration in sanctioning development proposals. However, no formal mechanism was designed to provide for this and implementation of the idea was left open to the wisdom of authorities.

Earliest effort for preservation of natural ecosystems started in the USA as early as 1872, when National Parks were established. Realization dawned over the adverse effects of water Resources and Highway developments.

It was not until the 1960, that this concern, as a result of population growth, urbanization, industrialization and the resultant pressure on natural resources, received a more public and global impetus. The harmful effects it had on environment are sustainability, continued throughout the sixties and eventually led to the evolution of the practice of Environmental Impact Assessment and Management.

Later Environmental Impact Assessments (EIA) emerged as a successful innovation of the 20th century for environmental conservation. About 5-6 decades ago there was no EIA but today, it is a formal process in many countries and is practiced all over the world.

EIA as a mandatory procedure originating in the early 1970, with the implementation of the National Environment Policy Act (NEPA)-1960 in the United States of America (USA). Other countries later followed the policy. Australia, Canada and New Zealand in 1973, and among the developing countries Columbia in 1974 and Philippines in 1978 considered EIA as a necessity.

The EIA process really took off after the mid-1980, when in 1989, the World Bank (WB) adopted EIA for major development projects, in which borrower country had to undertake EIA under the Bank's supervision.

History of EIA in India

The Indian experience with EIA began over 40 years back when in 1976-77 then Planning Commission directed the Department of Science and Technology (DST) to examine the river-valley projects from the environmental angle. This was subsequently extended to include those projects which required approval of the Government. However these were only administrative orders, and lacked any legislative support.

The Ministry of Environment and Forests (MoEF) Government of India was given the task of monitoring the Environmental Impacts of new projects. MoEF on January 27, 1994 under the Environmental (Protection) Act 1986 (EPA), promulgated EIA notification making Environmental Clearance (EC) compulsory for expansion, or modernization of any project or institution or setting up of new projects listed in the Schedule 1 of the notification. Since then many amendments have been made in the notification.

The environmental protection act came into force after the Bhopal gas tragedy (2-3 December 1984) it has 26 sections and it has been divided into four chapters. The objective of the act envisages to protect and improve environmental quality, control and reduce pollution from all sources.

Under this act the Union Government acquired the power to lay down standards for quality of environment, restrict areas for industrial operations. Provision for deterrent punishment to those who fail to comply or contrivance any provision or rules or act has been made in the Act. The punishment will be fine of Rs. 1 lakh or 5 years imprisonment or both.

At present the Environment Clearance (EC) process is required for 41 types of projects and covers aspects like screening, scoping and evaluation of upcoming projects.

22.4 CONCEPT OF EIA

The UN convention of 1991 on Environmental Impact Assessment (EIA) inferred that EIA is fundamental to any regulatory system to prevent, minimize environmental harm and promote sustainable development. An effective EIA provides an opportunity of public security and participation in decision making, together with this it introduces Independency and impartiality in balancing environmental and developmental needs. It alerts Governments and International Agencies to any trans-boundary harm so as to concern with other states also.

The Rio treatise (1992) broadened the concept of common concern and global responsibility to protection of environment. Despite the rapid economic growth the concern of environmental sustainability is the challenge of every country. Through effective legislation on environment issues and with National Instrument of Environmental Impact Assessment (EIA), the path of sustainable development can be achieved without much difficulty.

Stockholm declaration (1972) recognised the concept of viable environment, as a decent, healthy one and advocated the promotion of environment quality by active control on wrong practices in every member state.

Universally it is accepted that "all persons have the right to secure, healthy, and ecologically sound environment adequate to meet equitably the needs of present generation, so as not to impair the rights of future generations and their needs."

Basic needs irrespective of caste, creeds, language, religion or region include:

Freedom from pollution; preservation of soil, water air, sea, ice, flora and fauna and bio-diversity; high standard of health; adequate housing and healthy food and water; and sustainable use of natural resources.

It is the primary responsibility of government to fulfil the needs of all people. EIA plays its important role by guiding the State Governments to protect environment and save human race on this beautiful blue plant.

Obviously, the goal of sustainable development can only be achieved by imposition of strict liability rule on various governments with supervisory power of EIA.

22.5 BENEFITS OF EIA

Some of the benefits are direct, such as improved design or location of the project, or indirect such as better quality EIA work and increased awareness of the project personnels. In general the benefits of EIA include:

1. **Better environmental planning** and design of the proposed project, which can result in:
   (a) Selection of an improved technology.
   (b) Lower waste outputs.
   (c) Environmentally optimum location for a project.
   (d) Minimum risk involved.
2. **Economic Benefits:** Compliance with environmental standards results in:
   (a) Reduced damage to environment which avoids likelihood of penalties, fines and also of trust and credibility.
   (b) Savings in capital as well as operating costs.
   (c) Avoid undue costs of unanticipated impacts. Because changes which must be made late in the project are most expensive.
3. **Time Saving:** If all environmental concerns have been taken into account properly before submission of report for approval, then it is unlike that delay will occur in getting approval.
4. **Acceptance:** Proper EIA report generate confidence among stakeholders and there is increased project acceptance by the community.

The benefits can be achieved by a transparent EIA process, with provision of opportunity for public involvement of people who are directly affected by the implementation of the project.

Core Values of EIA

For a successful Environmental Impact Assessment certain core values have been prescribed by the Ministry of Environment and Forest (MoEF) as described below:

1. **Integrity:** The EIA process should be fair, objective, unbiased and balanced. It ensures that the project is in agreement with standard and good principles.
2. **Utility:** The EIA should provide a balanced approach and credible information for decision making process.
3. **Sustainability:** The EIA process should result in environmental safeguard. An environmentally sound development must have a regenerative capacity of the resources.

Main Features of EIA

The main features of EIA can be summarised as below:

1. Most important part of the EIA is to evaluate present status of various environmental parameters.
2. Evaluate possible changes in the character and atmosphere (ambience) of the proposed location of the project.
3. As already mentioned, the main job of EIA is to predict the negative and positive effects of the proposed project.
4. To find ways to avoid/mitigate negative effects.
5. To find ways to enhance positive effects.

22.6 THE STAKEHOLDERS OF EIA

A stakeholder is any individual, group, agency or organization affected by a project and/ or with concern or interest in a development project and its outcome, or in common resources by a development project. A stakeholder should be treated as a 'Partner in Development' and not as opponent of the project.

There are many potential stakeholders/partners in development projects. Some like the local public, are involved at all stages, throughout the life of a project, and they are often affected by it for years afterwards. Others become more or less important during various stages of the project development and implementation.

The effects of a project on stakeholders can be adverse or beneficial, direct or indirect sooner or later. The identity of the stakeholders, the timing of their involvement and the activities they may be encouraged and enabled to undertake depend largely on the type, size and location of a project as well as on the insightfulness, creativity, sensitivity and dedication to public involvement of the developer and the members of the Public Involvement Facilitation Team.

Generally speaking, those most directly affected by a project are clearly among the key stakeholders—they are at the greatest risk, they feel the impacts most intensely, they benefit the most from opportunities; hence, they should be the first to be involved. The poor, landless, vulnerable and marginalized people are among these stakeholders, and it is they who are often the most difficult to get involved.

EIA in itself is a huge process, and naturally involves a large number of people. Each of them have to play a role in the entire process of EIA. The major stakeholders and their contribution is discussed below:

(1) **Project Proponent:** The project proponents are anyone that want to start a project. They are the one who ask for an EIA, which is now necessary for project approval. Though they do not like the idea of EIA, they are obliged to ask for it under the national regulations.

(2) **Regulatory Institution:** Mostly these are Government bodies and are responsible for issuing the No Objection Certificate (NOC).

They oversee the entire process of EIA, check the EIA report and finally decide whether to accept or reject the project proposal. The institutions involved may be concerned Ministry, Central and State Pollution Control Boards, Water Control Board, Wildlife authority etc. and objection from even one of these may get the proposal rejected.

(3) **Project Affectors:** The project affectors include both the losers and beneficiaries. Often the most ignored stakeholder are the people living in the project area. These are generally rural people who are simple and get easily exploited by the proponents, whose main concern is profit earning and not the environment.

(4) **Social Activist Groups:** The social activists most of the time represent the affected people during the EIA process. They try to ensure that in EIA there is no foul play and vested interests are kept away.

(5) **Funding Agencies:** Big projects require huge amount of investment which is provided by various funding agencies like; State and Central Government, and International agencies like World Bank (WB), Asian Development Bank (ADB), International Monetary Fund (IMF), United Nation (UN) and also private institutions like, Ford foundation etc. These always keep a vigil on EIA process.

(**Note:** It is worth noting that even though all major, medium or even minor projects would have both negative as well as positive effects; the word **impact** in the context of Environmental Impact Assessment (EIA) has developed almost exclusively a Negative connotation.)

22.7 ENVIRONMENTAL IMPACT ASSESSMENT (EIA) PROCESS

The EIA-process is an interdisciplinary and multistep procedure to ensure that environmental considerations are included in decisions regarding project that may impact environment.

While EIA-process makes sure that environmental issues are raised when a project is planned, its recommendations may necessitate the redesign of some project components. It may suggest changes which alter the economic viability of the project. To be most effective EIA must be carried out to determine significant impacts early in the project cycle.

The way in which EIA is carried out is not rigid: it is a process comprising a series of steps. The stages of an EIA-process will depend upon the requirements of the donor agency or country. However, most EIA processes have a common structure. The EIA-process mainly constitutes eight steps as described below:

1. Screening

Screening is the first stage of EIA-process. In this first stage of EIA it is determined whether the proposed project requires an EIA and if it (EIA) is required, then what should be the required level of assessment. The output from the **Screening** process is often a document called an Initial Environmental Examination of Evaluation (IEE). The main conclusion will be a classification of the project according to their likely environmental sensitivity.

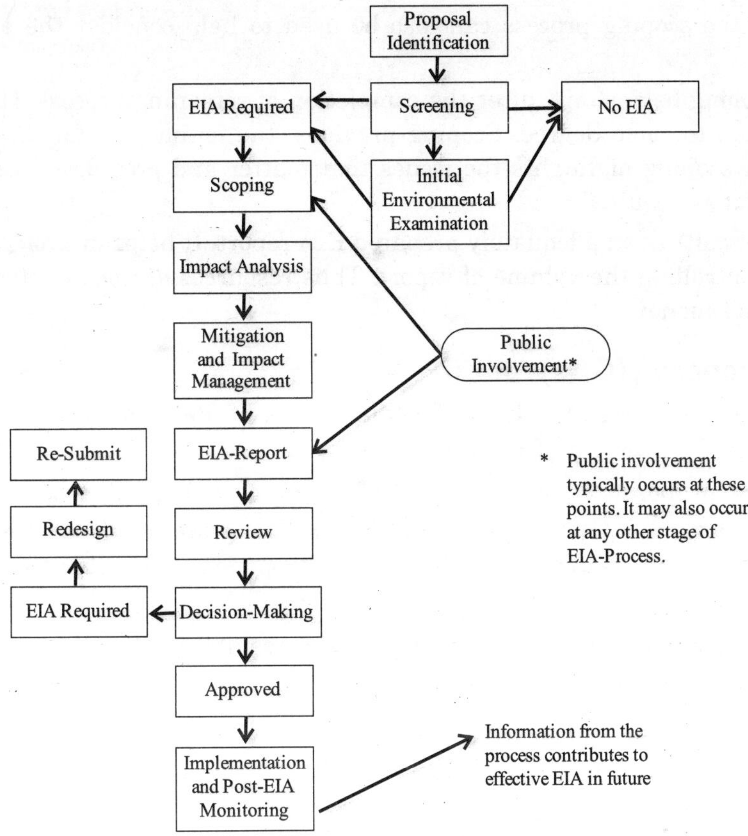

Flow chart of EIA–Process

2. Scoping

Scoping is a critical early step in preparation of EIA. The stage identifies the issues that are likely to be the most important (key issues) during the EIA and eliminates those that are of little or no concern. Typically, this process concludes with the establishment of Terms of Reference (TOR) for the preparation of an EIA (as discussed below).

The requirements and procedures established for this purpose differ from country to country. In many EIA systems, the involvement of public, as well as the competent authority and other responsible government agencies, is an integral part of the scoping process. In particular public input helps to ensure that important issues are not overlooked when preparing TOR or initiating the EIA study.

Briefly, the purpose of scoping is to identify

- The important issues to be considered in an EIA.
- The appropriate time and space boundaries of the EIA study.
- The information necessary for decision making.
- The significant effects and factors to be studied in detail.

Besides this, the scoping process can also be used to help consider the alternatives to proposed action.

Normally, scoping begins only after the completion of screening process. However, these stages may overlap to some degree. Scoping provides, the foundation for an efficient EIA-process. A proper scoping highlights the issues that matter and provides clear guide to the proponent on what is required.

Scoping also results in an adequately prepared EIA report. It helps in avoiding unfocussed problems, thus controlling the volume of report. Thus resources are not wasted and there is saving of time and money.

Terms of References (TORs) for EIA

As per guidelines of MOEF, Govt. of India Terms of References (TOR) for EIA are required for A and B1 category of projects.

The TOR are proposed by project proponent alongwith application seeking prior Environmental Clearance (EC) in the prescribed form. These are finalized by Environmental Appraisal Committee and conveyed to the project proponent. The objectives of the TOR are to:

(a) define what types of informations are to be presented in the EIA,

(b) delineate the relevant issues to be undertaken,

(c) define what studies will be performed,

(d) explain who will conduct the studies,

(e) state when the studies will be conducted, and

(f) outline the basic structure of EIA.

The purpose of preparing the TOR for understanding an EIA of a proposed project is to ensure that resulting EIA will be suitable for review and evaluation by concerned governmental agencies.

Time for developing TOR. Following the completion of scoping, a TOR should be development based entirely on the results of scoping. The most appropriate time for preparation of TOR is at the feasibility stage of the project cycle. The critical issues identified during the scoping exercise, to be carried out in EIA study, should be included in the TOR.

3. Impact Analysis

After the screening and scoping phase, the next stage of the EIA-process is the Impact Analysis which is the technical heart of the EIA-process. Here detailed assessment is undertaken to **forecast** the main potential impacts, both social and environmental.

This stage is broken down into three phases as:

- **Identification** : In this stage specific impacts associates with each phase of the project and activities are considered.
- **Prediction** : Here the nature, magnitude, extent and duration of the main impacts are to be forecasted.

- **Evaluation** : In this phase the importance/significance of residual impacts after mitigation is evaluated.

It is necessary to realize that impact identification and prediction can only be taken against an environmental baseline. The baseline information and relevant bio-physical and socio-economic factors begin during screening and scoping. Additional baseline data need to be collected to establish impact predictions.

Earlier the focus was primarily on natural or bio-physical environmental impacts. However, over time, consideration has been given to social, health and economic impacts also.

While performing Impact Analysis, priority is given to human health, welfare and safety. Effects on flora, fauna, ecosystems and bio-diversity alongwith soil, water and air are next to be considered. Other aspects to be considered are climate, landscape, use of natural resources, protection of important sites, heritage etc. Livelihood and well being of affected people needs to be given high priority.

4. Mitigation

One of the important stages of the EIA-process is 'mitigation' which generally refers to actions taken to lessen harmful impacts of any proposed project. Mitigation is the stage of the EIA-process when measures are identified to avoid, minimize or compensate for the effects (both direct and indirect).

Mitigation measures include those that 'restore, enhance, credit or replace damaged ecosystems'. Those mitigation actions aimed at compensating for unavoidable damage are referred to as "offsets."

However, the most effective mitigation strategy is to avoid environmental impacts wherever possible. Avoidance of impacts can only be achieved at the design and planning stage.

Mitigation Hierarchy: The mitigation hierarchy is a tool that guides users towards limiting as far as possible the negative impacts on environment and bio-diversity from development projects. It emphasizes best practice of avoiding and minimising any negative impacts, and then restoring sites no longer used by a project, before finally considering offsetting residual impacts.

Mitigation hierarchy generally has five steps as explained below:

(1) **Avoidance:** It is the first step of the mitigation hierarchy. It offers complete mitigation of any negative impacts expected from the project. Avoidance is often the easiest, cheapest and most effective way of reducing potential negative impacts to zero level, if considered in the early stages of the project. Obviously it is the most preferred forms of mitigation as it ensures no damage to the environment due to implementation of the project.

(2) **Minimize:** The second step in this hierarchy is to minimize the negative effects on environment. Here, the team recognizes that the environment impact cannot be completely avoided or eliminated; instead they take steps to ensure minimal damage is done to the environment.

(3) **Rectify:** Rectification of an impact implies that the negative impact on the environment has already happened and a damage control is required to be done. In a way, rectification

allows the managers to correct the mistakes that led to the adverse environmental impacts.

(4) **Reduce:** If rectification is not possible, the extent of impact is tried to be reduced through change in management practices and methodology.

(5) **Environmental offset:** When rectification and reduction both fail, action outside the development site is undertaken to compensate for the damage already done. In effect it means that the development authorities undertake environment conservation activities to compensate for what they do in order to achieve "no net bio-diversity loss," preferably a net gain of bio-diversity.

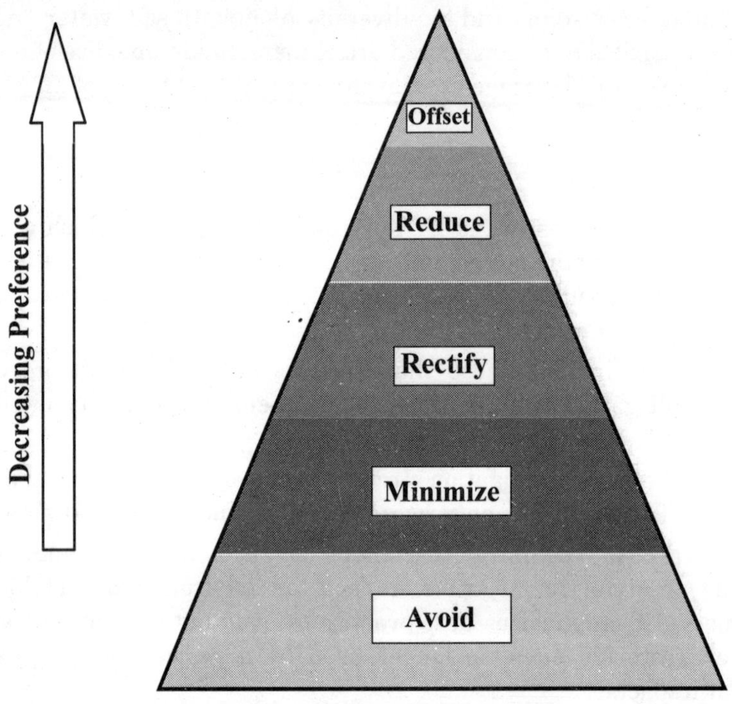

Mitigation Hierarchy

An environmental offset may be required as a condition of approval where a prescribed activity is likely to result in a significant residual impact.

22.8 ENVIRONMENTAL AUDIT

While EIA assesses the potential environmental effects of a proposed facility, the purpose of an environmental audit is the systematic scrutiny of environmental performance throughout a company's existing operations.

The International Chamber of Commerce (ICC) has defined environmental audit as, "A management tool comprising a systematic, documented, periodic and objective evolution of how well environmental organization, management and equipment are performing with the aim of helping to safeguard the environment."

Objectives of Environmental Audit

The overall objectives of an environmental audit are: to determine how well the environmental management information systems and equipments are performing, verify compliance with the relevant national, local or other laws, rules and regulations and minimize human exposure to risks from environmental problems.

Environmental audit needs for any project are internal as well external value. The external needs serve to achieve compliance standards and establish a report with regulatory bodies for implementation of environment management policies. Internal need serves the industry as well as self-evaluation tool for the process and technology.

The objective of environmental audit is to help pollution control, improve health and safety, conserve resources by waste prevention and reduction. Mainly assessing compliance with regulatory requirements and placing environmental information to the public.

Scope of the Environmental Audit

While the objective of the environmental audit is to test the adequacy of existing management systems, the scope of an audit can vary from just compliance to a more rigorous examination of the performance.

The technique is applied not only to operational environmental, health and safety management, but increasingly also to product safety and product quality management, and to areas such as loss prevention.

If the intention of auditing is to help ensure that these broad areas are managed properly, then all of these individual topics must be reviewed. Items which may be addressed in audits, including environment, health, safety and product safety are shown in table below.

Table 22.1: Items to be addressed in audits

Environment	*Safety*	*Occupational health*	*Product safety*
Site history	Safety policy/procedures	Employee exposure to air contaminants	Product safety programme
Process materials	Accident reporting	Exposures to physical agents *e.g.* noise, radiation, heat, etc.	Product quality control
Storage of materials above ground, below ground	Accident recording	Measurements of employee exposure	Product packaging, storage and shipping
Air emissions	Permit to work systems	Exposure records	Product recall-withdrawal procedures
Water discharges	Special procedures for:	Ventilation engineering controls	Customer information on product handling and quality
Solid wastes Liquid hazardous wastes	Confined space entry work on electrical equipment	Personal protective equipment	Regulatory compliance

Asbestos, PCBs	Breaking into pipelines, etc.	Information and training on health hazards	Labelling
Waste disposal on site off site	Emergency response Fire fighting	Medical surveillance programme	Specifications for Purchased materials products packaging
Oil/chemical spill permits, licenses	Job safety analysis safety training	Hearing conversation First aid	Material safety data Vendor qualification programme
Pollution control Contractors on site Past incidents	Safety communications promotion housekeeping regulations compliance	Regulatory requirements	QA testing and inspections Record keeping product literature process control

Audits can address one topic, or a whole range of issues. The greater the scope of the audit, the greater will be the size of audit team. The time spent onsite and the depth of investigation.

Although some companies have a regular annual audit cycle, audits are determined by need and priority. Thus, not all facilities or aspects of any company/organisation will be assessed at the same frequency or to the same level.

Benefits of Environment Audit

By proper and honest environmental audit, it can derive number of benefits. Actually it is a means for ensuring the continual improvement in environmental management. Following benefits have been identified of the environmental audit:

(1) It helps in safeguarding the environment and stop misuse of natural resources.

(2) It helps in verifying compliance with local, state, national and international environment laws.

(3) It identifies the areas of strength and weakness for improvement in performance.

(4) It assures safety of plant, environment and human population, by highlighting deficiencies.

(5) It results in loss prevention, in manpower development and marketing.

(6) It identifies potential cost saving from waste minimization and other activities.

(7) It provides an opportunity for management to give credit for good environmental performance.

(8) It facilitates exchange and comparison of information between different plants or the subsidiary companies of the parent company.

(9) It demonstrates company's commitment to environmental protection, to its employees, the public and the authorities.

(10) As a whole environment audit plays an important role in minimizing the environmental problems all over the place.

Note: Non-compliance with environmental laws and regulations can impose fines, penalties, legal liabilities and even criminal sanctions for responsible officials, persons.

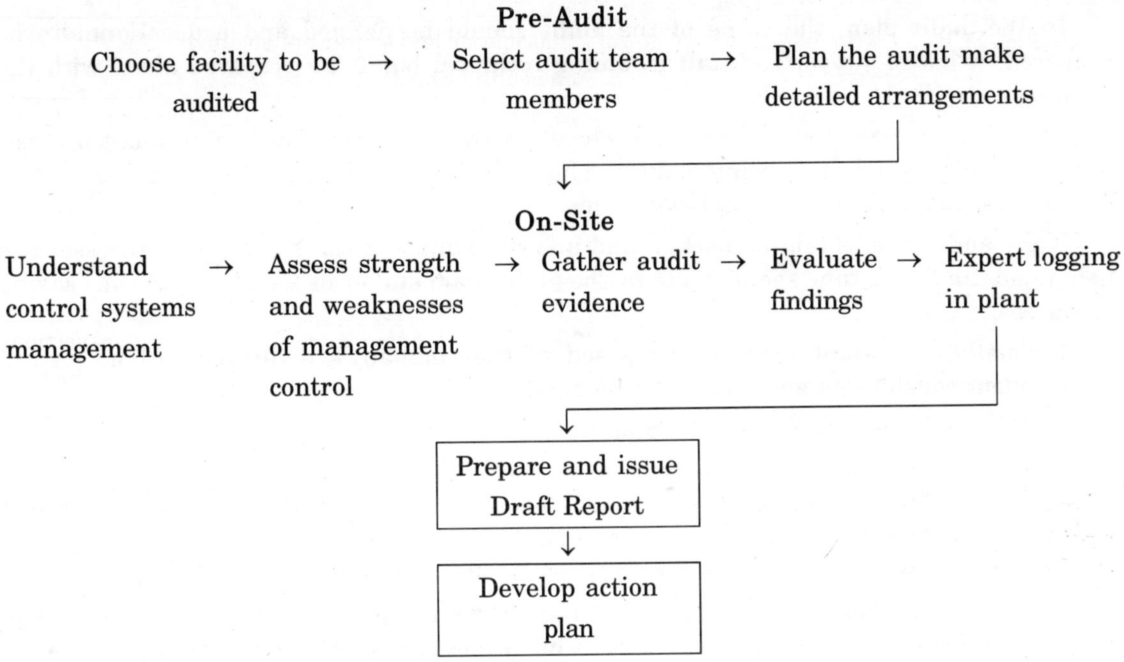

22.8.1 A Typical Audit Process

For an effective environmental audit, a team of knowledgeable person is formed who first assemble factual information prior to and during a site visit, analyze the collected information and compare them with the criteria for the audit, draw conclusions and report their findings.

Basic Steps in Environmental Auditing

Perhaps the most important step in establishing an audit programme is to decide the criteria against which the audit will be conducted. Also, it must be ensured that the management had been aware of these criteria. For an audit following criteria are normally used:

- Company policies and procedures on environmental matters.
- Applicable legislation and regulations.
- Good environmental management practice.

To achieve an effective environment audit 3-steps are recommended, these are briefly described below:

1. Pre-audit Steps

Pre-audit steps include: Planning of audit and selecting the personnel for the audit team. The composition of audit team will depend upon the nature of the project. Once proposed the availability of audit team member should be ascertained. Their travel and lodging managements should be made according to their status. Also responsibilities be assigned to members and supporting staff.

In the audit plan, the scope of the audit should be defined and a questionnaire be prepared. A list of supporting staff including technical hands be prepared along with the resources required.

If auditing is new, these is need for educating those who are involved in audit process, both the auditors and those being audited. This also applies to multinational companies who may not be conversant with local conditions.

If pre-audit step, obtaining background information about the site and its processes can help to minimize the time spent onsite by the audit team and focus its activities, thus saving resources.

Normally, the audit team is composed of the company's dedicated in-house staff, independent consultants and local stakeholders.

2. Onsite Steps (Activities at-Site)

This step is the backbone of environmental auditing. In this, first of all, an understanding of the controls that are existing is obtained. These include assessment of procedures and practices being followed, record keeping and monitoring. This is done by physical observation, through interviews and analyzing the response to questionnaire.

Here, auditors will evaluate the strength and weaknesses of internal controls. They will look into indicators such as - defined responsibility and competence of personnels, quality of documentation and records. Care should be taken whether the system is effective rather than only sophisticated.

The audit team must verify the records to confirm compliance with regulations. Evidence may be collected through enquiry, observation and testing. Once evidences are collected, the finding should be recorded. Wherever, any deficiency is found, it is noted as audit finding. Some of the findings may be resolved through some informal discussion with the management, for others, inclusion in the formal report will be appropriate.

Just before the end of the audit team's visit, reporting of audit finding is done at a meeting with the plant management. Each 'finding' and its significance is discussed with the plant personnel. Before the end of the exercise a written summary of findings is provided to the management.

3. Post-Audit Steps

After the activities at site, a draft report is prepared, which is presented to the plant management at the end of teams site visit. An action plan is then developed to address the deficiencies. Recommendations for corrective measures are included in the formal audit report. The plant on this basis of recommendations prepare their plan of action.

Once an audit programme is in place, future audits will include the past reports and judge the progress in the implementation of the recommendations made earlier as part of their evidence.

Some Special Types of Audit

Although the most widespread use of environmental auditing is to assess the environmental performance of a company's operations, there are some different types of environmental audits also. For example **pre-acquisition audit**, which is means to identify actual or potential problems, and taking these into account in the final negotiations of the deal.

A pre-sale audit is conducted when a company is being sold, then an audit is done as a part of the corporate policy. The purchasing company then knows the status of environmental issues and can take the remedial measures. Another special audit is the **issues audit** in which a particular issue is audited, for example the issue of waste, or water pollution or any other issue.

22.9 IMPACT ASSESSMENT METHODOLOGY

These are approaches developed to identify, predict and value changes of any action. It uses methods and techniques to quantify or qualify those changes. For quantitative impact identification following methods are being used :

22.9.1 Ad-hoc Method

In this method minimal guidance for total impact assessment while suggesting the broad areas of possible impacts and the general nature of these impacts. The broad areas considered are - impact on flora and fauna, water bodies (lakes, ponds, streams, groundwater) and forests. Specific environmental parameters are not considered within the impact area.

Ad-hoc method is useful when there is time-constraint and EIA rely exclusively on the expert opinion. Proper selection of experts can be beneficial for the EIA. However, when more scientific methods are available, ad-hoc method is avoided.

Delphi method is a popular ad-hoc method for EIA. The Delphi method is a forecasting process based on the results of multiple rounds of questionnaires sent to panel of experts, and the anonymous responses are aggregated and shared with the group after each round. Based on the "Group response" experts adjust their answers to reach the correct response through consensus.

The main advantage is that opinion of diverse set of experts can be sought. However, this method becomes time-taking and sometimes the results lack consensus.

22.9.2 Checklist Method of EIA

Like many other fields, checklists are also used in EIA. Checklist means a listing of potential environment impacts.

The checklist presents the specific list of environmental parameters to be investigated for possible impacts, or list of causes known to have caused the environmental concern.

This method also assesses the nature of impacts *i.e.* its type such as adverse/beneficial, short term/long term, no effect/significant impact, reversible/irreversible etc.

Checklists are the most common formal method used for impact identification. It indicates and explains the environmental features that need to be considered in any project or activity.

Checklist method is primarily for organising information or ensuring that no potential impact is overlooked. They are more formalised version of ad-hoc approach in which specific areas of impacts are listed and instructions are supplied for impact identification and evaluation.

Checklist is not supposed to avoid or repeat any required parameters.

Types of Checklists

(1) **Simple Checklists** : These merely list aspects to be considered in the analysis without any guidelines on how they are to be measured and interpreted. They point out issues that are likely to be affected, including those that may be less obvious/ important but may still be relevant. The checklist is subject to change and can be improved to suit local conditions as the experience with its use is gained.

(2) **Descriptive Checklist** : It adds to the simple checklist by providing additional background information on each aspect. They may identify the most important indicators to measure each component. The descriptive checklist provides information on measurement of environment parameters, impact and assessment.

(3) **Structured Checklist (Weighted Checklist)** : Such a list is strong in impact identification. They include a certain degree of interpretation and evaluation. In this list each impact is weighted. Each component is given a 'weight' according to its importance. The scale of 'weights' may be digital like (1,2,3,4) or alphabets (A,B,C,D).

Below is an example of a checklist for a Land Development Project

Table 22.2

SI.No.	*Parameter (1)*	*Impact on (2)*	*Weightage (3)**
1.	Local Economy	• Public fiscal balance • Employment	(A) or (4)
2.	Natural Environment	• Air, Water, Soil Quality • Wild Life • Flora • Natural Disasters	(B) or (3)
3.	Services (Public and Private)	• Water Supply • Medical care • Electric supply • Transport • Housing • Shopping • Security • Recreation	(A) or (4)
4.	Social	• Displacement • Social Interaction • Friendliness • Overall contentment	(B) or (3)
5.	Aesthetics	• Attractiveness • Landmarks	(C) or (2)

* The weightage is a guess and may be different for different persons.

The simple checklist will have columns (1) and (2) only. The structured checklist will have columns (1), (2) and (3). The third column will show the importance of each impact.

The scale can be choosen by the persons conducting EIA. The scale can be 1 to 4 or even 1 to 10. Each of the impact can be given weightage or each parameter collectively evaluated, as shown in the Table.

Advantages of Checklist Method : The checklist provides all possible relationships and impacts which gives opportunity to select specific assignments. It creates awareness among the stakeholders what they should be looking for when assessing the proposed project. Quantification of impact is possible by structured checklist to some degree. Even then it is the simplest method available for EIA.

Limitations of Checklist Method

In particular the descriptive checklists may be exhaustive, still lacking in quantitative information. Checklists also tend to compartmentalize the environment, whereas environmental systems comprise a complex web of interrelated components. It does not relate individual activities to environmental components affected by these activities.

22.9.3 Matrix Method of EIA

Though checklist method of EIA is the simplest, it is too primitive to be used for large scale projects. A step higher from the checklists is the matrices form of impact assessment in EIA. Contrary to checklist method, in matrix method quantitative aspect in the assessment of an impact has been introduced. Matrices are versatile and can be applied both to small as well as large-scale projects alike.

The matrix methodology incorporates both a list of project activities and a checklist of potentially impacted environmental characteristics. The two lists are then related in a matrix which identifies cause and effect relationship between specific activities and impacts. In a way the matrix present in both the attributes and activities.

Matrix methodology may either specify which actions impact which environmental characteristics or may simply list the range of possible actions and characteristics in an open matrix to be completed by the analyst.

Procedure : Matrix method identifies interactions between various project actions and environmental parameters likely to be affected by these actions. A checklist of environmental components is prepared alongwith a list of project activities.

A matrix of potential interactions is produced by combining these two lists placing one on horizontal x-axis and the other on vertical y-axis. The cells of matrix are filled in using experts' subjective judgement, or by using extensive or available data base.

Simple Matrix : This is simply a list of environmental aspects listed along the vertical axis, against which analyst determines whether an activity would have an **adverse effect**, **no effect** or **beneficial effect**. A simple mark is given under the appropriate column. Following is the example of land development project matrix.

Table 22.3

Sl.No.	*Activities*	*Effects*		
1.	Land Transformation and Compaction			
2.	Land Use			
3.	Water Resources			
4.	Air Quality			
5.	Noise and Vibration			
6.	Transport System			

Sometimes, the activities are listed along the horizontal axis with the environmental aspects on the vertical axis. Those pairs that have an interaction between themselves are marked.

An example of a Matrix for a Pulp Industry is shown below:

Table 22.4

Environmental	*Project Activities*							
Components	*Plant Constru-ction*	*Farm-ing*	*Use of Fertilizer & Pesticide*	*Transport of Raw Materials*	*Solid Waste*	*Efflu-ent Dis-charge*	*Emis-sion*	*Employ-ment*
Surface Water Quality			✔		✔	✔		
Air Quality				✔			✔	
Fisheries			✔			✔		
Wild Life	✔							
Land Use Pattern		✔						
Highway/Railway				✔				
Water Supply			✔			✔		
Agriculture		✔						
Housing								✔
Health					✔	✔	✔	
Socio-Economics								✔

One of the very famous Environmental Impact Matrix is the Leopold Matrix.

Leopold Matrix : This matrix analyses 100 specific actions and 88 Environmental/ Social aspects. Project activities are recorded on the horizontal (x-axis) and environmental/ social aspects on the vertical (y-axis). This was designed by Leopold in 1971. Out of these relevant actions and impacts are chosen for a particular project.

The cells of matrix are divided by a diagonal line. The top division is used to describe the magnitude of the impact that the activity will have on the environment or social aspect, and the bottom division is used to describe the significance of the impact.

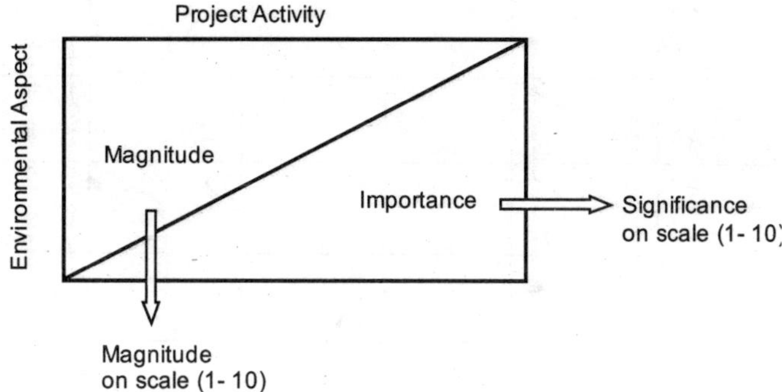

As shown, both the magnitude and importance/significance are rated on a scale of 1 to 10. This is subjective to the analyst and is based on baseline data and experts' opinion. An empty cell shows that there is no significant impact on the environmental/social aspect.

For example, if we consider the first row of the Table 22.4 considering surface water quality in the matrix the row will look like as follows:

Environmental Components	*Project Activities*							
	Plant Constru-ction	*Farm-ing*	*Use of Fertilizers & Pesticides*	*Transport of Raw Materials*	*Solid Waste*	*Efflu-ent Dis-charge*	*Emis-sion*	*Employ-ment*
Surface Water Quality			2 / 3		2 / 1	4 / 5		

Following is an example of Leopold Matrix for a Hydro-electric Project.

Proposed ⇒ *Action* / *Impacts* ⇓	*Immigra-tion of Labour*	*Dam Construc-tion*	*Trans-mission Lines*	*Reservoir Filling*	*Heavy Metal Discharge*	*Growth of Aquatic weeds*	*Relocation of Inhabitants*
Health	5 / 8	4 / 6		5 / 8	4 / 7	6 / 6	
Spawning of Fish		3 / 4		3 / 6	3 / 7	5 / 5	
Archaeological Artifacts	4 / 6			8 / 8			
Tourism			7 / 6	7 / 6			

D/s Water Pollution		7/7		7/8	2/4		
Social and Economic aspect							8/7
Forestry		4/2					
Fishery		2/5			2/5		
Navigation				6/5			
Aquatic Plants				6/5			
Leopold Aggregate	9/14	20/24	7/6	42/47	11/23	11/11	8/7

Note : The assigned values may vary with individuals.

Limitations of Leopold Matrix : The Leopold matrix has some glaring disadvantages:

(1) It does not describe spatial and temporal effects.

(2) It tends to be too simplified for a complex Environmental problem.

(3) It does not link various environmental aspects *i.e.* it does not describe secondary and tertiary impacts.

The Component Interaction Matrix :

For considering indirect impacts systematically which is not possible in Leopold Matrix, Component Interaction Matrix was developed in 1974 by Environment Canada. In this matrix both axes have environment components. In case two components were seen to be linked by secondary or tertiary interactions, they would be marked as 1, 2 ... etc. If they are not linked, they would be marked zero. An example is given below:

Activity / *Env. Aspect*	*Dam Construction*	*Reservoir Filling*	*Heavy Metal Discharge*	*Fisheries*	*Landscape Heterogeneity*
Dam Construction		1	0	1	1
Reservoir Filling	1		0	1	2
Heavy Metal Discharge	0	0		1	0
Fisheries	1	1	1		2
Landscape Heterogeneity	1	2	0	2	

Some other forms of Matrices :

The importance of matrices have been accepted around the world and they being used increasingly in EIA. More and more forms have been developed for better performance. These are briefly discussed here.

(1) **Modified Graded Matrix** : In this type each activity is assigned a relative value based on its priority level in the developmental activity. This method is particularly useful in indentifying activities that have a higher potential for causing damaged in the developmental site. This specifies the areas that need most attention.

(2) **Impact Summary Matrix** : This type of matrix can identify potential impact areas, specific mitigation measures and even list out the agencies which can help in the mitigation efforts. It gives an overview of the study that has been conducted. Basically it is the summary of the entire scoping stage.

(3) **Loran Matrix** : This matrix has 234 developmental activities and 27 impact against which they are assessed. The rest of the process of scaling and grading is same as in Leopold matrix. The idea behind is that even the smallest of developmental activity can show impact on the environment.

Note : The choice of matrix depends solely on the EIA-team. However, a combination of various types may be beneficial according to the nature and size of the project.

22.9.4 Mathematical Modelling

The use of modelling tools in EIA will surely increase the ability to present quantified results from field studies, which will contribute tremendously towards improving environmental practices and improve the quality of EIA-report.

Environment modelling is a method of simulating real-life situation with **mathematical equations** to forecast or predict their future behaviour. It is an analytical tool which enables the quantification of impacts which can affect the environment.

By mathematically representing a simplified version of an environmental system, reasonable alternative scenario can be predicted, tested and compared. Model Developments involve derivation of mathematical equation based on first principle of derivation. Following stages are involved:

(i) Carry out field monitoring
(ii) Collate existing data
(iii) Establish relation via mathematical equations (models)
(iv) Calibrate the equations (models)
(v) Verify the models
(vi) Adopt the models for prediction

Environmental mathematical models can make accurate predictions, only when the input data is reliable. Here it is to be realised that model output is not necessarily an end in itself but may be an input for assessing changes in economic, sociological and ecological terms.

Mathematical models require expert handling, mathematics and computer knowledge and also scientific understanding.

22.9.5 Expert Advice

An expert is someone who has a prolonged or intense experience through practice and education in the field of Environmental Science and Technology. He/She has acquired knowledge and skills through study and practice over the years and is capable of solving problems that have no pre-existing solutions.

Expert advice is sought for predictions which are inherently non-numeric and particularly suitable for estimating social and cultural impacts. In this local experience provides invaluable insight. Expert opinion is also likely to be needed to assess implications of any modelling predictions.

If the forecast is based on the opinion of several experts then the approach is called forecasting through the use of **panel consensus**. Such forecasting embodies collective wisdom of experts. When there are large difference in opinion, already discussed **Delphi Method** can be useful.

The greatest advantage of Experts' opinion method is in its simplicity as it does not require extensive statistical or mathematical calculations. However, it has some limitations also. Experts may have major difference of opinion, their own biases and intentions. If the experts are genuinely reliable then panel consensus can result in valuable forecasting.

22.10 ECONOMIC STUDY (PROJECT APPRAISAL)

A project plan which is optimal from both environmental and economic perspective will have a higher Benefit/Cost (B/C) Ratio and will obviously be sustainable.

It is important to realize that environmentally sound development brings long-term economic benefits. Unfortunately, short-term gains are often given priority.

The most commonly used method of project appraisal is cost-benefit or cost-effectiveness analysis. In cost-benefit analysis usually attempt is made to put monetary value on both costs and benefits to be derived, so that they can be expressed in the same units. However, it has not been found easy to incorporate environmental impacts (both positive and negative) into traditional cost-benefits analysis, primarily because of the difficulty in quantifying and valuing environmental effects.

A better EIA can provide some information on expected effects and quantify to some extent their importance. This information can be used in cost-benefit analysis. Cost-effectiveness analysis can be used to determine what is the most efficient and least-cost method of meeting environmental objectives.

Valuing the environment raises complex and controversial issues. Human health issues present similar problems. Some efforts have been made to translate these issues in monetary terms.

This would require to classify the benefits and losses into tangible and intangible groups. Tangible benefits/losses are those which can be measured in monetary terms. Intangible ones cannot be measured in monetary terms but they have significant impact on the environment and human health. For example, loss of property and services etc. are tangible, while loss of human life and environmental degradation intangible losses. Creation of employment is a tangible benefit, while improved health of the people may come under the intangible category.

22.11 THE FINAL REPORT OF EIA

The final EIA report is a compilation of important project components, including the project description, the prediction/assessment of its environmental and social impacts, mitigation measures and related management and monitoring plans.

The final report of an EIA is also referred as Environmental Impact Statement (EIS). The EIS "is the detailed analysis that serves to ensure that the policies and goals defined by National Environmental Policy (NEP) are infused into the ongoing programmes and actions of various agencies including the governments and other controlling agencies."

EISs are prepared for those projects that the proposing and controlling agencies view as having significant prospective environmental impacts. The report/statement should provide a discussion on significant environmental impacts and reasonable alternatives (including a non-action alternative) which would avoid or minimise adverse impacts or enhance the quality of human environment.

Experts preparing an EIA report must be able to appreciate that the final report will be circulated among a wide range of people, hence the EIA report should be concise, clear, to the point and should be supported by evidence that the EIA agency has made the necessary environmental analysis.

The World Bank (1991) has prescribed following points to be covered in the main text are summarised below:

(1) A description of programme including the time-scale of the proposal.

(2) A summary of the EIA-methodology, including the limits of the study.

(3) The policy, legal and administrative framework within the area where project is to be situated.

(4) A summary of the baseline data providing an overall picture of present conditions and physical, biological and ecological trends.

(5) A description of Governmental and Non-Governmental participation during EIA.

(6) Environmental Impacts : The most significant beneficial and adverse environmental impacts need to be clearly stated. Impact need to be quantified wherever possible and uncertainties in the results need to be highlighted, whether due to a lack of data, lack of knowledge or to critical but indeterminate assumptions such as future policy. The results of economic analyses need to be present in this section.

(7) Mitigation and enhancement measures that are proposed should be clearly mentioned. Impacts with no effective mitigation need be clearly identified and explained.

(8) The Environmental Action Plan : It has to be presented in two sections :

   (a) The first section covers the implementation of proposed mitigation measures, including both cost, training and institutional enhancement required to implement these.

   (b) The second part should cover monitoring requirements to measure predicted impacts and to determine the success of mitigation measures. A clear programme of implementation should be mentioned.

(9) The report should include Recommendations and Guidance to the decision makers.

(10) A statement of provision for auditing, who is to carry it out and when to start.

(11) Appendix : The appendices should include :

(a) Glossary of technical terms and units.

(b) List of team members who have prepared EIA.

(c) Records of public meetings and consultations.

(d) A catalogue of information, both data and written material, and their sources.

(e) Photographs supporting evidence.

Baseline Data : Baseline data is required to describe environmental and socio-economic status of project site and project impact area for pre-project condition. It consists of primary data (field tests, surveys, measurements) and secondary data (published information, unpublished information available with various agencies). Much of the required secondary data/information is often available within the various government agencies. Major difficulty with the unpublished information is determining which of it is important and then correctly interpreting it, while eliminating all the unimportant data.

The data is compiled for land environment, Water environment, Air environment, Noise environment and Socio-economic environment. Primary data related to the environmental attributes, like air, noise level, water quality and soil are collected from field studies, a structured questionnaire is used for collection of primary information on socio-economic aspects. Ecological information is collected from field studies as well as secondary sources. A summary of environmental attributes related parameters and source of information.

Format of Report : The format of the report should preferably follow a standard as recommended by the appropriate institution or required by legislation. The main report should preferably be of about 50 pages long and not more than 100 pages. An exceptionally complex study might require upto 150 pages. The main text should maximize the use of visual Aids-Maps, Tables, Charts etc.

Executive Summary : The quality of executive summary is particularly important as some decision makers may only read this part of the project report. The executive summary must include the most important impacts (particularly those that are unavoidable and irreversible), the key mitigation measures, proposed monitoring and supervision needs and the recommendations. The executive summary should be to the point and be about 2 to 5 pages.

Review of the Audit Report : The review of EIA report is held to judge the adequacy and quality of the report. It imparts authority and public confidence in the findings of EIA. The main purpose of the review is to determine whether the information included in the report is sufficient for decision making. Review can add value to the EIA process by acting as a quality assurance instrument and ensuring credibility of the process. According to UNEP (2002), the key objectives of EIA review are to:

(a) Assess the adequacy and quality of an EIA report.

(b) Take account of public comments.

(c) Determine if the information is sufficient for a final decision to be made.

(d) Identify the deficiencies in the EIA report.

Generic criteria listed by UNEP (2002) that can be used to conduct an EIA review include:

(i) Legal EIA requirements,

(ii) Environmental standards or guidelines,

(iii) Principles of good EIA practice and

(iv) Knowledge of the project and its typical impacts.

It must be realised that EIA is an aid to decision making and decision makers must keep in mind the welfare of the society as well as protection of environment.

The Final Decision : The decision makers are required to inform about the outcome of the EIA which may be, either of the following :

(a) The proposal is accepted

(b) The proposal is rejected or

(c) The proposal needs to be revised with comments and reviewed afresh.

Environment Impact Assessment (EIA)

(Laws, Rules and Notifications)

[EIA Notification, 2006 as amended alongwith, Official Memorandums/Circulars upto July 2018]

Summary

Environment Impact Assessment Notification, 2006

(1) Requirement of Prior Environmental Clearance (EC) : The following projects or activities shall require prior Environmental Clearance (EC) from the concerned regulatory authority, which shall hereinafter be referred to as the Central Government in the Ministry of Environment and Forests (MoEF) for matters falling under Category 'A' in the Schedule and State level the State Environment Impact Assessment Authority (SEIAA) for matters falling under Category 'B' in the said schedule [and at District level, the District Environment Impact Assessment Authority (DEIAA) for falling under Category 'B2' for mining of minor minerals in the said schedule], before any construction work or preparation of land by the project management except for securing land, started on the project or activity :

(i) All new projects or activities listed in Schedule to the notification;

(ii) Expansion and modernization of existing projects or activities listed in the Schedule to this notification with addition of capacity beyond the limits specified for the concerned sector, that is, projects or activities which cross the threshold limits given in the Schedule, after expansion or modernization.

(iii) Any change in product-mix in an existing manufacturing unit included in Schedule beyond the specific range.

(2) State Level Environment Impact Assessment Authority (SEIAA) is constituted by the Central Government comprising of 3 members including Chairman and Member Secretary, who is nominated by State Government or Union Territory (UT) Administration concerned. The other member will be an expert fulfilling the eligibility criteria.

(3) District Level Environment Impact Assessment Authority (DEIAA) will have four members including Chairperson and Member Secretary and shall be constituted by the

Central Government. The District Magistrate will be ex-officio chairman. The member secretary will be the Sub-Divisional-Officer (SDO) of the District Head Quarter. Other two members will include seniormost Divisional Forest Officer (DFO) and an expert.

The DEAC shall function on the principle of collective responsibility and Chairman shall endeavour to reach a consensus in each case otherwise majority shall prevail.

(4) Categorization of Projects and Activities : All the projects and activities are broadly categorised in two categories—Category A and Category B. The categorization is based on the spatial extent of potential impacts and potential impacts on human health and natural and man-made resources.

All category A projects or activities, including expansion and modernization of existing projects and activities and change in product mix, shall require prior environmental clearance from the Central Government in the Ministry of Environment and Forest (MoEF) on the recommendations of an Expert Appraisal Committee (EAC), constituted by Central Government.

All Category 'B' projects and activities including expansion and modernization of existing projects or activities will require prior environmental clearance from the State/UT Environment Impact Assessment Authority. The SEIAA shall base its decision on the recommendations of a State or Territory level Expert Appraisal Committee (SEAC). In the absence of duly constituted SEIAA or SEAC, a Category 'B' project shall be considered at the central level.

The 'B2' category projects pertaining to mining of minor mineral of lease area less than or equal to 5 Ha shall require prior environmental clearance from DEIAA. The decision will be based on the DEAC recommendations.

The EAC, SEACs and DEACs shall meet atleast once every month. The EAC and SEACs shall be reconstituted after every three years.

Application for Prior Environmental Clearance (EC) :

An application for seeking prior environmental clearance in all cases shall be made by the project proponent in the prescribed Form 1 and Form 1A, 1M if applicable, after the identification of prospective site(s) for the project and/or activities to which application relates. The project proponent shall furnish alongwith the application, a copy of the pre-feasibility report, in addition to Form 1, 1A or 1M; and in case of a construction project or activity, a copy of the conceptual plan shall be provided instead of pre-feasibility report.

Stages in the Prior Environmental Clearance (EC) :

The environmental clearance process for new projects will comprise of maximum of 4 stages, all of which may not apply to particular cases. These four stages in sequential order are:

- Stage I - Screening (only for B projects & activities)
- Stage II - Scoping
- Stage III - Public Consultation
- Stage IV - Appraisal.

In Stage I (screening), the 'B' projects or activities are scrutinized by SEAC for determining whether or not the project or activity requires further environmental studies for preparing

EIA. In case no further study is required, the B project or activity is labelled as B2. Those requiring an EIA report are termed as B1 category.

In Stage II, **Scoping** refers to the process to determine detailed and comprehensive Terms of Reference (TOR) addressing all relevant environmental concerns for preparation of EIA-Report in respect of project or activity for which clearance is sought. Standard TOR developed by Ministry in consultation with EAC or SEAC shall be deemed approval for the project or activities. The Appraisal Committees may stipulate additional TOR if found necessary.

The Stage III, **Public consultation** refers to the process by which the concerns of local affected persons or other stakeholders are taken into account. All category A and B1 projects or activities shall undertake public consultation. Category B2 projects, modernization of irrigation projects, expansion of roads and highways, building construction projects, area development projects and few others are exempted from this stage of EC process.

The public consultation ordinarily have two components:

(a) Public hearing at the site or in its close proximity district wise for ascertaining concerns of local affected persons as prescribed in the notification.

(b) Obtain written responses from persons having stakes in the environmental aspects of the project or activity.

The public hearing in all cases shall be conducted by the State Pollution Control Board (SPCB) or the Union Territory Pollution Control Committee (UTPCC) concerned in the specified manner and forward the proceedings to the regulatory authority.

If it is not possible to conduct the public hearing due to local situation, the authority may decide that the public consultation in the case need not include the public hearing clause.

For obtaining responses in writing from other concerned persons having a plausible stake in the environmental aspects of the project or activity, the concerned regulatory authority and the State Pollution Control Board (SPCB) or UTPCC shall invite responses by communicating them with the summary EIA report in the format provided in EIA notification. The responses are kept confidential. All the responses received as part of this public consultation process shall be forwarded to the applicant through the quickest available means.

After completion of the public consultation, the applicant shall address all the material environmental concerns expressed during this process, and make appropriate changes in the draft EIA and Environment Management Plan (EMP). The final EIA report, so prepared shall be submitted by the applicant to the concerned regulatory authority for appraisal. The applicant may alternately submit a supplementary report to draft EIA and EMP addressing all the concerns expressed during the Public Consultation exercise.

In Stage IV, **Appraisal** means the detailed scrutiny by the Expert Appraisal Committee or State level Expert Appraisal Committee of the application and other documents like the Final EIA report, outcome of the public consultations including public hearing proceedings, submitted by the applicant for grant of environmental clearance. The appraisal shall be made by Expert Appraisal Committee or State level Expert Appraisal Committee concerned in a transparent manner in presence of the applicant. On conclusion of this proceeding, the EAC or SEAC concerned shall make categorical recommendations to the regulatory authority concerned either for grant of prior environmental clearance on stipulated terms and conditions, or rejection of the application for prior environmental clearance, together with reasons for the same.

For the projects which are not required to undergo public consultation or submit EIA-report, appraisal is carried out on the basis of applications, other relevant validated information or even a site visit by EAC or SEAC.

The appraisal completed by EAC or SEAC must reach the competent authority within 60 days or within 15 days where public consultation is not necessary.

A procedure has been prescribed for appraisal in EIA Notification 2006.

Prior Environmental Clearance (EC) Process for Expansion or Modernization or Change of Product Mix in existing Projects :

All applications seeking prior environmental clearance for expansion with increase in the production capacity or with increase in least area or for modernisation of existing unit due to change in process or technology or involving a change in the product-mix shall be considered by EAC or SEAC who will decide on the due diligence necessary including preparation of EIA-report and public consultation and the application will be appraised accordingly for grant of environment clearance (EC).

Any change in configuration of the plant will not require EC, if there is no change in production and pollution load. Same will apply to any change in product mix. No change in pollution load will have to be certified by the concerned State Pollution Control Board as per the provisions in Notification.

Guidelines have also been provided for preparation of Distinct Survey Report for Sand Mining or River Bed Mining and mining of other minor minerals.

Grant or Rejection of Prior Environmental Clearance (EC) :

The regulatory authority shall normally accept the recommendations of the EAC or SEAC or DLEAC concerned. In case where it disagrees with these recommendations, the regulatory authority shall request reconsideration by these committees while stating the reasons for disagreement. An intimation of this decision shall be simultaneously conveyed to the applicant. The EAC, SEAC or DLEAC shall consider the observations of the regulatory authority and furnish its views on the same within stipulated period. Now the decision of the regulatory authority shall be final and conveyed to the applicant.

Deliberate concealment and/or submission of false or misleading information or data which is material to screening or scoping or appraisal or decision on the application shall make the application liable for rejection, and cancellation of prior environmental clearance granted on that basis. The decision of rejection is taken by the regulatory authority after giving a personal hearing to the applicant, and following the principles or natural justice.

Validity of Environmental Clearance :

The validity of Environmental Clearance will be ten years in the case of River Valley Projects, thirty years for mining projects and seven years for all other projects and activities. For construction projects it can be extended by three years. The period of validity can be extended by the regulatory authority in consultation with the Expert Appraisal Committee or State Level Expert Appraisal Committee or District Level Expert Appraisal Committee as the case may be.

Post Environmental Clearance Monitoring :

It shall be mandatory for the project proponent to make public the environmental clearance granted for their project alongwith the environmental conditions and safeguards. The Ministry of Environment and Forest (MoEF) and Environmental Impact Assessment Authority shall also place the environmental clearance in the public domain on government portals.

The copies of the environmental clearance shall be submitted by the project proponent to the heads of local bodies, Panchayats, and Municipal Bodies and relevant government offices.

Half Yearly Compliance Report :

It shall be mandatory for the project management to submit half-yearly compliance report in respect of the stipulated prior environmental clearance terms and conditions to the regulatory authority concerned on 1st June and 1st December of each year. All such compliance reports shall be public documents and shall be displayed on the website of concerned regulatory authority.

Preparation and Presentation of Environmental Impact Assessment (EIA) Report and Environment Management Plan (EMP) :

The Environmental Consultant Organisations which are accredited for a particular sector and the category of project for that sector with the Quality Council of India (QCI) or National Accreditation Board for Education and Training (NABET) or any other agency as may be notified by the Ministry of Environment, Forest and Climate Change from time to time shall be allowed to prepare the Environmental Impact Assessment Report and Environmental Management Plan of a project in that sector and category and to appear before the concerned Expert Appraisal Committee (EAC) or the State Expert Appraisal Committee (SEAC). The Ministry will also prepare a panel of national level reputed educational and research institutions to work as Environmental Consultant Organisations.

Integration of Environmental Condition in building bye-laws :

For construction of buildings permission being granted by the local authorities. The states adopting monitorable environment conditions in the building bye-laws shall not require a separate environmental clearance from the ministry of environment, forest and climate change for individual buildings.

After the State Government notifies the bye-laws and rules concurred by the Central Ministry, the Central Government may issue an order stating that no separate environmental clearance is required for buildings to be constructed in the areas concerned.

There are separate standards for Buildings Category '1' (5000 to $<$ 20,000 sq.m.) and other categories ($\geq$ 20,000 sq.m).

List of Projects or Activities Requiring Prior Environmental Clearance:

The projects are classified in eight groups.

Group 1 includes 7 types of projects:

1. Mining of Minerals
2. Slurry Pipelines

3. Oil Exploration
4. River Valley Projects
5. Irrigation Projects
6. Thermal Power Plants
7. Nuclear Power Projects.

Group 2 includes 2 types of projects:

8. Coal washeries
9. Mineral beneficiation

Group 3: Material Production

10. Metallurgical Industries
11. Cement Plants

Group 4: Material Processing

12. Petroleum Refining Industry
13. Coke Oven Plants, Coaltar Processing Units
14. Asbestos Milling and Asbestos Based Products
15. Chlor-alkali Industry
16. Soda Ash Industry
17. Skin/Hide Processing and Tanning Industry

Group 5:

18. Chemical Fertilizers
19. Pesticide Industry
20. Petro-chemical Complexes
21. Man-made Fibers Manufacturing
22. Petroleum Products and Petrochemical Based Processing
23. Synthetic Organic Chemical Industry
24. Distilleries
25. Integrated Paint Industry
26. Pulp and Paper Industry
27. Sugar Industry

Group 6: Service Sector

28. Oil and Gas Transportation Pipelines
29. Isolated Storage & Handling of Hazardous Chemicals

Group 7: Physical Infrastructure including Environmental Services

30. Air Ports.
31. All ship Breaking Yards
32. Industrial Processing Zones (EPZs) and Special Economic Zones (SEZs), Biotech Parks, Leather Complexes.
33. Common Hazardous Waste Treatment: Storage and Disposal Facilities (TSDFs).
34. Biomedical Waste Treatment Facilities.

35. Ports, Harbours, Breakwaters Dredging
36. Highways
37. Aerial Ropeways
38. Common Effluent Treatment Plants (CETPs)
39. Common Municipal Solid Waste Management Facility (CMSWMF)

Group 8: Building/Construction Projects/Area Development Projects and Townships

40. Building and Construction Projects
41. Townships and Area Development Projects

Note :

(1)General Condition (GC) :

Any project or activity specified in Category 'B' will be appraised at the central level as Category 'A', if located in whole or in part within 5 km. from the boundary of : (*i*) Protected areas (*ii*) Critically polluted areas (*iii*) Eco-sensitive areas (*iv*) Inter-state and International boundaries. Items in Group 1 and Group 7, the appraisal shall be made at central level even if located within 10 km.

(2) Specific Condition (SC) :

If any Industrial Estate/Complex/EPZs/SEZs/Biotech Parks/Leather Complex in Group 4 and 5 will not be required to take prior environmental clearance, so long as the terms and conditions are complied with. If not complied with the Estate/Complexes will be held responsible for violation throughout the life of Complex/Estate.

Categories of Projects

All projects and activities are broadly categorized as Category A and Category B projects on spatial extent of potential impacts on human health and natural resources. Projects under Category A require the prior environmental clearance from the MOEF and Category B projects require clearance of State Environmental Impact Assessment Authority (SEIAA).

The objective of this notification is to formulate a transparent, decentralized and efficient regulatory mechanism to:

(a) incorporate necessary environmental safeguards at planning stage;

(b) involve stakeholders in the public consultation process;

(c) identify development projects, based on impact potential instead of the investment criteria.

Schedule to this Notification gives a comprehensive list of projects or activities requiring prior environmental clearance. Here, we are mentioning only important *infrastructural projects* categorizing them in Category A and B:

Project or Activity	*Category A*	*Category B*
1. River Valley Projects	(a) ≥ 50 MW hydel generation (b) ≥ 10,000 ha of culturable command area	(a) < 50 MW ≥ 25 MW Hydel generation (b) ≤ 10,000 ha of culturable command area

2. Thermal Power Plant	(a) ≥ 500 MW (b) ≥ 50 MW (pet coke, diesel and all other fuels)	(a) < 500 MW (b) < 50 MW ≥ 5 MW (pet coke, diesel and all other fuels)
3. Airports	All projects	
4. All Ship Breaking yards.	All projects	
5. Industrial Estates/parks/ complexes/arras, Export processing zones (EPZs), Special Economic Zones (SEZs), Biotech parks, Leather complexes	(a) If atleast one industry falls under category A, entire area shall be treated as category A. (b) Industrial estates with area greater than 500 ha and housing atleast one category B industry.	(a) Industrial estates housing atleast one category B industry and area < 500 ha. (b) Industrial estates of area > 500 ha and not housing any industry belonging to category A or B.
6. Ports, Harbours	≥ 5 million TPA of cargo handling capacity	< 5 million TPA of cargo handling capacity and/or ports/harbours ≥ 10,000 TPA of fish handling capacity
7. Highways	(a) New National Highways (b) Expansion of National Highways greater than 30 km involving land acquisition and passing through more than one state.	(a) New State Highways (b) Expansion of National/ State Highways greater than 30 km involving land acquisition.
8. Building and Construction Projects		≥ 20,000 sq. mtrs. and < 1,50,000 sq. mtrs. of built up area
9. Townships and Area Development Projects		Covering an area ≥ ha and/ or built up area ≥ 1,50,000 sq. mtrs.

QUESTIONS

1. Define and explain 'Environmental Impact Assessment'.
2. Explain the importance of EIA.
3. Explain the objectives of EIA study.
4. Discuss the 'History of EIA'. Also discuss the 'History of EIA in India'.
5. Explain the concept of EIA.

6. Explain different benefits of EIA.
7. Write short notes on:
   (a) Core values of EIA (b) Aims of EIA
8. Who are the main stakeholders in an EIA process? Explain their role in the EIA study.
9. Explain the necessity of involving stakeholders in the EIA study.
10. Discuss EIA process in detail.
11. Write short notes on:
    (a) Scoping (b) Flow chart of EIA process
    (c) Terms of References for EIA
12. Explain the steps involved in EIA study.
13. Discuss mitigation measures considered in EIA study.
14. Explain Environmental Audit of a company explaining its objective and scope.
15. What are the benefits of Environmental audit?
16. Discuss various steps involved in an Environmental Audit.
17. Explain Environmental impact assessment methodology in detail.
18. Discuss different methods used for quantitaive impact identification.
19. Explain project appraisal economic study. Also explain cost-benefit analysis for the project appraisal.
20. Discuss about the final report of EIA. Explain its components.
21. Explain as to what baseline data should be collected for an EIA study.
22. Enumerate the environmental impacts expected from a water resource development project.
23. Explain the procedure for obtaining the 'environmental clearance' from MoEF.
24. Explain the salient features of MoEF notification dated September 14, 2006.

23

Chapter

Environmental Management and Environmental Management Plan

23.1 INTRODUCTION TO ENVIRONMENTAL MANAGEMENT

As discussed earlier, environment is "the sum total of all conditions and influences that affected the development and life of organisms." Therefore, environmental management is "the reducing or minimisation of the impact of human activities on the environment, thus, an endeavour to avoid overuse, misuse and abuse of environmental resources."

Environmental management can be defined as under:

1. It is the process of balancing the socio-economic, technological, cultural and ecological forces in the development and identify allocation of resources in order to fulfill the needs and aspirations of present and future generations of mankind.
2. Environmental management is an inter-disciplinary approach to resource conservation and re-cycling and it acts as a regulatory force on human greed in resource exploitation and resource wasting.
3. It is proper resource use and resource management.
4. Environmental management involves: (*i*) judicious and gainful utilisation of natural resources for socio-economic development, and (*ii*) maintenance of environmental quality by pursuing ecological principles.

The **main tasks** of environmental management are:

(i) environmental planning considering all aspects discussed in this chapter,

(ii) environmental status evaluation,

(iii) environmental impact assessment,

(iv) minimizing adverse environmental impact of activities, products and services,

(v) limiting any adverse impact within the laws/prescribed norms and their monitoring,

(vi) safety of environment while working with hazardous materials,

(vii) management of disposal of waste from the construction sites,

(viii) considering positive environmental contribution particularly after completion of construction, and

(ix) mechanism to review concerns of interested parties.

Thus, we must make attempt to identify the priority areas and the strategies to solve the problems that should lead to sustainable development. In managing the environment, our basic philosophy should be of harmony with nature. As human activities of modern 'economic and technological man' have disturbed the harmonious relationships between the environment and man. Environmental management is, thus, the process to improve the relationship between man and environment so that quality of both may be improved. This improvement of relationship may be achieved through check on destructive activities of man, conservation, protection regulation and regeneration of nature.

Environment management must relate the demands and activities of man in such a way that the ability of environment to sustain future development remains unimparted. Environmental Management is, therefore, related to the rational adjustment of man with nature involving judicious exploitation and utilisation of natural resources without disturbing the ecological balance and ecosystem equilibrium.

Objectives of Environmental Management

Environmental management has following broad objectives:

(i) To restrict and regulate the exploitation and utilisation of natural resources mainly non-renewable.

(ii) To regenerate degraded environment and to renew natural resources.

(iii) To control environmental degradation and pollution.

(iv) To reduce the impacts of extreme event and natural disasters.

(v) To make optimum utilisation of natural resources and also re-cycle and re-use waste materials.

(vi) To assess the impacts of proposed projects and activities on the environment.

The environmental managers should always keep in mind that the earthly resources are finite and the environment is a natural gift and is a property shared by all types of life.

Tools for Environmental Management

For the purpose of environmental management, following tools are used to achieve sustainable development by ensuring installation and operation of major industries and infrastructural projects, which are environmental friendly, causing the least adverse environmental impacts :

1. Evironmental Impact Assessments (EIAs).
2. Life Cycle Assessments (LCAs).
3. Environmental Audits.
4. ISO Certifications.

These tools have been discussed at appropriate places in the book.

23.2 INTRODUCTION TO ENVIRONMENTAL MANAGEMENT PLAN (EMP)

Environmental Management Plan (EMP) consists of all mitigation measures for each item-wise activity to be undertaken during the construction, operation and the lifecycle to minimize adverse environmental impacts as a result of the activities of the project. It also delineate the environmental monitoring plan for compliance of environmental regulation. It also states the steps to be taken in case of emergency, such as accidents at the site including fire.

The Environmental Management Plan (EMP) is the key to ensure a safe and clean environment. The desired results from the environmental mitigation measures proposed in the project may not be obtained without a management plan to assure its proper implementation and function. The EMP envisages the plans for the proper implementation of mitigation measures to reduce the adverse impacts arising out of the project activities during pre-construction, construction and operation stage.

Objectives of EMP

Environmental Management Plan (EMP) has following broad **objectives:**

1. To restrict and regulate the exploitation and utilisation of natural resources, mainly non-renewables.
2. To regulate degraded environment and to renew natural resources.
3. To control environmental degradation and pollution.
4. To reduce the impacts of extreme events and natural disasters.
5. To make optimum utilisation of natural resources by re-cycling and re-using waste materials.
6. To assess the impacts of proposed project and activities on the environment.
7. To formulate laws and regulations for the implementation of environmental protection and conservation programmes to maintain the environmental quality.

The environmental managers should always keep in mind that the earthly resources are finite and the environmental system is a closed system, and that the environment is a natural gift and is a public property shared by all types of life (plants and animals) in the biosphere. In other words, since the environment belongs to man and all living organisms, its planning should involve entire biological world.

In order to minimise the adverse impacts of development activities various environmental management plans have to be proposed. **Following EMPs are required to be submitted by the project developers to the Ministry of Environment & Forests (MoEF) for obtaining environmental/forest clearance of projects.**

1. Re-settlement and Rehabilitation Plan for Project Affected People.
2. Plan for Protection and Rehabilitation of Flora, Fauna, Forests and Wildlife.
3. Catchment Area Treatment Plan.
4. Rehabilitation and Re-location of Archaeological, Religious and Historical Monuments.
5. Compensatory Afforestation Plan.

6. Disaster Management and Emergency Response Plan.
7. Other Plans:
   (i) Biodiversity Conservation and Management Plan.
   (ii) Fisheries Development Plan.
   (iii) Public Health Delivery System.
   (iv) Provision of LPG Depots and Energy Conservation Measures.
   (v) Restoration of Muck Disposal Sites.
   (vi) Landscaping and Restoration of Construction Area.
   (vii) Green Belt and Voluntary Afforestation.
   (viii) Health Management/Control of Water-borne Diseases.
   (ix) Solid Waste Management and Sanitary Management.
   (x) Air Pollution Management.

23.3 ISSUES COVERED IN EMP

The environmental impacts and their mitigation measures of proposed project in the light of the construction activity of project including housing, infrastructure, dumping of excavated material, excavation in borrow area etc. Need to be worked out. The main issues which need to be properly addressed, so that their impacts do not continue after the construction phase, are discussed hereunder:

23.3.1 Impact and Mitigation Measures on Flora and Fauna (Biodiversity Management)

Baseline studies and impact on flora, fauna and forest due to project activities are conducted for the project influence area (7 to 10 km area around the project), project immediate affected area (500 m on all sides of project sites) and the directly affected area, which includes the land required for carrying out various project activities. Various mitigation measures to minimise the impacts are taken under the Biodiversity Management Plan.

The major direct impacts on natural vegetation cover occurs at the project site, areas selected for quarrying, muck disposal activities and along the access roads. At the major construction sites, dust is likely to create air pollution. Indirect impact on forest cover would occur as a result of increased demand for forest products.

Based on the impact assessment study, a plan for protection and rehabilitation of flora, lental fauna, forests and wildlife is prepared. Biodiversity management plan should provide for interventions to enhance the quality and the management of the forests, coming in the project influence area (especially in the hydro-power projects), even if the projects impacts are not significant. The plan should include the afforestation work (including compensatory afforestation as required by law), densification, medicinal plant plantation, and assisted natural regeneration in the area. Wildlife monitoring should be established in association with the Forest Department and hunting/poaching should be strictly banned.

Measures should be suggested in the plan to improve the forest cover and wild habitat, and should be strictly implemented as per approved Compensatory Afforestation Plan and Catchment Area Plan. Compensatory Afforestation should be carried out to compensate the loss of trees. It should be carried out in double, the area of forestland acquired as regulated by Forest (conservation) Act, 1980. Additional measures by the project should include roadside plantation, green belt development and Catchment Area Treatment (CAT) Plan so as to add to the natural resource to the area.

Indirect impacts of a construction project are due to various construction activities, such as generation of dust due to earthwork, excavation, transportation of construction materials, quarry, crusher and blasting operations, air pollution due to movement of construction vehicles, equipments and machineries etc. These impacts are for short term and limited to construction period only. The necessary measures should be taken to control the dust during construction period.

For the control of dust vehicles should be covered, approach roads should be kept clean and clear of dust and should be moist or watered, all plants, machinery and equipment should be handled so as to minimise generation of dust.

In order to *control soil and moisture conservation* following engineering works should be carried in forest areas:

- Vegetative check dams.
- Stone check dams.
- Spurs
- Gully plugging.
- Crate wire check dams
- Water percolation tanks.

The noise level of the construction site may increase due to various construction activities, which may cause disturbance to the fauna in the area. In order to control the generation of noise, the machineries, vehicles and equipments used, should strictly confirm to CPCB standard. All these vehicles, equipment and machinery should be fitted with exhaust silencers. These should be maintained regularly and sound proof gadgets should be used. Blasting should be carried out as per the statutory laws, regulation and rules pertaining to acquisition, transport, storage, handling and use of explosives.

Main mitigation measures for quarry activities are:

- The extraction of material should be done from the identified quarry site only.
- The quarry area should be reclaimed back. The pits formed should be backfilled by construction waste and site should be stabilized.
- Stockpiles should be utilised for re-development of quarry areas.
- Grasses and shrub species should be planted.

Regarding biodiversity, awareness programmes should be held for the stakeholders to develop concern for conservation of biodiversity in the area.

23.3.2 Water Quality Issues

In any hydro project water quality related studies should be carried out covering River flow measurement, water usage survey, Aquatic ecological studies, Pollution load studies and Assessment of water-borne diseases. These studies intend to identify the critical stretches of the river (within the study stretch) for flow requirements considering aquatic habitats, cultural values, water quality and waste assimilation. The dam construction will block the movement of the species. This impact should be studied and the fish management programme should be undertaken, if required, accordingly.

Generally effects of construction on water quality in a river system are primarily from the discharge of waste water (both construction of waste water and sanitary waste water from workforce housing areas) to the adjacent river and run-off form quarries and construction areas.

Discharge of waste water from construction areas that may affect water quality include discharge of water used to wash concrete mixing and hauling equipment, and other on-site discharges of water to the river system. Dissolved or suspended particles may also increase significantly as a consequence of run-off from the construction area.

Deposition of suspended solids in the river channel downstream form the construction area, may cause changes in the river channel, if significant amounts of sediments is contributed to the river (for example in muck disposal), may cause increase in concentration of heavy metals, and also may cause problems with spawning areas of fish in the river.

Discharge of sanitary waste form work camps and other human sanitary facilities can affect Biochemical Oxygen Demand (BOD) in the river as increasing fecal coliform bacteria concentrations, an indicator of potential disease risk to the downstream users. Suitable handling and treatment of sanitary wastes will significantly reduce the potential for adverse effects to water quality from this source.

Run-off from construction areas, such as excavated areas, quarries, access road construction, spoil disposal areas, flushing of oils, greases and other hydrocarbon lubricants from maintenance areas and potential flushing of other hazardous materials used at the project site for construction purposes. Containment of such materials and installation of erosion barriers will significantly reduce the potential effects to water quality. Maintenance of the erosion control measures and containment facilities must be accomplished throughout the construction period.

23.3.3 Construction Related Issues

Management of construction related issues is the responsibility of project proponent (through contractor, as and where relevant and described in the EMP). The environmental issues during construction stage generally involve safety and public health issue. The construction agency is required to comply with the laws with respect to Environment, protection, Pollution prevention, Forest conservation, Re-settlement, safety, Building and Other Construction Workers Act, and any other applicable law.

The EMP constitutes the Emission and Dust Management Plan, Construction/Labour Camp Management, Borrow Area Management Plan, Public Health and Safety, Green Belt

Development, Solid Waste and Muck Disposal Plan/Slope stabilization, drainage and bioengineering measures to protect slopes/erosion of soil.

The EMP involving construction related issues include measures to control air and noise pollution. Regular water sprinkling on construction sites, haul and unpaved roads particularly near habitation will be undertaken to control fugitive dust. Trucks carrying soil, sand and stone should be duly covered to avoid spilling. Plants, machinery and equipment should be handled so as to minimize generation of dust. To ensure that the sewage from labour camps do not pollute the river-water, sufficient number of community latrines and septic tanks need to be constructed.

23.3.4 Catchment Area Treatment (CAT) and Other Environmental Enhancements

Catchment area can be defined as the area of the drainage basin, which drains the water, sediment and dissolved material along a stream channel. Catchment area is the area draining into dam through different local streams. Catchement area treatment is an important mitigation measure in case of hydro-power development, as the life of the reservoir depends on the type of catchment from which it collects the drainage. A Catchment Area Treatment (CAT) plan is formulated with the main objective of arresting soil erosion in the upstream catchment. According to the topographic factors, soil type, climate and vegetation cover in the catchment area, various afforestation measures are required to be undertaken with the aim to check soil erosion, prevent/check siltation of reservoir and maintain storage capacity.

Soil erosion in catchment area disturbs the ecological balance by destroying the vegetated cover, dislocation wildlife, removing precious topsoil, modification of stream morphology. The removal of top fertile soil from catchment also adversely affects the agricultural production.

For clearance of a hydropower project for environmental forest angle by MOEF, a CAT plan for protection of total river basin is required to be submitted. A well-designed CAT plan is framed to address the issue of degradation and improve the environment of the area. It gives maximum emphasis through plantation; soil and water conservation measures, as discussed under biodiversity management.

The *broad objectives* of the CAT plan are:

1. To prevent degradation of land and to enhance its productivity through optimum land-use pattern.
2. To improve ecological balance amongst land, water, plants and animal life.
3. To reduce siltation of the reservoir/arresting soil erosion in the free drainage catchment soil by taking up adequate and effective soil conservation measures in erosion prone areas.
4. Rehabilitation of degraded slopes.
5. Rehabilitation of degraded forest areas through afforestation and facilitation natural re-generation.

23.3.5 Land Acquisition and Social Environment

A Social Impact Assessment (SIA) for the project area includes consultations with stake-holders, information on socio-economic and cultural features of the population and baseline data on land acquisition impacts. This information together with consultations with communities in the project area regarding their concerns, and subsequent verification to ensure that negative impacts are mitigated and that people receive benefits from the project.

To address land acquisition impacts, project specific re-settlement and rehabilitation policy which goes beyond the requirements of the National Re-settlement and Rehabilitation Policy (NRRP) of 2007 should be developed.

Social environment aspects have been discussed in the chapter on "Re-settlement and Rehabilitation (R&R)."

23.3.6 Muck Disposal Management Plan

Construction of projects involve excavation of earth and rock generating muck in large quantities. The muck, thus, need proper disposal. With the objective to protect the disposal areas from further soil erosion and develop the surrounding areas in harmony with the environment. The muck disposal plan should give the quantification of muck, identifies location and activities, wherein muck is generated and qualifies muck generated from each activities with relevance to disposal options.

The identified locations/sites of muck disposal is in conjunction with various characteristics *viz.*, landscape, cost effectiveness, nearness to source of generation, ground water/blockage to surface water, relief and scope of afforestation and erosion control/sediment arrest.

The muck disposal management plan identifies landscaping measures for disposal of muck, modes of transportation for muck disposal and specifies selection for use of biofertilizer method for vegetative growth on muck spread, delineates muck disposal options for each site implementation and development of landscape. Plantation should be carried out at the muck disposal sites for the stabilization of the slopes, landscaping and improving the aesthetic value to the area.

The waste material dumped at spoil pit should be mechanically compacted and properly levelled with suitable safe slopes. A retaining wall of suitable height should be constructed at the muck disposal sites. In order to re-store the area, all these dumping sites need to be rejuvenated by means of turfing and vegetation growth. Once the dumping completed these areas should be developed into terraces and re-stored by laying of the top soil on the top, digging of pits and planting of plant sapling. These areas may be developed as parks, gardens and viewpoint sites for tourists. In between the spoil tips drains or channels should also be provided for drainage of water.

23.3.7 Green Belt Development Plan

Although the forest loss due to various project appurtenances is compensated as a part of compensatory afforestation, it is proposed to develop green belt around the perimeter of various project appurtenances, select stretches along the periphery of water spread area in case of hydro projects, etc. General considerations involved while developing the green belt are:

1. Local/native tree species should be planted.
2. Planting of trees should be undertaken in appropriate encircling rows around the project site.
3. Generally fast growing trees should be planted.
4. It may be useful to have shrubbery infront of the trees so as to give complete coverage.

23.3.8 Quarry and Borrow Area Management

Project authorities/contractors are required to take quarry material only from licensed quarries. In the case of existing quarries, it should be ensured that all actions in these quarries are in accordance with the environmentally sound and legally acceptable manner. In case, additional quarries and dedicated crushing plants are established, it is necessary to ensure that all actions are in accordance with the environmental and legal requirements.

In the case of borrow areas, it is necessary to specify a detailed arrangement including the agreement with the owner of the land. It is also necessary to comply with the provisions of taxes, lavies, royalties etc. of the state government.

Quarry and borrow area rehabilitation programme is carried out to re-instate the quarry/borrow pit sites to a safe and secure area. Securing borrow pits/quarry sites in a stable condition should be fundamental requirement of the rehabilitation process. This could be achieved by filling the quarry/borrow pit with suitable materials to approximately the ground level/access road level.

Quarries and borrow pits may be backfilled with rejected construction wastes and should be given a vegetative cover. Plantation be done extensively for slope stabilization. During execution of work, care should be taken to facilitate water percolation and allow natural vegetation growth; re-establishment of previous natural drainage flows, improvement of site appearance. For re-storation of such sites care should be taken to use green manure, fertilizer and pesticides as per requirement.

23.3.9 Mitigation Measures for Approach Road Construction

The approach roads are constructed as a part of the access to the construction site. In a hilly environment, construction of roads sometimes disturb the scenic beauty of the area. In addition, landslides are often triggered due to road construction because of the loosening of rocks by water trickling from various streams. Steeply sloping banks are liable to landslide which can largely be controlled by provision of suitable drainage. The basic principle is to intercept and divert as much water as possible, before it arrives at a point, where it becomes a nuisance.

The other hazard is that of surface erosion of the bank, which is best controlled by vegetation. Engineering solutions, such as surface drainage, sub-surface drainage, toe-protection and rock-bolting can be used. Landslides can be stabilized by several methods: engineering or bio-engineering measures alone or a combination of these.

In hilly terrain, road construction often generates significant quantity of wastes (muck) due to the stripping of the rocks to make way for the roads. The stripped muck is generally cleared by dumping the material along the slopes. These dumped material finally flow down to the valleys and ultimately finds its way in the river. However, it is advisable that stripped material should be collected and dumped in the designated muck disposal area which will have check dams to prevent the muck to flow down into the river. After disposal operation is complete at the dump site, dumpyard should be contoured and vegetated.

Drainage of water from hill slopes and road surface is very important. All artificial drains should be linked with the existing natural drainage system. Surface drains should have gentle slopes and should be constructed such that soil is not eroded and carried away by high velocity flows.

23.3.10 Public Health Delivery System

The increase in water fringe area provides suitable habitats for the growth of vectors of various diseases and they are likely to increase the incidence of water-related diseases. Malaria is one such disease. The anti-malarial operations can be co-ordinated by Primary Health Centre (PHC) in the nearby villages and Hospital at Tehsil/District headquarters in association with the project proponents. In this regard, following precautionary measures should be taken:

1. Site selected for labour camp should not be in the path of natural drainage.
2. Adequate drainage system to dispose storm water should be provided.
3. Adequate vaccination and immunization facilities should be provided for workers at construction site.
4. Labour camps and resettlement sites should be atleast 2 km away from a main water body or quarry areas.

It is recommended that necessary medical facilities should be developed at the project site so as to serve labour population migrating in the area as well as the local population.

During the construction period, technical staff and workers are not only exposed to physical strain of work but also to the physical effects of the environment in which they are working. The health facilities to be provided at construction sites and labour camps have been discussed in detail in the 'Health Management' section in this book.

23.3.11 Landscaping and Re-vegetation

For 'Landscaping and Re-vegetation Plan' project should follow the biodiversity management plan, muck disposal management plan, green belt development plan, and borrow area and quarry area management plan. All areas disturbed by construction activity, including temporary access roads should be landscaped to reflect natural contours, restore suitable drainage paths and encourage re-establishment of vegetation.

Spoil heaps and excavated slopes should be re-profiled to stable batters and grassed to prevent erosion. Re-establishment of vegetation should be commenced at the earliest possible opportunity.

Landscape should be properly planned alongwith the buildings and other structures of the project. Ideally, a professional landscape architect should be consulted on planning the landscape.

Project authorities should ensure that, at the end of the construction phase, the construction areas have been re-graded to conform to the natural topography and that appropriate grasses and shrubs have been planted to start the re-vegetation process. Necessary measures need to be taken for restoration of the aesthetic environment of the area through landscaping and re-vegetation.

23.4 ENVIRONMENTAL MANAGEMENT SYSTEM (EMS)—ISO:14000

Environmental management system provides a structural process for the achievement of continual improvement, as it is a tool which enables the organisation to achieve and systematically control the level of environmental performance that it sets itself. EMS is an integral part of an organisation's overall management system. The basic concept of EMS is to manage all activities so as to eliminate or atleast minimise the degradation of environment in which we live and work.

EMS, therefore, enables an organisation to address environmental problems in a structured and systematic way and comes out with mitigation measures to handle such environmental problems. 180:14001 standard is mainly designed to protect environment. It provides an overall framework for EMS integrating it with overall business policy. It places emphasis on prevention rather than corrective action.

Environmental Management system provides:

1. Structured and systematic approach.
2. Control of environmental aspects.
3. Compliance with legislation.
4. Improvement in environmental performance.
5. Integrated with quality system.
6. Continual improvement.
7. Integration with different agencies.
8. Confidence to its interested parties.

The **elements/sub-elements required** under the EMS-14000 are:

1. Environment Policy.
2. Environment Planning:
   (i) Various environmental issues
   (ii) Legal and other requirements

(iii) Objectives and Targets

(iv) Environmental management programmes

3. Implementation and operation:

(i) Structure and Responsibility.

(ii) Training, Awareness and Competence.

(iii) Communication.

(iv) EMS Documentation.

(v) Document Control.

(vi) Operational Control.

(vii) Emergency Preparedness and Response.

4. Checking and Corrective Action:

(i) Monitoring and Measurement.

(ii) Non-conformance and Corrective and Preventive Action.

(iii) Records.

(iv) EMS Audit.

5. Management Review.

The ISO-14000 Family of Standards and Guides

The various standards and guides published by ISO under ISO-14000 series are listed in the following table :

Srl. No.	*ISO No.*	*Title of Publication*
1.	ISO-14001 : 1996 (Revised in 2004)	Environmental Management System (EMS)—Specification with guidance for use.
2.	ISO-14004 : 1996	EMS-General guidelines on principles, systems and supporting techniques.
3.	ISO-14010 : 1996	Guidelines for environmental auditing—General principles.
4.	ISO-14011 : 1996	Guidelines for environmental auditing—Audit procedures.
5.	ISO-14031 : 1999	Environmental management—Environmental performance evaluation guidelines.
6.	ISO-14040 : 1997	Environmental management—Life cycle assessment—principles and framework.
7.	ISO-14042 : 2000	Environmental management—Life cycle impact assessment.

23.5 INSTITUTIONAL AND IMPLEMENTATION ARRANGEMENT

The organisations executing the project, for taking care of environmental and social aspects, a separate department should be established at Corporate Office with a separate cell/ section at project headquarter supported by environmental engineers/scientists, a biodiversity specialist and a social scientist. The Environmental Management Cell (EMC) at project, level should be made responsible to handle all issues, related to different environmental attributes from investigation level to execution at project level.

23.5.1 Activities within the Project Area

The activities within project area, having major impact on environment will be land acquisition, felling of trees, clearing of land for construction work, establishing of construction camps, construction of project facilities and operation of the project on completion of work.

Implementation of management measures for impacts related to forest compensatory afforestation and tree felling is the joint responsibility of the project authorities and State Forest Department. Muck/quarry area redevelopment plan and landscaping should be implemented by the project authorities. Wherever contractors are involved, they should be responsible to implement mitigation measures and project authorities should monitor the implementation programme.

Functions of Corporate Environment Department

The functions of Corporate Environment Department include:

1. Obtaining statutory and non-statutory clearances, *viz.* Site/Environment Clearances, Forest Clearance, Wildlife Clearance and No Objection Certificate from concerned agencies/departments.
2. Co-ordination with the project on environmental matters and providing required assistance/inputs.
3. Compliance reporting to statutory/monitoring agencies.
4. Co-ordination and interaction with MOP, MOEF, State Forest Deptts., State Pollution Control Board (SPCB) etc.
5. Monitoring of all environmental studies/implementation of environmental conditions.
6. Policy related issues, *viz.* National Environment Policy, forestland transfer, apportionment, CAT etc.
7. Reply of Parliament/Assembly Questions and other quarries of government related to environment and other VIP references.

Functions of Project Environment Management Cell

1. Planning, implementation and monitoring of all environmental and social mitigation and compensation measures outlined in BMP, R&R policy and Rehabilitation Action Plan (RAP).
2. Implementation of community development initiatives under CSR.

3. To ensure that relevant environmental mitigation measures are reflected in the project contract.
4. Monitor mitigation measures that are implemented by contractor to ensure compliance with Contractor's contract.
5. Prepare regular reports for submission to SPCB and MOEF.
6. Implementation and monitoring environmental mitigation measures.
7. To work in close association with the Administrator and Commissioner of R&R for land acquisition and implementation of RAP.
8. Hiring of consultants and NGO, where required.
9. Conducting internal and external audits to ensure compliance with the EMP and procedures of the organisation.

Functions of Construction Contractors

1. Meeting effluent standards before discharging into surface water.
2. Managing waste at construction sites.
3. Muck disposal at pre-decided locations.
4. Location of construction camps at proper place—away from Forest Area and Settlements.
5. Arrange fuel for the construction camp—no tree cutting for fuel and hut making for the camp.
6. Manage erosion and sedimentation in construction area.
7. Managing onsite traffic-provide diversion signs/flags/guard along approach road.
8. Control noise emissions.
9. Control dust and fugitive emission.
10. Managing, preventing and developing emergency plans for chemical pollution incidents.
11. Implementing a health program for all persons engaged in undertaking construction works.

Functions of State Forest Department

1. Implementing Catchment Area Treatment (CAT) plan.
2. Implementing Compensatory Afforestation Plan.
3. Wildlife Conservation Programme.
4. Develop Herbal Garden in consultation with State Medicinal Plant Board (SMPB).
5. Facilitate and guide *Vanpanchayats*.

Functions of Administrator–R&R

1. Approval of award passed by Land Acquisition Officer.
2. Oversee the disbursement of compensation for land acquired.
3. Formulation of Re-settlement and Rehabilitation Plans/schemes.

4. Minimize displacement of persons and identify non-displacing or least displacing alternatives in consultation with the project authorities.
5. Hold consultation with the PAPs and ensure the interest of adversely affected PAFs of STs and weaker sections.
6. Verification and approval of PAFs eligible for various entitlements.
7. Issuance of individual certificates to the landless people and those loosing houses.
8. Review the implementation of RAP and community development works.

Functions of NGO (Engaged by the Project)

1. Develop rapport with PAFs and between PAFs and Project authorities.
2. Verification of PAFs.
3. Consultations with the VDAC and local community during the implementation of the RAP. (Rehabilitation Action Plan).
4. Preparation of micro plans and assist the PAFs in receiving the rehabilitation assistance.
5. Motivate and guide PAP for productive utilization of the compensation and assistance amounts.
6. Assist the PAFs in getting benefits from the appropriate local development schemes.
7. Forward the grievances of the PAFs to the grievance redressal mechanism.
8. Assist the PAFs in usage of modern techniques in agriculture and animal husbandry.
9. Assess the level of skills and efficiency in pursuing economic activities, identify needs for training and organize training programmes.
10. Participate in the monthly review meetings with the Project Social Group and VDACs.
11. Carry out other responsibilities as required from time to time.

Functions of Van Panchayats

1. Protection of Plantation sites in the project area.
2. Regulate usage of Van Panchayat resource.
3. Cultivation of herbal species.
4. Develop nurseries for plantation under the Project area.

23.5.2 Activities in the Project Influence Area

The major activities which are involved in the project influence area are:

- Biodiversity Management.
- Catchment Area Treatment (CAT) Plan.
- Compensatory Afforestation.
- Wild Life Management.
- Awareness Programme.
- Income Generation Activities.

- Control of Communicable Diseases.

Biodiversity management is the joint responsibility of the project and the State Forest Department. CAT plan, Compensatory afforestation and Wildlife management should be implemented by the State Forest Department.

The Project is responsible to undertake environmental awareness programme in the area involving Vanpanchayats, Gram Panchayats, Schools and Women's group. Income Generation activities should be the responsibility of project. Project should also organise health camp and periodic checkup in the area in consultation with the Public Health Centre (PHC). The large influx of labourers in the project is a major problem and requires monitoring by the District Administration.

23.6 MITIGATING MEASURES

While studying the adverse impacts of projects, it is necessary to find out the extent of impact, analyse it and also find out mitigating/controlling measures. If it is not possible to eliminate the adverse impact completely, efforts should be made to find out alternatives and if even then it is not admissible, atleast minimise the impact. Some of the common mitigation measures taken in this regard are discussed in this section.

23.6.1 Preventing Air Pollution

In order to prevent air pollution following mitigation measures must be taken:

1. Planting of trees purify the air. Trees breathe carbon dioxide and release oxygen. Forests act as vast sinks for carbon dioxide and tend to reduce global warming.
2. Curtail vehicular emission by using them to minimum possible, using fuel efficient engines, keeping vehicles in maintained condition. Conserve fuel, monitor the fuel emission.
3. While executing a project an EMP should be prepared for formation, implementation and monitoring of environmental protection measures.
4. Emission levels of pollutants from chimneys should conform to the pollution control standards, prescribed by the central or state pollution control boards, for which necessary arrangements must be made. Generally used air cleaning devices are: Electrostatic precipitators, Cyclone separators, Wet collectors, Gravity settlers, Bag filters, Wet scrubbers etc.
5. Adequate dust suppression measures, such as regular water sprinkling on construction sites, haul and unpaved roads, particularly near settlement may be undertaken to control fugitive dust.
6. Workers may be provided with mask to prevent breathing problems.
7. Trucks carrying sand, soil and stone may be duly covered to avoid spilling.
8. Air quality monitoring may be conducted at construction sites.

23.6.2 Prevention of Noise Pollution

Areas of high level noises in construction industry are: movement of compressed air into the atmosphere, the combustion fuel in the burner, crushers, blowers, exhaust fans, pug mills. Employees working near plant area outside cabin are exposed to excessive noise. Excessive noise is irritating, mental fatigue and reduce the efficiency. Following are some of the common methods used for noise control:

1. Replacing noise producing machinery with quiet alternatives.
2. Interrupting the path of sound with insulating materials.
3. Use of baffles, sound proof materials.
4. Mounting the noise and vibration source on flexible supports.
5. Doing proper maintenance and lubrication of equipments.
6. Protecting the workers by providing earplug, earmuff, helmets, control cabins etc. Thus, shielding the receiver.
7. Trees should be planted in plenty in colonies, hospitals, school campuses etc. to protect the people from noise pollution from outside.
8. Noise producing activities may be prohibited during night hours.
9. Blasting should be carried out as per the statutory laws, regulation and rules pertaining to acquisition, transport, storage, handling and use of explosives.
10. Noise level monitoring must be conducted during construction phase.

23.6.3 Prevention of Water Pollution

1. The water quality required to be maintained to desired standard.
2. Discharge of waste from construction/labour camp into water bodies must be strictly prohibited. Adequate drainage system to dispose stormwater drainage from the labour colonies should be provided. Labour camps will be located atleast 200 m away from water bodies and settlements.
3. Silt fencing may be provided near water bodies to avoid spillage of construction material.
4. Water quality monitoring should be conducted during construction phase.
5. Proper water and water management plan must be prepared for the colony area.
6. Provision of temporary drainage arrangement due to construction activities should be made.
7. Construction methodologies with minimum or no impact on water quality may be adopted and disposal of construction wastes at designated sites may be provided.
8. Construction activity maybe prohibited during rainy season near water bodies.

23.6.4 Prevention of Impacts on Aquatic Ecology

1. Where dam construction block the local movement of the species, scientific management of the existing stock needs to be adopted for the conservation of species.

2. Fish management programme should be undertaken in consultation with the concerned specialist organisation.
3. Ban on fishing should be enforced in the affected stretch of river during lean season and critical period of spawning season.
4. Discharge of untreated sewage into the river must be strictly prohibited.

23.6.5 Mitigation Measures from Impacts of Construction Camp

1. Construction of camps should be located outside forest.
2. Construction of camps should be located atleast 500 meters away from habitation.
3. Adequate supply of fuel in the form of kerosene or LPG may be provided to the labours to avoid felling of trees for cooking etc.
4. Adequate sanitary facilities should be provided within the camp.
5. Reference to the illegal cutting trees, hunting and other prohibited activities in community areas to be included in the contract document.
6. Periodic health checkups and awareness programmes on communicable diseases may be conducted in consultation with State Public Health Department.
7. In every Construction Camp, first-aid facility and ambulance should be provided.
8. During construction workers may be hired from local communities also to avoid social/resources conflict.
9. It should be ensured that after completion of construction, area of the camp be cleared of the debris and other wastes deposited. The land should be restored back to its original form/condition.
10. Discharge of waste from construction/labour camp into water bodies must be strictly prohibited. Adequate drainage system to dispose storm water drainage from the labour colonies should be provided.

23.6.6 Prevention of Impacts on Health

1. For maintaining community health, programme may be framed, such as:
   (a) Augment existing health programme.
   (b) Health education for local project workers and community.
   (c) Vaccination programmes.
2. Specific HIV/AIDS prevention activities.
3. Safety goggles, helmets, earplugs, masks and other PPEs must be provided to the workers.
4. All workers employed on mixing of asphaltic material, cement, lime mortars, concrete etc. may be provided with protective footwear and protective goggles. Workers involved in welding work may be provided with welder's protective eye shields.
5. Necessary steps must be taken to prompt first-aid treatment of all injuries likely to sustain during the course of work.
6. Public Health Delivery System be provided, as discussed earlier.

7. Malaria control measures which aim at destroying the habitat and interrupting the life cycle of mosquitoes by mechanical or biological or chemical means needed to be implemented.

23.6.7 Mitigation Measures for Impact on Biodiversity

A. Pre-Construction Phase

1. **Acquisition of Forestland.**
   (a) Obtain forest clearance from MoEF.
   (b) Strict implementation of approved compensatory afforestation plan in accordance with Forest (Conservation) Act, 1980 and State Govt. Forest Policy.
   (c) Strict compliance of MoEF stipulated condition(s) of forest clearance.
2. **Felling of Trees (other than Forestland).**
   (a) Compensatory afforestation should be carried to compensate the loss of trees. Double the number of trees should be planted in lieu of trees felled.
   (b) Compensation should be given as per R&R policy.
   (c) Compensation for fruit bearing trees should be compensated as per R&R policy.
3. **Clearing of Project Sites for Construction.**
   (a) Compensatory afforestation will be carried out as per measures given in para (1) and (2) above.
   (b) Vulnerable species found in the area should be developed in separate herbal garden to be developed at suitable place in consultation with forest department.

B. Construction Phase

1. **Impact on Fauna.**
   (a) Poaching/hunting should be strictly banned in the forest area.
   (b) Plantation of tree species which are major sources of fodder for wildlife in consultation with forest department in degraded and open areas.
   (c) Awareness programme on environment and wildlife conservation should be provided to work force. Forest Act and Wildlife Act should be strictly adhered to.
2. **Quarry Activities.**
   (a) Extraction of material should be done from the identified quarry site only.
   (b) The quarry area should be re-claimed back. The pits formed should be backfilled by construction waste and site should be stabilized.
   (c) The topsoil (150 mm) from all areas should be preserved in stockpiles. Stockpiles should be utilized for re-development of quarry areas.
   (d) Grasses and shrub species should be planted.
3. **Soil Erosion.** Already discussed.
4. **Muck Disposal.** Already discussed.

C. Operation Phase

1. **Loss of Flora.**

   (a) Proper protection measures should be taken for the plantation work carried under the project. Vanpanchayats should be involved in afforestation activity and monitoring of the plantation work.

   (b) Fencing of plantation area should be done.

   (c) Watchman should be employed to take care of plantation for minimum of 3 years.

2. **Loss of Fauna.** Awareness programmes should be held for the stakeholders to develop concern for conservation of biodiversity in the area.

23.6.8 Use of Fly Ash

In the past, fly ash produced from coal combustion was simply entrained in flue gases and dispersed into the atmosphere. This created environmental and health concerns that prompted laws which have reduced fly ash emissions. Worldwide, most of the fly ash produced from coal power stations is disposed off in landfills and also ponds.

The recycling of fly ash has become an increasing concern in recent years due to increasing landfill costs and current interest in sustainable development. The ways of fly ash utilization include the following applications:

1. Fly ash is used as a replacement for some of the portland cement content of concrete.
2. Embankments and other structural fills (usually for road construction).
3. Grout and flowable fill production.
4. Waste stabilization and solidification.
5. Mine reclamation.
6. Stabilization of soft soils.
7. Road subbase construction.
8. Brick production.
9. Mineral filler in asphaltic concrete.
10. Agricultural uses: soil amendment, fertilizer, cattle feeders, soil stabilization in stock feed yards, and agricultural stakes.
11. Other applications include roofing tiles, PVC pipe, railway sleepers, highway sound barriers, sign posts, paving stones, park benches, etc.

23.7 ENVIRONMENTAL MONITORING

Environmental monitoring is one of the most important component for sustainability of any construction project and is an integral part of any environmental assessment process. Monitoring will determine during the construction period whether the mitigation measures undertaken are achieving their objectives. A monitoring plan is framed to monitor the implementation of activities provided in EMP. However, a third party monitoring for the implementation of the EMP may also be conducted by an agency or a committee of experts who will visit the project site, generally, twice a year to oversee and evaluate the EMP implementation and suggest improvements.

Environmental monitoring is essential for:

1. Ensuring that impact do not exceed the established legal standards.
2. Checking the implementation of mitigation measures in the manner described in the EIA report.
3. Providing early warning of potential environmental damage.

An Environmental Monitoring Programme is designed with the following objectives:

(a) Assess the changes in environmental conditions, if any, during construction and operation of the project.

(b) Monitor the effective implementation of mitigation measures.

(c) Warning of any significant deterioration in environmental quality so that additional mitigation measures may be planned in advance.

From the monitoring point of view, the important parameters are: water quality, ecology, air quality, noise, soil erosion and siltation, incidence of water-related diseases:

S.No.	*Monitoring Component*	*Parameters*
1.	Air quality	SPM, SO_2, NOX at different locations.
2.	Water quality	Physical, chemical and bacteriological parameters and heavy metals at different locations (surface water and groundwater) including pH, temperature etc.
3.	Noise and Vibration levels	At different locations in d_B (A).
4.	Meteorology	Wind speed and direction, rainfall, temperature, relative humidity, solar radiation.
5.	Effluent from septic tanks	pH, BOD, COD, TSS, TDS (Total dissolved solids).
6.	Water-related diseases	Identification of water-related diseases, adequacy of local vector control and curative measure etc. at labour camps and colonies, three times a year.
7.	Soil	Parameters related to agricultural and forestation potential. Soil erosion rates, stability bank embankment etc.
8.	Ecology	Status of afforestation programme, green belt develop programme. Existing terrestrial and aquatic flora and fauna in 10 km radius.
9.	Socio-economic	Socio-economic and demographic features and labour details.

23.8 ENVIRONMENTAL AUDITING

The term "Audit" is usually associated with the professional of finance and accounting. Auditing refers to the examination and assessment of certain type of performance. In the case of EIA and EMP, an audit should assess the actual environmental impact, the accuracy of prediction, the effectiveness of environmental impact mitigation and enhancement measures, and the functioning of monitoring mechanisms.

Environmental auditing should compare monitoring results with data generated during the pre-project period. Comparisons can be made with similar projects or against standard norms. Relating actual impacts with predicted impacts will help in evaluating the accuracy and adequacy of EIA and EMP.

An *internal audit mechanism* is required to keep a watchful eye on its environmental effects, and to review and suggest ways and means for continuously upgrading its processes and practised to ensure reduced adverse environmental effects.

Environmental audit is, in fact, an organisation own management plan to help in the evaluation of the environmental protection measures, periodically at regular intervals. The final action plan prepared by the audit team would normally contain a specific implementation schedule. The management will finally study and implement the suggested measures considering their financial impacts. The environmental audit will, thus, help the management to:

1. Adopt the suggested waste treatment systems.
2. Recover and reuse materials.
3. Implement the action plan for controlling environmental pollution.
4. Take measures in accordance with the various environmental related laws.
5. Plan and move towards the goal of sustainable development.

QUESTIONS

1. What is meant by "environmental audits", and why such audits are necessary?
2. Discuss the importance of regular environmental audits in the functioning of a good environmental management system.
3. How does the ISO certification under ISO 14000 services of standards help the general well-being of the environment?
4. Explain as to how ISO certification helps in regards to sustainable development in a society.
5. Define environmental management. Enumerate the tasks of environment management. List objectives of environment management.
6. Discuss 'Environmenal Management Plan (EMP)' in detail.
7. What are the objectives of EMP? Give a list of tools for environmental management.
8. What are the water quality issues, which should be addressed in EMP?

9. What is the 'Catchment Area Treatment'? Discuss CAT plan.
10. Write short notes on:
    (a) Muck disposal plan.
    (b) Construction related issues to be covered in EMP.
    (c) Mitigation measures for quarry activities.
11. Explain the following in brief:
    (a) Green belt development plan.
    (b) Quarry and borrow area management.
    (c) Landscaping and re-vegetation.
12. What do you understand by 'mitigation measures' in 'environmental management'? Explain important mitigation measures for: (*a*) preventing air pollution, (*b*) preventing noise pollution, and (*c*) preventing water pollution in the construction of industrial project.
13. Write short notes on:
    (a) Use of fly ash.
    (b) Mitigation measures for impact on biodiversity.
14. What is 'environmental monitoring'? Why it is essential?
15. What are the implementation measures which should be made by the executing project organisation for taking care of social and environmental aspects?

24
Chapter

FIELD WORK (PROJECT WORK)

24.1 INTRODUCTION

In the book, we have discussed various environmental issues and related problems. In order to understand the environment, its advantages, problems, impacts and the state of seriousness ourselves, the students are required to take up field work (project work). This will help them to verify and validate the various descriptions and ideas presented in the book. The field work can be either undertaken individually or as a group activity.

By conducting field work/survey students can get an opportunity to understand himself how thigs are moving in the real environment. Thus, students will then begin to appreciate the nature and complexity of the environmental problems being faced.

24.2 CRITERIA FOR THE SELECTION OF FIELD WORK

The students should choose a field project which can be completed in prescribed time. It is always better to do a thorough job with a small project, instead of taking a big project and leave it incomplete. Following are some of the criteria which should be considered while selecting a project. Some of the criteria considered for the purpose:

1. Interest in the topic.
2. Usefulness of topic in real life.
3. Specialist knowledge is needed to be acquired.
4. If the project sound to be interesting, judging the earlier reports.
5. Easy accessibility to the project location, and necessary facilities are available in the field.
6. If the facilities are available to carry out the work each week.
7. For having more practical understanding about the ecosystems, natural resources etc. in a specific area.

8. Study of a polluted site or waste management systems in an urban, rural, industrial, agricultural or municipal areas.
9. Study of water resources, flora and fauna in a particular area.
10. Project (field work) maybe selected as per the discipline of student. For example:

(a) Botany / Zoology students may carry out study of plants, insects, birds etc in a given area; tracing the origin of different plants, trees, vegetables, and fruits available in an area; study of any seed bank preserving indigenous species.

(b) Science students can take up testing of water quality, air quality and resource mapping of the area.

(c) Engineering students can take up designing products with waste material, improving the design of solar energy products.

(d) Philosophy students can study about environmental ethics.

(e) Students of humanities and management can study about the effective ways of communicating to people about mobilizing them for preserving our environment.

(f) Economics students can take up the study of alternative economic models keep sustainable development in view.

24.3 PREPARATION FOR PROJECT WORK

It is always better to work as a volunteer or project trainee with an organisation, which is already working in the area or site. This organisation could be an NGO, a government department, or a company that is working for an environmental project or works related to environment.

For carrying out the project one should prepare himself in following steps :

1. Define objectives in detail and revise it as project proceeds.
2. Formulate the questions which are expected to be answered by you through the project.
3. Conduct a literature survey on the topic.
4. Keep notes of the relevant informations.
5. For interviewing a person go prepared with a set of questions. If permitted, you may record the interview.
6. Obtain permission for entering the project site. For this purpose always carry a letter of introduction from your institution and your identity card.

24.4 HOW TO WRITE A REPORT?

1. A report should be written, after detailed planning, considering the time frame available. Follow guidelines set by your institution or your project guide for formatting the report. The general format of the report should have following sections :

   (i) Executive Summary

   (ii) Contents

   (iii) Acknowledgement

(iv) Preface
(v) Introduction. The ecosystem or problem in brief
(vi) Objective of the study
(vii) Background
(viii) Observation and Results
(ix) Conclusions and recommendations
(x) References
(xi) Photographs

2. Keep the language simple. If your language abilities are weak, get the draft corrected by a professional.
3. Get the draft approved by the organisation and your guide.
4. Final version of the report should be thoroughly checked for errors and style of report.
5. Follow guidelines, if any, on the size, number of copies etc.
6. Generally a viva-voice examination is conducted on the project, for which you should be prepared.
   (a) A good project report should be used in job interviews.
   (b) You may write an article on your work for the appropriate journal, magazine, newspaper or website.

24.5 VISIT TO STUDY ENVIRONMENTAL ASSETS

After going through the book, and having knowledge about the nature, biotic and abiotic environment, students are expected to see the nature themselves. For this purpose they should start the study from the nearby natural environment, this will enable them to understand the availability of natural resources and functions of ecology, their importance from economic, aesthetic and recreational angle. Environment provide multiple benefits for mankind at the global, national, regional and local level. Environment has different significance for different groups such as rural, urban, tribal etc.

Degradation of the ecosystem is a serious threat to the mankind. Through the visit to the environmental ecosystems, the students can know the real threats/causes resulting to the environmental degradation. Students through these visits may try to learn various mitigation measures to save environment. During the study, students are required to have documentaion of (i) what they see, and (ii) findings after survey (going through records, asking from users and other stakeholders etc.).

In the study report, students are expected to :

(a) Describe the ecosystem as they see to it, with regards to structural nature, its quality, geographical features, flora and fauna. Also mention the environmental assets of the ecosystem.
(b) Describe the functioning of ecosystem, links between different species, their habitat, food chains, food webs.

(c) Use of natural resources be examined and decide whether utilisation is sustainable or not.

(d) If the ecosystem is degraded, go into details about when and how it was degraded, measures for its mitigation are required to be found and necessary suggestions should be made.

(e) For the study of socio-economic environment describe (*i*) population density, (*ii*) SC/ST, (*iii*) literacy, (*iv*) employment and occupation (*v*) economic status (income, landholding, house holding) (*vi*) life style (use of gadgets etc.) (*vii*) cultural aspects, (*viii*) Health aspects.

24.5.1 Study of River Ecosystem

As mentioned in the earlier part of this chapter, the study is conducted and a report is prepared. The report for the river eco-system should include the following aspects:

1. The use of water: drinking and domestic use, fishing etc.
2. Use of nearby area *e.g.* grazing of cattle, installation of pumps of irrigation purposes. Land use pattern around the aquatic ecosystem.
3. Observation about the river front in the rural as well as in an urban area *e.g.* ways to use the water, quality of water, reason for pollution (if so).
4. If river flows through forest area : its catchment, different animals who tracks at the edge, name the important wildlife animals, fishes. Observe/enquire if the fish catch has decreased, increased or remained same during about last 10 to 20 years.
5. Plants used for food for men, animals etc.
6. Observe/enquire about the flow of water round the year, speed of moving the flow, deep/shallow water.
7. Aquatic plants and animal life.
8. Status of flood, frequency, impact, preventive steps required to be taken to mitigate its impact.
9. How the river ecosystem is linked to your own life and how can you enhance the public awareness to keep the river clean.

24.5.2 Forest Assets

For the study of forest following should be recorded :

1. Identify the type of forest and its qualitative aspects.
2. Land use and land cover pattern *i.e.* homestead land, grazing land, marshes, etc. should be shown on the map of the area.
3. Prepare a list of natural resources, goods and services available in the study area,
   (i) Goods include food, fuelwood, fodder, water etc.
   (ii) Services include water regime, climate control, oxygen production, carbon dioxide removal etc. It should also be documented as to who uses the forest ecosystem's natural resources and to what extent? Whether this use is sustainable. Whether the use of these resources are for personal use, for marketing or for both?

4. The proportion of income of local people comes from sale of fruit, fodder, wood, medicinal plants etc.
5. Drainage pattern *i.e.* catchment area, ground water strata, springs.
6. Soil characteristics *i.e.* soil erosivity in catchment area.
7. Migrant behaviour of human and livestock population.
8. Geology of the area *i.e.* rocky, slopy strata, minerals.
9. Floral diversity, species diversity index.
10. Presence of economically important species *viz.* non-wood forest producing species including medicinal plants, timber, fuel wood etc. Presence of rare, endangered and threatened flora species.
11. Faunal diversity: The study area should be divided into different strata based on vegetation and topography. Observe wildlife status in the study areas. The data should be presented in the report in the categories of : mammals, avifauna, herpetofauna (reptiles and amphibions), and butterflies.
12. Water body inundating forest area.
13. Presence of national park, sanctuary, biosphere, reserve forest etc. in the area.
14. Presence of rare, endangered and threatened faunal species.
15. Existence of barriers and corridors on migration routes for wild animals, if any in the study area should be covered as a part of the study.
16. Whether the ecosystem is overused due to high number of people depend on it; or the greed of a few people, or both.
17. Whether the ecosystem is protected and/or efforts are being made/to be made to restore it.

24.5.3 Grassland

Steps for the study of grassland are :

1. Identify the type of grassland *e.g.* Himalayan, terrestrial, semi-arid or area developed for grazing or grass collection.
2. Describe the qualitative aspects of the grassland *e.g.* biotic and abiotic features *e.g.* nature of soil, plant and animal species.
3. Discuss with local people as to how they use grassland for grazing cattle, cutting fodder and collecting fuelwood.
4. Assess the quantity of grass available and needed by the domestic herbivores dependent on this grassland. How many animals can be supported by this grassland (cattle, sheep, goats etc. and their proportion).
5. Whether the utilization level is sustainable?
6. Whether the grassland burned frequently why the local people burn the grass.

24.5.4 Visit to Hill Slope/Mountain

The hills are generally fragile ecosystems and are easily degraded. These are used by different types of user groups. The report on field visit of hill slopes/mountains should cover the following aspects :

1. Utilization patterns by different user groups.
2. Identify the type of hill *i.e.* steep or gradual slope and top having peak or plateau.
3. Describe soil characteristics and vegetation characteristics. Whether the soil cove is intact or degraded and extent of degradation.
4. Whether it has any water courses such as springs, nala courses etc.
5. Extent of grazing by the cattles/domestic animals. Identify the wild and domestic animals present.
6. Describe vegetation pattern (tree cover, shrub, grass cover or simply rocks). Identify and prepare a list of plants that grow on it.
7. If the hill slopes are eroded, what are the reasons for it and what measures can be suggested for its mitigation.
8. Observe and describe the food chains.
9. With reference to this hill/mountain ecosystem, describe various cycles such as water cycle, nitrogen cycle, energy flow and detritus cycle.
10. In the forest ecosystem, identify the forest type and its dominant tree species and describe the utilisation pattern of forest cover and grassy slopes, and who uses it.
11. What are the natural resources available in the area? How they are being managed? If required, what measures should be taken for their erosion?
12. Have there been landslides or floods in the valley? What was its impact?
13. How common life is linked to the ecosystem?

24.6 VISIT TO STUDY A LOCAL POLLUTED SITE

As discussed in the book, pollution occurs from variety of sources and effects our environment, which adversely impact our life and health. Polluted sites may be urban, rural, agricultural and industrial areas. For the purpose of study any of these polluted site can be selected for its visit. The pollutants mainly affect the air, water, soil, and biodiversity *i.e.* plant and animal life.

24.6.1 Air Polluted Sites

Air polluted sites inclue urban areas having traffic congestion, industrial areas due to release of gaseous products during manufacturing processes, rural areas burning *parali* or using pesticides/chemical fertilizers, and noise producing areas (as noise pollution is also considered as a part of air pollution).

The study report of air pollution site should cover following details :

1. Compare the air quality level with ambient levels, especially with reference to SPM, NO, SO_2 and CO.
2. Find out the reasons for deviation.
3. Find impact of air pollution on soil, vegetation, material and human health.
4. Suggest remedial measures to control air pollution.
5. Create awareness by asking people from the area as to what are the effects of the pollution on their lives and how these can be reduced. Also suggest methods to make more people aware of this problem and remedial measures.

24.6.2 Water Polluted Sites

Observe the river/lake/tank and considering its level of pollution, categorise it as unpolluted, slightly polluted, moderalely polluted or severely polluted. The study report of the water pollution site should cover the following details :

1. Name of the river. Compare the existing water quality with that of BIS water quality standards.
2. Find out the water contaminants and their types, such as oxygen-demanding wastes, disease causing agents, synthetic organic compounds, plant nutrients, inorganic chemicals and mineral substances, suspended solids etc.
3. Find out if the garbage is dumped on the bank.
4. If there is any industry which is discharging its waste water in the water body.
5. Are there any sources of water contamination from the surface run-off from nearby agricultural land on which chemical fertilizers and pesticides are used.

24.6.3 Solid Waste Polluted Site

Solid wastes are of 3 types, namely:

(i) *Garbage i.e.* bio-degradable organic waste which produces foul odours and health hazards, insect problems;

(ii) *Rubbish* which includes variety of materials like plastic, paper, textile, broken glass, crockery, metal and

(iii) Others such as dust, sand, construction debris, abandoned appliances etc.

The problem of solid waste is increasing due to increase in population and increasing consumerism. Solid waste pollution problem can be studied at a dumping area. The problems are of two types:

(a) Due to rubbish, which is due to over-utilisation of non-biodegradable disposable consumer goods, and the lack of awareness about the management of waste at the household level. These are dumped in the form of heaps which ultimately takes the shape of mountain.

(b) The garbage is a source of various diseases. Garbage, being an organic waste, leads to a large population of flies, cockroaches and rats that are responsible for the spread of disease.

(c) The wastes, since so many years are being dumped into oceans, rivers or on land. All these methods contribute to the contamination of soil, river waters or ground water under the dumping site. It also foul up the air and spread diseases.

The students are required to prepare the study report giving the details of:

(i) Type of waste, quantum of waste

(ii) Population of the area from where the waste is generated for dumping.

(iii) Can this waste be reduced, or utilized by way of segregating and then:

  (a) recycling

  (b) used for electricity generation

(c) producing the manures,

(d) filling of pits by rubbish or by any other method

(iv) Awareness programmes

(v) Impacts of waste pollution and management of waste scheme be drawn.

24.7 STUDY OF PLANTS, ANIMALS AND BIRDS

For the purpose of study one can visit a national park or sanctuary to see several plants, mammals, reptiles, avifauna, butterflies, birds etc.

While carrying out study of plants, animals, insects, birds etc. one should carry a pair of binacular, reference books which can help to identify plants, insects, birds, reptiles and mammals, a field guide is also helpful for this purpose.

(A) Study of Plants (Flora)

1. Identify and list the common plant species available at the study site. These should include trees, shrubs, herbs, climbers, grasses etc.
2. Also identify and prepare separate list of rare species available in the area, for help of field guide or a botanist may be taken.
3. The identified and listed plants should be documented mentioning their characteristic features.
4. The characteristic features should also be described for the flowers, fruits and seeds.
5. Who are using these plants and how?
6. Also mention about the sustainability or over-harvesting of these plants.

(B) Study of Animals (Fauna)

1. Identify and list all the faunal species which are seen at the study site.
2. Document the major identification features of the common animals of each category, such as, mammals, birds (avifauna), reptiles, butterflies, amphibians and insects.
3. Migration routes of animals, existence of barriers and corridors for wild animals should be covered in the study.
4. List of resident and migratory avifauna.
5. Classify the level of abundance at the site, such as abundant, common, rare and very rare.
6. Mention the role of each species played in the food chain and energy pyramid.
7. Observe and record the habits of each of the selected fauna species, such as feeding behaviour, habitat, nesting for birds and the breeding behaviour.

24.8 VISIT TO STUDY SIMPLE ECOSYSTEMS

Students are expected to study different simple ecosystems, such as a forest, grassland, semi-arid/desert, aquatic eosystems (lake, river or seacoast). In this study one should identify the local landscape pattern which are unmodified as well as those which are modified ecosystems (*e.g.* farmland, grazing land, urban land, industrial land).

The plants and animals have several exciting features that are identified. Observe and document their abundance or rarity, their habitat, their behaviour and their links with other faunal species.

For the purpose of carrying out the study of an ecosystem:

1. Mention about its special features, and its abiotic and abiotic characteristics.
2. Whether the goods and services of a particular ecosystem are misused or overused. How is the ecosystem being degraded? Degradation of the ecosystem is indicated by deforestation, pollution of a water body, and soil erosion.
3. The impact of the degradation can be indicated.
4. How this degradation process be prevented by sustainable use of goods and services of the ecosystem.
5. Observe and document the water cycle, carbon cycle, oxygen cycle, nitrogen cycle, and energy flow.

24.8.1 Visit to a Forest Ecosystem

A nearby reserved forest, national park or wildlife sanctuary may be selected for visit. For this, concerned forest officials should be contacted, and available brochers/literature should be referred. Following aspects should be studied and covered in the report:

1. Identify the type of forest.
2. Study the connection between abiotic and biotic features of the ecosystem. Observe the difference in types of vegetation and relate this to abiotic features, such as soil, temperature rainfall and topographic patterns.
3. Observe the abundance of different species (flora as well as fauna), and record in the report, as discussed earlier while discussing the "study of plants, animals and birds."
4. Observe and record the names of animals found in the forest, and classify them as mammal, reptiles, amphibians, insects and birds. Also classify them as harbivores and carnivores. Also indicate the relative abundance of herbivores over carnivores.
5. Identify the layers of forest *i.e.* ground, undergrowth, trunk, branches, canopy.
6. Document the micro habitat for species in different levels of forest.
7. Worms and insects, such as ants, termites etc. live on the ground with the fallen leaves. Lizards and woodpeckers live in the middle layer on branches and tree trunks. Birds live in the canopy of the tree tops, these are fruit-nector-dependent birds. Some birds are insectivore birds, who also live in the canopy. Many insects live under the ground. The study report should cover all these types of commonly found fauna.
8. Observe and document how an ant colony works together in the ecosystem.
9. Observe and document the abundance of different animal species.
10. Observe and interpret the temperature and moisture control functions of the forest.
11. Document the species density and also the floral species diversity.
12. Observe and document the detritus cycle.
13. Observe and document as to how forests influence the water cycle.

14. Observe and record the food chains, food webs and food pyramid and biomass distribution of the ecosystem. (The food chains, food webs and food pyramids have been discussed in detail in the book in the chapter on "Ecosystems"). Observe what all the animals (fauna) are feeding on and construct many food chains as possible. It can also be observed that a single species can play a role in several food chains. Such chains form a foodweb.

Document what you have seen about the food web in the area observed in the ecosystem. Some of the examples of food chains are :

- Flower → Butterfly → Spiders
- Flower → Sunbirds → Birds of prey
- Flowers → Bees → Bee caters
- Fruits → Parakeet → Birds of prey
- Fruits → Langurs → Leopards
- Leaves → Monkeys → Leopard
- Grass → Chital → Tiger

15. Observe and study the wildlife signs. There are some animals which even when not seen, leave behind several signs, such as footprint for their identification. These signs in the forest area can tell us which species live there and also indicate about their movements with regard to their day-to-day activities.

24.8.2 Visit to a Grassland

1. Observe and record the variety of plant and animal life on the grassland. Document the food used by each identified animal.
2. Indicate abundance of grassland species.
3. Describe as to how the grassland would look in different seasons. Describe the changes in colour and the condition of grasses in growing phase, flowering phase, dry phase and dying phase.
4. Identify and make a checklist of common grassland birds by referring from the some good books on Indian birds.
5. Observe as to how the ants, beetles and grasshoppers live, collect food and behave in the grassland.
6. Observe and document the behavioural patterns of all the animals, birds and insects in the grassland. Also describe the relationship of animals, birds and insects in the grassland as a habitat.

24.8.3 Visit to a Desert/Semi-Arid Area

Following points should be covered while preparing a study report on a desert or semi-arid area.

1. Observe and document the specialised nature of vegetation in a desert or semi-arid landscape. Document the number of animal species that are seen in the study area. The animal species in the desert or semi-arid area are very few in comparison to other types of ecosystems.

2. Identify and document the birds and insects which are easily seen in the ecosystem.
3. Record in the study as to how each species is using its habitat. Also record as to what these species feed on in this harsh environment of the ecosystem.
4. Observe and document the birds of prey that use this ecoystem.
5. Rare birds like the great Indian bustard are also available in a few areas.
6. Rare mammals, such as the wolf are also available in such ecosystem.
7. Observe and document the typical species, such as dung beetles that roll dung into a ball, in which they lay their eggs so that their yound get food.

24.8.4 Visit to an Aquatic Ecosystem

An aquatic ecosystem may be a pond, lake, river or seacoast.

1. Study of the pond ecosystem

A pond ecosystem changes rapidly during the year. The study of pond ecosystem should cover the monsoon phase (when the pond changes from a dry to a wet state *i.e.* a growing phase), at the peak of the monsoon (a mature aquatic phase, which is full of life). Once the rain stops, the pond begins to shrink and its periphery becomes dry and is colonsied by terrestrial plants, like grasses and herbs. When the pond shrinks, its aquatic flora and fauna die, giving way to land flora and fauna. Thereafter, it may remain only in the form of a ditch or depression. This process is then repeated year after year.

For the study of the ecosystem following actions are taken :

1. For studying the pond ecosystem, make observations at different occasions, such as before, during and after the monsoon and record seasonal changes in the plant (flora) and animal (fauna) life.
2. Observe the different vegetation zones in the pond *i.e.* grasses on the periphery, floating vegetation and underwater plants.
3. Observe the seasonal field variations on the pond *i.e.*
   (i) soon after the pond fills with water in the monsoon, observe the algae and microscopic animals,
   (ii) in fully active phase, observe the submerged and emergent vegetation alongwith fish, frogs, snails, worms and other aquatic insects,
   (iii) in the shrinking phase, observe the drying aquatic plant life with dead and dying plants, and the plants growing on the mud exposed in the pond,
   (iv) in the dry phase observe the overgrown grasses and shrubs with hidden dormant animals in the mud which are not seen.
4. Observe and document that as what species use the pond.
5. Check the water quality in different stages with the help of microscope and note down in the study report.
6. Observe the periphery of the pond and describe with regards to its soil, flora and fauna. Similarly describe the pond's floor with regards to soil, flora and fauna.
7. Also record in the study report about the food chain of the pond ecosystem.

2. Study of Lake Ecosystem

While carrying out study of a lake ecosystem following steps should be taken :

1. Observe and make a checklist of all the visible aquatic flora and fauna.
2. Observe the abundance (after estimating the count of the population) of different species in the aquatic ecosystem.
3. Observe and document the food chains of aquatic flora and fauna.
4. Observe and document the way in which different water birds use the various habitats both on the shore and in the water.
5. Observe and document as to which of aquatic birds shares its habitats with other species.
6. Observe and document that which species specialises in which specific food and feeds at different depths.
7. Highlight about the length of legs of different wading birds and its relationship with the depth at which they feed.
8. Observe and document that the length of the beaks of aquatic birds indicate the depth of mud or sand into which they can penetrate.

EXERCISE

(DEFINING, DISTINGUISHING, SHORT ANSWER QUESTIONS, EXPANSIONS, FILL-IN-BLANKS, MCQs, TRUE-FALSE)

A. DEFINITIONS

Define the following:

1. Sustainable development
2. Biosphere
3. Ecosystem
4. Ecology
5. Predation
6. Biotic and abiotic components of an ecosystem
7. Succession
8. Bio-diversity
9. Forest
10. Deforestation
11. Flood
12. Minerals
13. Soil degradation
14. Fossil fuel
15. Biomass and biogas
16. Tidal power
17. Food and food resources
18. Crop
19. Environmental pollution
20. Water pollution
21. Air pollution
22. Biofuel
23. Toxicants
24. Extinction of species
25. Ground water
26. Tropical rain forest
27. Noise pollution
28. Municipal solid waste
29. Disaster and crisis
30. Disaster management
31. Cyclone
32. Landslides
33. Tsunami
34. Acid rain
35. Watershed management
36. Ethics
37. Consumerism
38. Ozone depletion
39. Global warming

40. Green house gases
41. El Nino
42. Involuntary re-settlement
43. Noise
44. Environmental impact assessment
45. Environmental management
46. Environmental monitoring
47. Environment
48. Radioactivity
49. Ecological balance
50. Waterlogging
51. Energy
52. B.O.D.
53. C.O.D.
54. Municipal solid waste
55. Biomedical wastes
56. Wetland
57. Hotspot
58. Population
59. Community
60. Ecosystem
61. Biome
62. Tundra
63. Crop rotation
64. Swamps
65. Arid
66. Desert
67. Species richness
68. Species evenness
69. Agriculture
70. Biodiversity coldspots
71. Threatened species
72. Endangered species
73. Rainwater harvesting
74. Silviculture
75. Apiculture
76. Olericulture
77. Viticultre
78. Floriculture
79. Pomology
80. Tillage
81. Farming
82. Organic farming
83. Sprinkler irrigation
84. Drip irrigation
85. Manures
86. Fertilizers
87. Pesticides
88. Smog

B. DISTINGUISHING

Distinguish between the following:

1. Nature and Environment
2. Physical environment and Biological environment
3. Troposphere and Stratosphere
4. Evaporation and Condensation
5. Ecology and Ecosystem
6. Producers and Consumers
7. Food chain and Food web
8. Biotic and Abiotic components of an ecosystem
9. Terrestrial and Aquatic ecosystems
10. Marine and Freshwater ecosystems
11. Species richness and Species diversity
12. River and Lake ecosystems
13. Endangered and Endemic species
14. National park and Wildlife sanctuary

15. Grassland and Desert ecosystems
16. National park and Biosphere reserves
17. Atmosphere and Biosphere
18. Sound and Noise
19. Producers and Decomposers
20. Renewable and Non-renewable resources
21. Renewable and Non-renewable resources of energy
22. Biodegradable and Non-biodegradable pollutants
23. Local and Regional environmental challenges
24. Primary and Secondary air pollutants
25. Continuous noise, Intermittent noise and Impulse noise
26. Climate change and Global warming
27. Thermal power plant and Hydro-electric power plant
28. Dense forest and Open forests
29. Deforestation and Afforestation
30. Water erosion and Wind erosion
31. Natural and Man-made air pollution
32. Hazardous and Non-hazardous industrial solid wastes
33. Recycling and Reuse
34. Disaster and Crisis
35. Cyclone and Tsunami
36. Primary and Secondary air pollutants
37. Garbage and Rubbish
38. Ocean and Tidal energy
39. Species richness and species evenness
40. Biodiversity hotsposts and Biodiversity coldspots
41. Sanctuaries and National parks
42. Manures and Fertilizers

C. SHORT ANSWER QUESTIONS

Write short notes on:

1. Environmental degradation
2. Overpopulation
3. Depletion of resources
4. Global warming
5. Green house effect
6. Depletion of ozone
7. Shape of the earth
8. Water table
9. Desert
10. Effect of altitude on temperature
11. Latitude and time
12. Photosynthesis
13. Aquifers
14. Geysers
15. Glaciers
16. Population dynamics
17. Flora
18. Fauna
19. Biomass
20. Poaching of wildlife
21. Sanctuaries
22. National Parks
23. Biosphere reserves
24. Levels of biodiversity
25. Habitat loss
26. Land resources
27. Soil resources
28. Animal resources
29. Wildlife resources
30. Impacts of deforestation

31. Causes of deforestation
32. Surface water
33. Underground water
34. Flood proofing
35. Flood plain zone
36. Drought
37. Rainwater harvesting
38. Recycling of water
39. Land as a resource
40. Land degradation
41. Landslides
42. Soil erosion
43. Salination
44. Wind energy
45. Wind mill
46. Global gas
47. Tidal power
48. Animal resources
49. Water logging
50. Pesticide related problems
51. Fertilizer related problems
52. Garbage and rubbish
53. Incineration
54. Composting
55. Acid rain
56. Ozone depletion
57. Kyoto Protocol
58. Earth Summit
59. Wastelands
60. Child labour
61. Women welfare
62. Child welfare
63. Indoor air pollution
64. Green belt development plan
65. Quarry and borrow area management
66. Landscaping and revegetation
67. Fauna
68. *In-situ* conservation
69. *Ex-situ* conservation
70. Noise pollution and human health
71. India as a mega-diversity nation
72. Multi-disciplinary nature of environmental studies
73. Reasons for depletion of water table
74. Impact of urbanisation on environment
75. Tundra
76. Aquatic organism
77. Swamps
78. Biodiversity measurement
79. Realms
80. Biomes
81. Bio-geographic provinces
82. Bird's migration
83. World heritage site
84. Sustainable mining
85. Fossil fuels
86. Sustainable agriculture
87. Smog
88. Earthquake magnitude
89. Richter scale
90. National Green Tribunal

D. EXPANDING THE ABBREVIATIONS

Expand the following abbreviations:

1. OTEC
2. BIS
3. ISI
4. NDMA

5. TWS
6. HIV
7. AIDS
8. RAP
9. SIA Study
10. R&R
11. EIA
12. NRRP
13. PIC
14. VDAC
15. EPA
16. UNEP
17. CPCB
18. FSI
19. GSI
20. MoEF
21. NEERI
22. SPCB
23. TERI
24. WII
25. ZSI
26. ECBC
27. BEE
28. CDM
29. EMP
30. ISO
31. BOD
32. DO
33. PNG
34. FRI
35. SPM
36. CNG
37. MAB
38. CFCs
39. pH value
40. dB
41. COD
42. NGT
43. NERF
44. BARC
45. EMS
46. EPA
47. FAO
48. FWP
49. MSW
50. TB
51. UV radiation
52. BSI

E. FILL IN THE BLANKS:

1. Environment friendly products are given ISO ______ certification.
2. The UN Conference on Environment and Development was held at ______.
3. World Environment Day is celebrated on ______.
4. Environment is derived from the French word ______ which means encircle or surround.
5. All the planets revolve around the ______ in the same direction.
6. Earth's Polar diameter is ______ km.
7. Earth's Equatorial diameter is ______ km.
8. Earth rotates on its own axis that is tilted at an angle of ______ to the vertical.
9. The organisms who feed directly on producers are called ______.

10. The sequence of eating and being eaten in an ecosystem is called a ______.
11. Pyramid of ______ is always upright.
12. Species restricted only to a particular area are called ______.
13. Loss of habitat in installments leading to small scattered patches is known as ______.
14. Illegal killing of prohibited endangered animals is called ______.
15. Herbivores are primary consumers that eats only ______.
16. The species which have not been seen continuously for the last 50 years or more are known as ______ species.
17. ______ resources are inexhaustible resources which can be generated within a given span of time.
18. ______ resources cannot be generated.
19. Plants use ______ gas for photosynthesis.
20. Deforestation means ______ of forests.
21. ______ dam is the highest dam on river Bhagirathi in Uttarakhand.
22. The ______ dam on river Satluj in Himachal Pradesh is the largest dam in terms of capacity.
23. About ______ % of the earth's surface is covered by water.
24. ______ conditions are created when annual rainfall is below normal and less than evaporation.
25. In water logged soils the plant roots do not get ______ for adequate respiration.
26. Ocean tides are produced by gravitational forces of ______ and ______.
27. Inadequate drainage and poor quality irrigation water often lead to ______ and ______ of soil.
28. The point where first movement occurs during earthquake is called ______.
29. Power plants utilize only ______ of the energy provided by fossil fuel for their operation.
30. Noise levels considered as threshold of pain are ______.
31. Sound frequency is expressed in ______.
32. CO has affinity for haemoglobin ______ times more than oxygen.
33. Sulphur dioxide during coal burning is produced due to oxidation of ______ contained in the coal.
34. The main pollutants emitted by Thermal Power Plants are ______ and ______.
35. Smog is mainly formed by mixture of ______ and ______.
36. pH value for drinking water should be between ______ and ______.
37. Dissolved oxygen in drinking water should be minimum of ______ mg/l.
38. BOD for drinking water should not be more than ______ mg/l.
39. Noise level in urban residential area should not be more than ______ dB in day time and ______ dB in night time.

40. The 3-R approach or resource use stands for Reduce, Reuse and ______.
41. The land area from which water drains under gravity to a common drainage channel is called a ______.
42. Atmosphere emissions of NO_2 and ______ cause acid rains.
43. Ozone layer acts as a natural sunscreen which protects life on the earth against ______ rays.
44. ______ Act provides for setting up of National Parks and Wildlife sanctuaries.
45. The World Environment Day is celebrated on ______.
46. Nodal ministry for natural disasters is Ministry of ______.
47. Nodal ministry for epidemics is Ministry of ______.
48. Nodal ministry for drought is Ministry of ______.
49. Avalanche is a mass of ______ that slides down a mountain slope.
50. When a population increases by a fixed percentage it is called ______ growth.
51. When birth plus immigration in a population are just equal to deaths plus emigration, it is known as ______.
52. GIS stands for ______.

Answers

1. ISO-14000
2. Rio de Janeiro
3. 5th June
4. environment
5. sun
6. 12714 km
7. 12756 km
8. 23.5°
9. Herbivores
10. Food-chain
11. Energy
12. Endemic
13. Habitat fragmentation
14. Poaching
15. Plants
16. Extinct
17. Renewable
18. Non-renewable
19. Carbon dioxide
20. Cutting
21. Tehri
22. Bhakra
23. 67%
24. Drought
25. oxygen
26. sun and moon
27. Salinity, water-logging
28. Epicenter
29. one-third
30. 140 dB
31. Hertz (Hz)
32. 210
33. Sulphur
34. Fly ash and SO_2
35. smoke and fog
36. 6.5, 8.5
37. 6
38. 2
39. 55, 45
40. Re-cycle
41. Watershed
42. SO_2
43. Ultra Violet (UV)
44. Wildlife (Protection)
45. June 5
46. Home
47. Health
48. Agriculture
49. snow
50. Exponential
51. zero population growth
52. Geographic Information System

F. MULTIPLE CHOICE QUESTIONS

Choose correct answer in respect of following:

1. The biotic component of environment consists of:
   (*a*) all the non-living things around us
   (*b*) all the living things around us
   (*c*) all the living as well as the non-living things around us
   (*d*) none of the above
2. The abiotic component of the environment consists of:
   (*a*) all the non-living things around us
   (*b*) all the living things around us
   (*c*) all the living as well as the non-living things around us
   (*d*) none of the above
3. The environment consists of:
   (*a*) biotic world (*b*) abiotic world
   (*c*) both (*a*) and (*b*) (*d*) none of the above
4. The science that deals with the relationship of various organisms with their environment is known as
   (*a*) economics (*b*) geology
   (*c*) ecology (*d*) anthropology
5. The objective of environmental studies is to:
   (*a*) raise consciousness about the environmental conditions.
   (*b*) create environmental ethics in a human community.
   (*c*) teach environmentally appropriate behaviour.
   (*d*) all of the above
6. Environmental studies involves:
   (*a*) Biology and geology (*b*) Physics and chemistry
   (*c*) Anthropology and economics (*d*) All of the above
7. Impacts of industrial development are:
   (*a*) natural resource depletion
   (*b*) deforestation which in turn leads to floods
   (*c*) decrease in non-renewable resources
   (*d*) All of the above
8. UN Conference on environment in 1972 was held at:
   (*a*) Rio de Janerio (*b*) Stockholm
   (*c*) Geneva (*d*) Johannesberg

9. World population day is celebrated on:
   (*a*) March 21 (*b*) April 22
   (*c*) July 11 (*d*) September 28
10. The organisms which feed on dead organisms, wastes of living organisms are called:
   (*a*) Chemotrophs (*b*) Decomposers
   (*c*) Detritivores (*d*) Carnivores
11. Gross primary productivity is the highest in:
   (*a*) Grasslands (*b*) Open oceans
   (*c*) Wet tropical forests (*d*) Agro-ecosystem
12. An ecosystem consists of:
   (*a*) the abiotic component (*b*) the biotic component
   (*c*) both (*a*) and (*b*) above (*d*) none of the above
13. Biosphere is a term used to represent the:
   (*a*) entire atmosphere
   (*b*) entire hydrosphere
   (*c*) entire lithosphere
   (*d*) a small zone of Earth, where lithosphere, hydrosphere and atmosphere come in contact with one another.
14. The entire biosphere lies:
   (*a*) below the surface of the Earth
   (*b*) above the surface of the Earth
   (*c*) partly below and partly above the surface of the earth
   (*d*) none of the above
15. The organisms which feed on dead organisms and excreta of living organisms are known as:
   (*a*) producers (*b*) carnivores
   (*c*) consumers (*d*) decomposers
16. The organic species which are classified as producers, include:
   (*a*) plants and bacteria producing their own food
   (*b*) animals that graze on plants
   (*c*) animals that consumes each other (*d*) none of the above
17. The organisms which feed on plants or on animals or both, are known as:
   (*a*) Producers (*b*) Consumers
   (*c*) Decomposers (*d*) None of these
18. The organisms who eat plants, dairy foods, as well as meat, are known as:
   (*a*) herbivores (*b*) carnivores
   (*c*) omnivores (*d*) none of these

19. Man is a constituent of:
    (*a*) Ecosystem (*b*) Biosphere
    (*c*) Environment (*d*) All of the above
20. The maximum energy is received by the:
    (*a*) producers (*b*) primary consumers
    (*c*) secondary consumers (*d*) decomposers
21. The word ecology is derived from two Greek words, which mean:
    (*a*) study of life (*b*) study of home
    (*c*) study of life and land (*d*) study of life at home
22. The study of interaction between the living organisms and their environment is called:
    (*a*) Ecosystem (*b*) Ecology
    (*c*) Phytosociology (*d*) Autecology
23. Which one of the following is a possible producers in an ecosystem?
    (*a*) Plants and some bacteria capable of self nourishment
    (*b*) Animals
    (*c*) Human beings
    (*d*) Fish
24. The perfect equilibrium existing in the biosphere between the various organisms is known as:
    (*a*) environmental balance (*b*) ecological balance
    (*c*) ecological cycle (*d*) none of them
25. The biotic and abiotic components of an ecosystem represent its:
    (*a*) structure (*b*) function
    (*c*) type (*d*) none of them
26. Role of decomposers in a food chain is:
    (*a*) to clean the environment
    (*b*) to feed on dead organic matter
    (*c*) to act as scavengers for removing excreta
    (*d*) all of the above
27. In a shorter food chain, the availability of food energy is:
    (*a*) less (*b*) more
    (*c*) equal to each other (*d*) may be more or less
28. Which one of the following food chains directly depends on solar energy:
    (*a*) grazing (*b*) detritus
    (*c*) paracitic (*d*) all of the above

29. In a food chain, humans are:
(*a*) primary consumers (*b*) secondary consumers
(*c*) primary and secondary consumers both
(*d*) producers

30. Detritus feeders do not obtain their energy directly from:
(*a*) sun (*b*) dead organic matter
(*c*) dead plants (*d*) dead animals

31. A food chain usually starts with:
(*a*) producers (*b*) consumers
(*c*) decomposers (*d*) detritoves

32. Which of the following statements is false?
(*a*) Energy is wasted at each stage of a food chain
(*b*) Vegetarian people consume less energy than the non-vegetarians
(*c*) Inorganic nutrients are recycled in an ecosystem
(*d*) Energy is recycled in an ecosystem

33. Fraction of sunlight used by plants in photosynthesis is about:
(*a*) 20.0% (*b*) 2.0%
(*c*) 0.2% (*d*) 0.02%

34. Hydrological cycles involve movement of:
(*a*) hydrogen (*b*) water
(*c*) oxygen (*d*) nitrogen

35. Biogeochemical cycles represent the:
(*a*) biological nature of the process (*b*) chemical nature of the process
(*c*) geological nature of the process (*d*) all of the above

36. Biogeochemical cycles refer to cycling of:
(*a*) nutrients in the ecosystem (*b*) energy in the ecosystem
(*c*) water in the ecosystem (*d*) all of the above

37. Which among the following is a gaseous type of cycle?
(*a*) phosphorus cycle (*b*) carbon cycle
(*c*) sulphur cycle (*d*) water cycle

38. The largest reservoir of nitrogen on Earth is:
(*a*) atmosphere (*b*) oceans
(*c*) fossil fuels (*d*) none of them

39. During photosynthesis trees produce:
(*a*) Oxygen (*b*) Carbon dioxide
(*c*) Nitrogen (*d*) Carbon monoxide

40. Forests prevent soil erosion by binding soil particles in their:
    (*a*) Stems (*b*) Leaves
    (*c*) Roots (*d*) Buds
41. Wood pulp is used for making:
    (*a*) Lumbar (*b*) Chipboard
    (*c*) Paper (*d*) Plywood
42. Deforestation rate is alarming in:
    (*a*) Temperate countries (*b*) Tropical countries
    (*c*) Polar region (*d*) None of them
43. Major causes of deforestation are:
    (*a*) Shifting cultivation (*b*) Fuel requirements
    (*c*) Raw material for industries (*d*) All of these
44. Major consequences of deforestation are:
    (*a*) Destruction of natural habitat of wild species
    (*b*) Disturbances in hydrological cycle
    (*c*) Soil erosion
    (*d*) All of these
45. Ground subsidence occurs due to:
    (*a*) Withdrawal of more groundwater than its recharge
    (*b*) More recharge of groundwater than its withdrawal
    (*c*) Equal rates of recharge and withdrawal
    (*d*) None of the above
46. The Satluj-Yamuna Link (SYL) canal dispute is between:
    (*a*) Punjab and Haryana (*b*) Karnataka and Tamil Nadu
    (*c*) Delhi and U.P. (*d*) All of the above
47. Over-grazing results in:
    (*a*) Productive soils (*b*) Soil erosion
    (*c*) Retention of useful species (*d*) All of these
48. Biomass energy can be obtained from:
    (*a*) Energy plantations (*b*) Petro crops
    (*c*) Agricultural and urban waste biomass (*d*) All of these
49. Which of the following types of coal has maximum carbon and calorific value?
    (*a*) Anthracite (hard coal) (*b*) Bituminous (soft coal)
    (*c*) Lignite (brown coal) (*d*) Wood coal
50. Nuclear energy can be generated by:
    (*a*) Nuclear fusion (*b*) Nuclear fission
    (*c*) Both of these (*d*) None of these

51. The minimum time needed for the formation of one inch of top soil is:
(*a*) 10 years (*b*) 50 years
(*c*) 100 years (*d*) 200 years

52. Which of the following is responsible for desertification?
(*a*) Deforestation (*b*) Overgrazing
(*c*) Mining (*d*) All of these

53. Industrial wastes may contain toxic:
(*a*) Chemicals (*b*) Phenols
(*c*) Acids (*d*) All of these

54. Dissolved oxygen in water comes from:
(*a*) Photosynthesis of aquatic plants (*b*) Atmosphere
(*c*) Both the above (*d*) None of these

55. Thermal pollution can be controlled by:
(*a*) Cooling ponds (*b*) Spray ponds
(*c*) Cooling tower (*d*) All of these

56. Which of the following human activity has caused serious irreversible environmental imbalances?
(*a*) Agriculture (*b*) Industrialisation
(*c*) Urbanisation (*d*) Mining
(*e*) All of the above

57. Which of the following human activity has caused maximum environmental pollution having regional and global impacts?
(*a*) Agriculture (*b*) Industrialisation
(*c*) Urbanisation (*d*) Mining

58. EIA is abbreviated for:
(*a*) Energy Impact Assessment (*b*) Ecological Impact Assessment
(*c*) Environment Impact Assessment (*d*) Emission Impact Assessment

59. Which of the following gases has maximum contribution to enhanced greenhouse effect?
(*a*) CFC's (*b*) CH_4
(*c*) CO_2 (*d*) N_2O

60. Cattle, sheep and termites are responsible for the release of the following greenhouse gas:
(*a*) Methane (*b*) Carbon dioxide
(*c*) Nitrous oxide (*d*) All of these

61. The most important agents for ozone depletion are:
(*a*) Methane (*b*) CFC's
(*c*) Nuclear fallout (*d*) Nitrous oxide

62. Maximum depletion of ozone occurs on:
(*a*) Equator (*b*) North pole
(*c*) South pole (*d*) Tropics

63. The natural resource, among the following, which is not a renewable resource, is:
(*a*) air (*b*) water
(*c*) fossil fuels (*d*) forests

64. The natural resource, among the following, which is a renewable resource, is:
(*a*) fossil fuels (*b*) metallic minerals
(*c*) non-metallic minerals (*d*) forests

65. The perpetually available natural resources, like solar energy, wind energy, tidal energy, are usually considered as a part of:
(*a*) non-renewable resources (*b*) renewable resources
(*c*) neither (*a*) nor (*b*) (*d*) both (*a*) and (*b*)

66. The total quantity of water available on Earth:
(*a*) is in a finite quantity
(*b*) is decreasing over time
(*c*) is variable, depending upon the climatic factors
(*d*) can be increased by reducing consumption

67. The Satluj-Yamuna Link (SYL) canal dispute exists in India between the States of:
(*a*) Karnataka and Tamil Nadu (*b*) Delhi and U.P.
(*c*) Punjab and Haryana (*d*) Haryana and Himachal Pradesh

68. The Indus-Water Treaty was signed between:
(*a*) India and Pakistan (*b*) India and China
(*c*) India and Bangladesh (*d*) None of the above

69. The major quantity of water stored by dam reservoirs is being used for:
(*a*) Irrigation (*b*) Drinking water supply
(*c*) Recharging ground water (*d*) Generation of hydro-power

70. Mining does not cause:
(*a*) water pollution (*b*) air pollution
(*c*) land subsidence (*d*) noise, vibrations and mining accidents
(*e*) none of the above

71. Which one of the following is not an ecological service, being provided by the forests?
(*a*) production of oxygen and consumption of carbon dioxide
(*b*) providing habitat to biodiversity
(*c*) providing peaceful congenial environmental, and absorbing air and noise pollution
(*d*) conservation of soil
(*e*) regulation of hydrological cycle and controlling floods
(*f*) none of the above

72. The major cause for land degradation in our country is:
   (*a*) soil erosion
   (*b*) water-logging
   (*c*) pollution of soil
   (*d*) none of the above

73. The majority of soil erosion in India is caused by:
   (*a*) water erosion
   (*b*) wind erosion
   (*c*) both (*a*) and (*b*) cause almost equal erosion
   (*d*) none of them

74. Degradation of soil can be reduced by:
   (*a*) adopting proper and sustainable agricultural practices
   (*b*) large scale afforestation
   (*c*) adopting and constructing engineering structures
   (*d*) all of the above

75. The engineering construction(s) which may help in reducing land degradation is (are):
   (*a*) construction of bench terraces
   (*b*) construction of contour bunds
   (*c*) construction of channel traces
   (*d*) all of the above

76. The best engineering technique to be adopted for reducing land degradation in low rainfall areas and in permeable soils at slopes of upto about 4% in agricultural lands is:
   (*a*) contour bunding
   (*b*) channel terraces
   (*c*) bench terracing
   (*d*) none of the above

77. The best advised engineering structure to be constructed for reducing land degradation in high rainfall area and in less previous soil is:
   (*a*) contour bunds
   (*b*) channel terraces
   (*c*) bench terraces
   (*d*) none of the above

78. The desert regions have expanded primarily due to:
   (*a*) change of natural forces creating dry conditions
   (*b*) human activities like over-grazing, mining, destruction of forests, unsustainable farming methods, etc.
   (*c*) both (*a*) and (*b*)
   (*d*) none of the above

79. The impact of traditionally adopted shifting cultivation is:
   (*a*) deforestation
   (*b*) soil erosion
   (*c*) depletion of soil nutrients
   (*d*) all of the above

80. Fertilizers and pesticides used in modern agriculture cause serious pollution of:
   (*a*) air atmosphere
   (*b*) ground water
   (*c*) surface water
   (*d*) all of the above

81. The fossil fuel which causes maximum environmental pollution due to its use in generation of thermal power is:

(*a*) Coal (*b*) Oil

(*c*) Natural gas (*d*) None of the above

82. The fossil fuel which causes minimum pollution due to its use in thermal power generation is:

(*a*) Coal (*b*) Oil

(*c*) Natural gas (*d*) None of the above

83. Declaration of Human Rights and Environment was drafted the United Nations at:

(*a*) Stockholm (*b*) Geneva

(*c*) Rio de Janerio (*d*) Johannesberg

84. HIV can pass from infected persons to others by:

(*a*) Contaminated syringe (*b*) Saliva

(*c*) Sweat (*d*) All of the above

85. Which one of the following is a non-biogradable pollutant?

(*a*) organic substances of plant origin

(*b*) organic substances of animal origin

(*c*) crude oil (*d*) DDT

86. Which one of the following is a biodegradable pollutant

(*a*) heavy toxic metals like lead or mercury (*b*) plastics

(*c*) nuclear wastes (*d*) none of the above

87. Which one of the following pollutions has acquired regional and global dimensions?

(*a*) water pollution (*b*) air pollution

(*c*) land pollution (*d*) none of the above

88. Which one of the following air pollution effects has not yet acquired global dimensions?

(*a*) acid rain (*b*) ozone depletion

(*c*) global warming (*d*) none of the above

89. The primary air pollutant responsible for acid rain is:

(*a*) carbon monoxide (*b*) sulphur dioxide

(*c*) carbon dioxide (*d*) ozone

90. The region most affected by acid rain is:

(*a*) India and China (*b*) Europe and North East America

(*c*) Africa (*d*) South America

91. The acid rains are more prominent in:

(*a*) temperate regions (*b*) tropical regions

(*c*) arid regions (*d*) equally prominent in all regions

92. A rainfall is generally classified as acidic, if its pH is less than about:
(*a*) 5.6 (*b*) 6.5
(*c*) 7 (*d*) 7.5

93. An ordinary rain will generally be:
(*a*) slightly acidic (*b*) slightly alkaline
(*c*) neither acidic nor alkaline (*d*) may be both acidic or alkaline

94. The important gaseous pollutants contributing to acid rain are:
(*a*) SO_2 and NO_x (*b*) CO_2 and H_2S
(*c*) NO_x and O_3 (*d*) None of the above

95. Acid rains lead to destruction of:
(*a*) Forests (*b*) Monuments
(*c*) Productive soil (*d*) All of the above

96. The fossil fuels responsible for causing acid rains are used in:
(*a*) Industries (*b*) Automobiles
(*c*) Domestic cooking and heating (*d*) All of the above

97. The ozone depletion in the stratosphere (outer atmosphere) is chiefly caused by the release of:
(*a*) chlorofluorocarbons (CFCs) (*b*) carbon dioxide (CO_2)
(*c*) methane (CH_4) (*d*) none of them

98. The chlorofluorocarbons (CFCs) are released into the environment by humans, as they are largely used:
(*a*) in refrigeration and air-conditioning
(*b*) as cleaning solvents in factories
(*c*) in aerosol sprays
(*d*) as blowing agents to create certain kinds of plastic foams
(*e*) all of the above

99. Depletion of ozone layer in the outer atmosphere is likely to increase the incidence of:
(*a*) skin cancer (*b*) lung cancer
(*c*) bronchitis (*d*) none of them

100. Pollution by depletion of ozone layer in stratosphere, threatening the environment, is caused due to reaction of ozone with:
(*a*) carbon monoxide (*b*) chlorine
(*c*) sulphur dioxide (*d*) nitrous oxide

101. The most significant environmental issue, causing serious diseases in humans, is:
(*a*) acid rain (*b*) ozone hole
(*c*) global warming (*d*) none of these

102. Harmful U.V. radiations emanating from the sun are preventing from reaching the Earth by the presence of ozone in the:

(a) thermosphere (b) mesosphere

(c) stratosphere (d) troposphere

103. Consider the following statements :

1. Biodiversity is nomally greater in the lower latitudes as compared to the higher latitudes.
2. Along the mountain gradients, biodiversity is normally greater in the lower altitudes as compared to the higher altitudes.

Which of the statement given above is/are correct?

(a) 1 only (b) 2 only

(c) Both 1 and 2 (d) Neither 1 or 2

104. Consider the following :

1. Carbon dioxide
2. Oxides of nitrogen
3. Oxides of sulphur

Which of the above is/are the emission(s) from coal combustion at themal power plants?

(a) 1 only (b) 2 and 3 only

(c) 1 and 3 only (d) 1, 2, and 3

105. Regarding "carbon credits," which one of the following statement is not correct.

(a) The carbon credit system was ratified in conjunction with the Kyoto Protocol.

(b) Carbon credits are awarded to countries or groups that have reduced greenhouse gases below their emission quota.

(c) The goal of the carbon credit system is to limit increase of carbon emission quota.

(d) Carbon credits are traded at a price fixed from time to time by the United Nations Environment Programme.

106. Which of the following can be threats to the biodiversity of a geographical area?

1. Global warming
2. Fragmentation of habitat
3. Innovation of alien species
4. Promotion of vegetarianism

Select the correct answer using the codes given below :

(a) 1, 2 and 3 only (b) 2 and 3 only

(c) 1 and 4 only (d) 2, 3 and 4

107. In which one among the following categories of protected areas in India are local people not allowed to collect and use the biomass?

(a) Biosphere Reserves

(*b*) National Parks

(*c*) Wetlands declared under "Ramsar Convention"

(*d*) Wildlife Sanctuaries

108. Vultures which used to be very common in Indian country side some years ago are rarely seen non-a-days. This is attributed to

(*a*) The destruction of their nesting sites by new invasive species

(*b*) A drug used by cattle owners for treating their diseased cattles

(*c*) Scarcity of food available to them

(*d*) A widespread persistent and fatal disease among them

109. Consider the following statements :

Chlorofluoro carbons (CFCs), known as ozone-depletion substances, are used

1. In the production of plastic foams
2. In the production of tubeless tyres
3. In cleaning certain electronic components
4. As pressurizing agents in aeroson cans

Which of the statements given above is/are correct?

(*a*) 1, 2 and 3 only (*b*) 4 only

(*c*) 1, 3 and 4 only (*d*) 1, 2, 3 and 4

110. The National Green Tribunal Act, 2010 was enacted in consonance with which of the following provisions of the constitution of India?

1. Right to healthy environment, construed as a part of Right to life under Article 21.
2. Provision of grants for raising the level of administration in the Scheduled Areas for the welfare of Scheduled Tribes under Article 275 (1).

Which of the statements given above is/are correct?

(*a*) 1 only (*b*) 2 only

(*c*) Both 1 and 2 (*d*) Neither 1 nor 2

111. With reference to the wetlands of India, consider the following statements :

1. The country's total geographical area under the category of wetlands is recorded more in Gujarat as compared to other States.
2. In India, the total geographical area of coastal wetlands is larger than the inland wetlands.

Which of the statements given above is/are correct?

(*a*) 1 only (*b*) 2 only

(*c*) Both 1 and 2 (*d*) Neither 1 nor 2

112. Consider the following statements :

Which one among the following industries is the maximum consumer of water in India?

(*a*) Engineering (*b*) Paper and Pulp

(*c*) Textiles (*d*) Thermal power

113. With reference to the food chains in ecosystems, which of the following kinds of organisms is/are known as decomposer organism(s)?

1. Virus
2. Fungi
3. Bacteria

Select the correct answer using the code given below :

(*a*) 1 only (*b*) 2 and 3 only

(*c*) 1 and 3 only (*d*) 1, 2 and 3

114. Which of the following is/are unique characteristic(s) of equatorial forests?

1. Presence of tall, closely set trees with crowns forming a continuous canopy.
2. Coexistence of a large number of species.
3. Presence of numerous varieties of epiphytes.

Select the correct answer using the code given below :

(*a*) 1 only (*b*) 2 and 3 only

(*c*) 1 and 3 only (*d*) 1, 2 and 3

115. Acid rain is caused by the pollution of environment by

(*a*) Carbon dioxide and nitrogen

(*b*) Carbon monoxide and carbon dioxide

(*c*) Ozone and carbon dioxide

(*d*) Nitrous oxides and sulphur dioxide

116. Which of the following is the correct sequence of ecosystems in the order of decreasing productivity?

(*a*) Oceans, lakes, grasslands, mangroves

(*b*) Mangroves, oceans, grasslands, lakes

(*c*) Mangroves, grasslands, lakes, oceans

(*d*) Oceans, mangroves, lakes, grasslands

117. In India, the problem of soil erosion is associated with which of the following?

1. Terrace cultivation
2. Deforestation
3. Tropical climate

Select the correct answer using the code given below.

(*a*) 1 and 2 only (*b*) 2 only

(*c*) 1 and 3 only (*d*) 1, 2 and 3

118. What are the benefits of implementing the Integrated Watershed Development Programme?

1. Prevention of soil runoff.
2. Linking the country's perennial rivers with seasonal rivers.
3. Rainwater harvesting and recharge of ground water table.

4. Regeneration of natural vegetation.

Select the correct answer using the code given below :

(*a*) 1 and 2 only (*b*) 2, 3 and 4 only

(*c*) 1, 3, and 4 only (*d*) 1, 2, 3 and 4

119. Which of the following are some important polutants released by steed industry in India?

1. Oxides of sulphur
2. Oxides of nitrogen
3. Carbon monoxide
4. Carbon dioxide

Select the correct answer using the code given below :

(*a*) 1, 3 and 4 only (*b*) 2 and 3 only

(*c*) 1 and 4 only (*d*) 1, 2, 3 and 4

120. Which one of the following is the best description of the term "ecosystem"?

(*a*) A community of organisms interacting with one another

(*b*) That part of the Earth which is inhabited by living organisms

(*c*) A community organisms together with the environment in which they live

(*d*) The flora and fauna of a geographical area

121. With reference to "fly ash" produced by the power plants using the coal as fuel, which of the following statements is/are correct?

1. Fly ash can be used in the production of bricks for building construction.
2. Fly ash can be used as a replacement for some of the Portland cement contents of concrete.
3. Fly ash is made up of silicon dioxide and calcium oxide only and does not contain any toxic elements.

Select the correct answer using the code given below :

(*a*) 1 and 2 (*b*) 2 only

(*c*) 1 and 3 (*d*) 3 only

122. Which one of the following is the national aquatic animal of India?

(*a*) Saltwater crocodile (*b*) Olive ridley turtle

(*c*) Gangetic dolphin (*d*) Gharial

123. Which one of the following regions of India has a combination of mangrove forest, evergreen forest and deciduous forest?

(*a*) North Coastal Andhra Pradesh

(*b*) South-West Bengal

(*c*) Southern Saurashtra

(*d*) Andaman and Nicobar Islands

124. With reference to "Agenda 21", sometimes seen in the news, consider the following statements :

1. It is a global action plan for sustainable development.
2. It orginated in the World Summit on Sustainable Development held in Johannesburg in 2002.

Which of the statements given above is/are correct?

(*a*) 1 only (*b*) 2 only

(*c*) Both 1 and 2 (*d*) Neither 1 and 2

125. Biological Oxygen Demand (BOD) is a standard criterion for

(*a*) Measuring oxygen levels in blood

(*b*) Computing oxygen levels in forest ecosystems

(*c*) Pollution assay in aquatic ecosystem

(*d*) Assessing oxygen levels in high altitude regions

126. How is the National Green Tribunal (NGT) different from the Central Pollution Control Board (CPCB)?

1. The NGT has been established by an Act whereas the CBCB has been created by an executive order of the Government.
2. The NGT provides environmental justic and helps reduce the burden of litigation in the higher courts whereas the CPCB promotes cleanliness of streams and wells, and aims to improve the quality of air in the country.

Which of the statements given above is/are correct?

(*a*) 1 only (*b*) 2 only

(*c*) Both 1 and 2 (*d*) Neither 1 nor 2

127. Which of the following is/are the possible consequence(s) of heavy sand mining in riverbeds?

1. Decreased salinity in the river
2. Pollution of groundwater
3. Lowering of the water table

Select the correct answer using the code given below :

(*a*) 1 only (*b*) 2 and 3 only

(*c*) 1 and 3 only (*d*) 1, 2 and 3

Answers

1. (*b*)	2. (*a*)	3. (*c*)	4. (*c*)	5. (*d*)
6. (*d*)	7. (*d*)	8. (*b*)	9. (*c*)	10. (*c*)
11. (*c*)	12. (*c*)	13. (*d*)	14. (*c*)	15. (*d*)
16. (*a*)	17. (*b*)	18. (*c*)	19. (*d*)	20. (*a*)
21. (*b*)	22. (*b*)	23. (*a*)	24. (*b*)	25. (*a*)
26. (*d*)	27. (*b*)	28. (*a*)	29. (*c*)	30. (*a*)

31. (*a*)	32. (*d*)	33. (*d*)	34. (*b*)	35. (*d*)
36. (*a*)	37. (*b*)	38. (*a*)	39. (*a*)	40. (*c*)
41. (*c*)	42. (*b*)	43. (*d*)	44. (*d*)	45. (*a*)
46. (*a*)	47. (*b*)	48. (*d*)	49. (*a*)	50. (*c*)
51. (*d*)	52. (*d*)	53. (*d*)	54. (*c*)	55. (*d*)
56. (e)	57. (*b*)	58. (*c*)	59. (*c*)	60. (*a*)
61. (*b*)	62. (*c*)	63. (*c*)	64. (*d*)	65. (*b*)
66. (*a*)	67. (*c*)	68. (*a*)	69. (*a*)	70. (*e*)
71. (f)	72. (*a*)	73. (*a*)	74. (*d*)	75. (*d*)
76. (*a*)	77. (*b*)	78. (*b*)	79. (*d*)	80. (*d*)
81. (*a*)	82. (*c*)	83. (*b*)	84. (*d*)	85. (*d*)
86. (*d*)	87. (*b*)	88. (*a*)	89. (*b*)	90. (*b*)
91. (*a*)	92. (*a*)	93. (*a*)	94. (*a*)	95. (*d*)
96. (*d*)	97. (*a*)	98. (*e*)	99. (*a*)	100. (*b*)
101. (*b*)	102. (*c*)	103. (*c*)	104. (*d*)	105. (*d*)
106. (*a*)	107. (*b*)	108. (*b*)	109. (*d*)	110. (*a*)
111. (*a*)	112. (*d*)	113. (*b*)	114. (*d*)	115. (*d*)
116. (*c*)	117. (*b*)	118. (*c*)	119. (*d*)	120. (*c*)
121. (*a*)	122. (*c*)	123. (*d*)	124. (*a*)	125. (*c*)
126. (*b*)	127. (*b*)			

G. TRUE OR FALSE

Whether following are 'True' or 'False':

1. Biotic and abiotic components of an ecosystem influence each other and are linked through energy flow and material cycling.
2. Landslides, drought, frost or volcanoes can result in development of a barren area for succession to start.
3. Environment in urban areas has high energy demands and dense population.
4. Environment includes all biological organisms, air, water and soil surrounding us, but not the material structures like buildings.
5. Human encroachment into forest area is one of the main reactions for attack by wildlife on humans.
6. Endangered animals cannot be prevented from becoming extinct by providing them special protection.
7. When ecological surveys are carried during a tour, it is known as ecotourism.
8. The commercial value of a true timber is much more than its ecological services value.

9. Maximum global biodiversity is found in tropical rain forests.
10. Soil erosion helps to maintain soil fertility.
11. Sludge left over in the biogas plant cannot be used as a fertilizer.
12. Hydropower also causes environmental pollution.
13. Ideal location for installation of wind-mills are coastal regions, open grasslands, hilly regions.
14. Networking of rivers is being proposed at national level to deal with the problem of floods.
15. Small dams are environmentally more sustainable than big dams.
16. Mining of Uranium exposes local people to radioactive hazards.
17. The major source of salinisation of soil is excessive irrigation.
18. Solar heat cannot be used to operate street lights, water pumps, television etc.
19. SO_2 does not affect respiratory passage.
20. Washing of coal cannot remove sulphur.
21. Sound can propagate without any medium.
22. Landslides occur when coherent rock of soil masses move down slope.
23. Solid waste material degraded by microorganisms are called non-biodegradable.
24. Groundwater contamination is irreversible.
25. Waste heat in water does not affect the survival of sensitive species.
26. *Talaab*, *Bawri*, *Johar* and *Hauz* are examples of ancient rainwater harvesting technologies.
27. As per UN declaration of Human Rights, right to housing is a basic human right.
28. Due to ozone depletion there will be increased incidence of skin cancer and eyecataract.
29. Wastelands can be formed due to activities like deforestation, overgrazing, mining and erroneous agricultural practices.
30. Total fertility rate and infant mortality rates are lower in developing countries.
31. Life expectancy over the globe has improved over the past century.

Answers

1. True	2. True	3. True	4. False	5. True
6. False	7. False	8. False	9. True	10. False
11. False	12. False	13. True	14. True	15. True
16. True	17. True	18. False	19. False	20. False
21. False	22. True	23. False	24. True	25. False
26. True	27. True	28. True	29. True	30. False
31. True				

❑❑❑

INDEX

A

B

C

D

E

F

G

H

I

K

L

M

N

O

P

Q

R

S

T

U

V

W

ENVIRONMENTAL STUDIES

Prof. M.P. Poonia, S.C. Sharma, Santosh Kumar

FEEDBACK FORM

1. Which topics from your examination point of view are inadequately or not covered in this book?

____

____

2. Is there any misprints/ mistakes/ factual inaccuracy in this book? If so, please specify.

____

____

3. What is your assessment of this book as regards the presentation of the subject-matter, expression, precision and price in relation to the other recommended books available on this subject?

____

____

4. Which competing books you regard as better than this? Please mention their author and publishers.

____

____

5. Any further suggestions/comments you would like to make for the improvement of the book?

____

____

PS: If needed, please attach a separate sheet.

Name : ______________

College : ______________

Address : ______________

Email : ______________

Mobile : ______________

Please mail the feedback form at:

KHANNA BOOK PUBLISHING CO. (P) LTD.

4C/4344, Ansari Road, Daryaganj, New Delhi-110002

You can also email this feedback form at contact@khannabooks.com or Whatsapp at +91 – 99109 09320

BEST FEEDBACK WILL BE REWARDED SUITABLY